43. Colloquium der Gesellschaft für Biologische Chemie
9.–11. April 1992 in Mosbach/Baden

DNA Replication and the Cell Cycle

Edited by
E. Fanning, R. Knippers and E.-L. Winnacker

With 97 Figures and 13 Tables

Springer-Verlag
Berlin Heidelberg New York
London Paris Tokyo
Hong Kong Barcelona
Budapest

Prof. Dr. ELLEN FANNING
Institut für Biochemie
Karlstraße 23
8000 München 2, FRG

Prof. Dr. med. ROLF KNIPPERS
Fakultät für Biologie
Universität Konstanz
7750 Konstanz, FRG

Prof. Dr. ERNST-L. WINNACKER
Institut für Biochemie der LMU
im Max-Planck-Institut für Biochemie
Am Klopferspitz
8033 Martinsried, FRG

ISBN-13:978-3-642-77042-5 e-ISBN-13:978-3-642-77040-1
DOI: 10.1007/978-3-642-77040-1

Library of Congress Cataloging-in-Publication Data: Gesellschaft für Biologische Chemie. Colloquium (43rd: 1992: Mosbach, Baden-Württemberg, Germany).
DNA replication and the cell cycle / 43. Colloquium der Gesellschaft füt Biologische Chemie 9.–11. April 1991 in Mosbach/Baden; edited by E. Fanning, R. Knippers, and E.-L. Winnacker.
p. ... cm. Includes bibliographical references and index.
ISBN-13:978-3-642-77042-5
1. DNA replication – Congresses. 2. Cell cycle – Congresses. I. Fanning, E. (Ellen), II. Knippers Rolf, and Winnacker, Ernst L. IV. Title

Softcover reprint of the hardcover 1st edition 1993

Typesetting: Fa. M. Masson-Scheurer, 6654 Kirkel 2
31/3145-5 4 3 2 1 0 – Printed on acid-free paper

Contents

Contributors

You will find the addresses at the beginning of the respective contribution

Introductory Remarks.
The Initiation of Eukaryotic DNA Replication and Its Control

R. Knippers[1] and J. Ruff[1]

1 Introduction

Interactions of peptide growth factors with their receptors on the cell surface induce quiescent mammalian cells to enter the cell cycle. The signals received are transmitted to their intracellular targets mainly by a series of phosphorylation events with $p34^{cdc2}$ and related protein kinases as primary participants as well as cyclin proteins that regulate the activity of the kinases (see E. Nigg and coworkers, this Vol., p 147). But other enzyme systems may also participate in signal transmission as, for example components of the ubiquitin-dependent protein degradation pathway (see S. Jentsch et al., this Vol., p. 177).

Targets of the signal-transmission pathways include transcription factors which are activated to transcribe a family of genes, the "immediate early" or "primary response" genes (Herschman 1991; Hipskind et al., p. 185, this Vol.). The products of these genes may directly or indirectly activate other sets of genes encoding proteins required for genome replication, mitosis and cell division. An example is given by P. Plevani and G. Lucchini (p. 199, this Vol.) who describe, for yeast cells, the cell cycle-dependent expression of genes encoding DNA polymerases and other enzymes required for DNA replication (for a short review see Andrews 1992).

Another group of targets includes the products of some tumor suppressor genes such as the $p110^{Rb}$ and the p53 protein. These proteins are believed to act normally as negative proliferation controls. Their neutralization by external signals may relieve this block (for review see Hamel et al. 1992; see also C. Finlay and R. Quartin, p. 231, and V. Rotter et al., p. 237, this Vol.).

In addition, external signals stimulate the expression of rRNA genes (R. Voit et al., this Vol. p. 203) or enhance the activity of several key components involved in translation initiation (Proud 1992), e.g. the activity of the 40S ribosomal subunit through the S6 protein kinase (S. Ferrari and G. Thomas, this Vol., p. 171).

An important outcome of these events concerns the amount and the activity of proteins interacting with replication start sites on genomic DNA as a first step in the assembly of the replication apparatus.

Some replication functions are known to be directly activated by phosphorylation of defined amino acid side chains, as we shall see later (I. Moarefi et al., p. 157; B. Stillman, p. 127, this Vol.).

[1] Fakultät für Biologie Universität Konstanz, D-775 Konstanz, FRG.

43. Colloquium Mosbach 1992
DNA Replication and the Cell Cycle

However, many important reactions that eventually lead to the initiation of replication are not yet known, even though considerable advances have been made in recent years in the study of eukaryotic replication.

We know most about the elongation phase of the eukaryotic replication cycle. Much of this information was obtained by straight forward biochemistry (U. Hübscher et al., p. 63) and, most notably, by the study of a particularly useful model system, simian virus 40 (SV40). As will be described in this volume by B. Stillman (p. 127), this virus engages the replication machinery of the host cell for the replication of its genome, but it encodes its own initiator protein which has evolved to serve the special purposes of the virus, fast and unconstrained replication.

In contrast to viral DNA replication, the initiation of cellular DNA replication is tightly controlled. Therefore, the initiation of SV40 replication may not be an appropriate model for eukaryotic replication initiation. Despite much recent progress (summarized on pp. 93 and 113 of this volume by M. DePamphilis and P. Held and their respective coworkers), the reactions at the start of an eukaryotic replication cycle are still elusive.

An important lesson from recent studies on eukaryotic DNA replication is that a remarkable similarity to the process of DNA replication in bacteria exists. This may not be surprising. Once invented by evolution, a reaction as complicated as DNA replication is unlikely to be drastically changed with the appearance of new types of organisms. And it appears to be a reasonable first assumption that similar general principles may determine bacterial and eukaryotic initiation reactions. Therefore, a brief inspection of the events at the bacterial origin of replication may be useful as a basis for further considerations.

2 Initiation of Replication in Bacteria

The genomes of gram-negative bacteria possess one fixed starting point of replication, a DNA element of about 245 base pairs, termed *oriC*. One important feature of the bacterial replication origin is a group of specific binding sites for the initiator protein, the dna A protein. Another important structural element is a set of 13-base-pair blocks of AT-rich DNA (Fig. 1).

The key factor is the initiator dna A protein, whose supply is carefully regulated by the controlled expression of its gene; it appears that the amount of active dna A protein in the cell is one critical signal for the induction of replication (von Meyenburg and Hansen 1987; see also E. Boye et al., p. 15 this Vol.).

Some 20–30 dna A protein molecules, activated by bound ATP, bind specifically to the four recognition sequences within the *oriC* sequence, forming a large protein core around which the DNA is wrapped in a negative superhelical turn (Bramhill and Kornberg 1988). Most importantly, the nucleoprotein complex induces an untwisting or "melting" of the adjacent AT-rich 13-mer region (Gille and Messer 1991), which may have an intrinsic low free energy requirement for DNA unwinding. The separation of DNA strands enables a DNA helicase, encoded by the replication gene *dna B*, to enter and to enlarge the unwound region as a prerequisite for an establishment of replication forks followed by the synthesis of RNA primers and their elongation by DNA polymerase (Fig. 2).

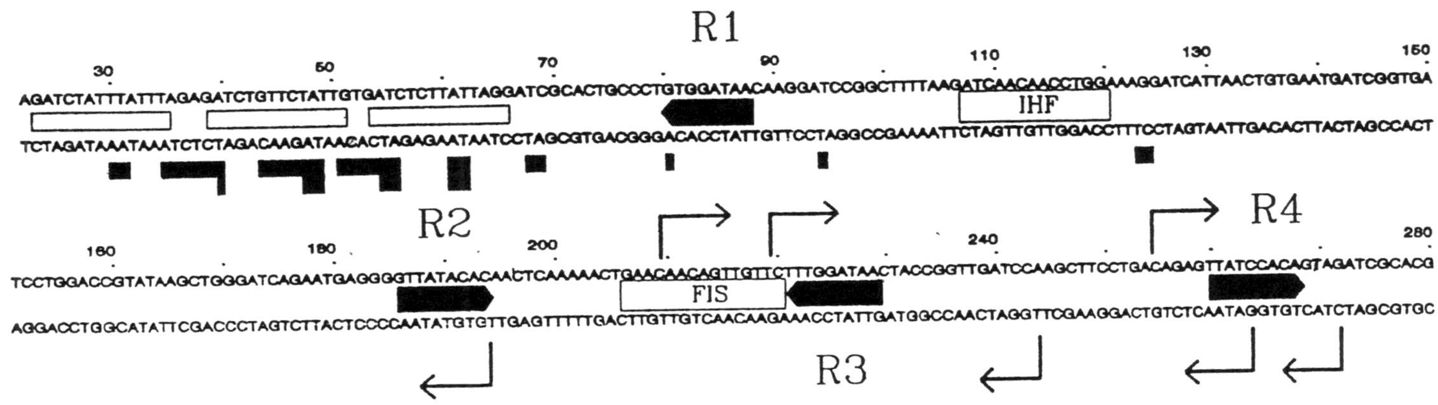

Fig. 1. Structure of the *Escherichia coli* origin of replication. The sequences of the indicated structural DNA elements and the distances between them are essential for origin function. *R1, R2, R3* and *R4* Binding sites for the bacterial initiator, the Dna A protein; *IHF* and *FIS* binding sites for auxiliary proteins; *stippled rectangles* three 13-bp repeat regions (the 13-mers) where DNA unwinding is initiated during formation of the "open complex" (experimentally determined unwound regions are indicated by *boxes* below the sequence); *arrows* location and 5' to 3' direction of the first primers for bidirectional DNA strand synthesis. (Gille and Messer 1991)

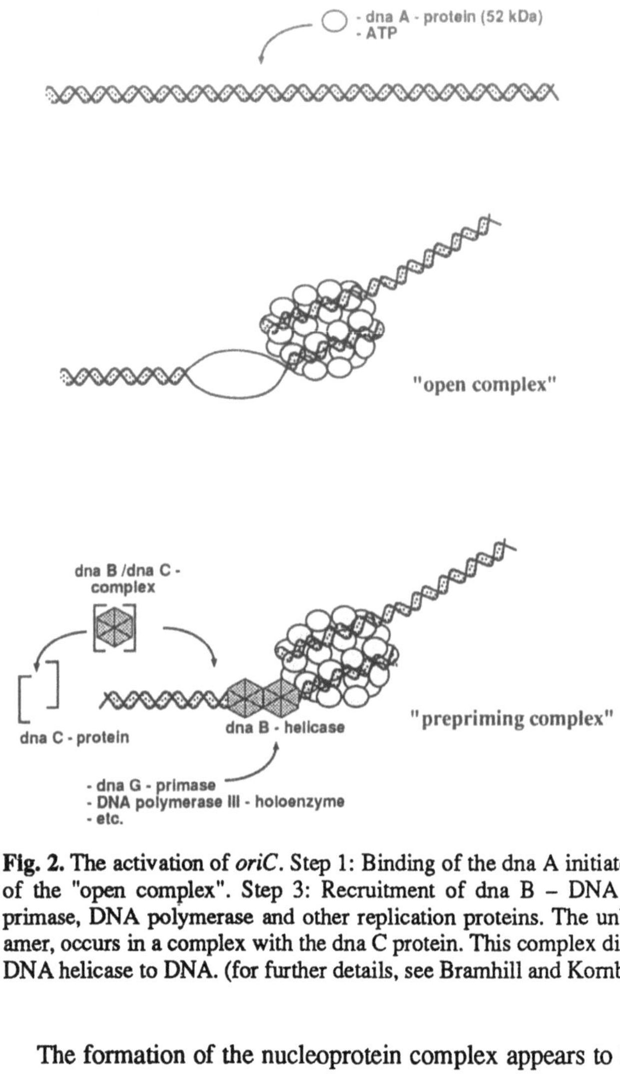

Fig. 2. The activation of *oriC*. Step 1: Binding of the dna A initiator protein. Step 2: Formation of the "open complex". Step 3: Recruitment of dna B – DNA helicase and association of primase, DNA polymerase and other replication proteins. The unbound DNA helicase, a hexamer, occurs in a complex with the dna C protein. This complex dissociates upon binding of the DNA helicase to DNA. (for further details, see Bramhill and Kornberg 1988)

The formation of the nucleoprotein complex appears to be facilitated by auxiliary proteins, such as FIS and IHF, which bind to particular sites within the core region of the origin and induce an extensive bending of the bound DNA (Gille et al. 1990).

An additional aspect should be noted: transcriptional activation. Transcription serves to facilitate the initiation of replication, probably through an RNA-mediated structural transition (R loop) near the origin (Skarstad et al. 1990). This may assist in the unwinding of the AT-rich section of the origin.

In more general terms, the process of "prepriming" proceeds in distinct steps: the specific binding of an activated initiator protein to DNA sites in the origin sequence; induction of localized untwisting of the bound DNA; and, finally, the recruitment of a DNA helicase. A similar general scheme of replication initiation is also used by bacteriophage λ, as will be described by R. McMacken (this Vol. p. 35).

However, it is important to note that this scheme of replication initiation does not apply to all DNA genomes. A variety of alternative replication-mechanisms exist.

Well-known examples include adenovirus and bacteriophage Φ 29 (see: M Salas, this Vol., p. 27) which initiate the replicatin of their genome via a protein covalently bound at the DNA termini. Another example is replication via an RNA intermediate by reverse transcription as in hepadnaviruses (see H. Schaller et al., this Vol., p. 5).

3 A Model System: Simian Virus 40

However, in broad outline, the replication initiation mechanisms of some eukaryotic viruses appear to be functionally related to the bacterial system. As already mentioned, a well-studied example is SV40 which uses cellular replication elongation factors for multiplication, but encodes its own initiator protein.

Like bacterial dna A protein, the viral initiator, T antigen, is activated by bound ATP (or a non-hydrolyzable ATP-analogue), to form a nucleoprotein complex, consisting of two closely adjacent protein hexamers, at a specific site in the SV40 origin. Binding alone induces conformational changes in the bound sections of DNA. But extensive unwinding occurs when the hydrolysis of ATP enables the protein to exert its intrinsic DNA helicase function (Fig. 3; for review see Fanning and Knippers 1992).

We note some obvious differences between the T antigen-induced activation of the viral origin and the dna A protein-induced activation of the bacterial origin. The SV40 origin is enclosed by the double-hexameric T antigen complex, whereas the bacterial origin DNA is wrapped in a superhelical turn around the dna A protein core.

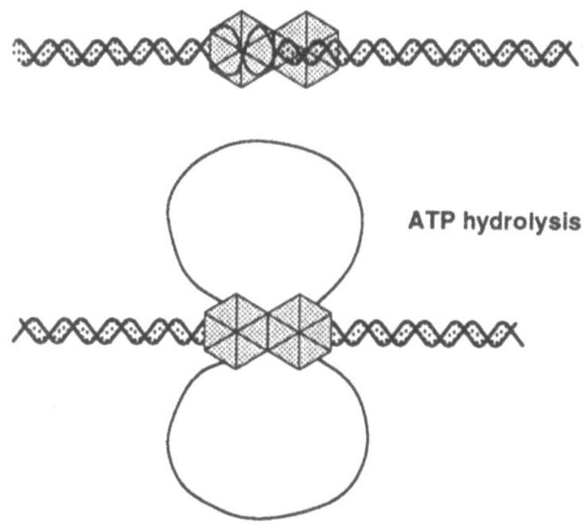

Fig. 3. SV40 T-antigen catalyzed unwinding of double-stranded DNA. The viral genome carries a "minimal" origin of 65 bp which includes a set of specific binding sites for two closely spaced T-antigen hexamers. Bound T antigen induces structural changes in the bound DNA, including the untwisting of a 17-bp-long AT-rich stretch in the minimal origin (Borowiec and Hurwitz 1988). Upon ATP hydrolysis, the DNA double strand is unwound and appears to be reeled through the stationary double hexameric complex. (Wessel et al. 1992)

An additional difference is that T antigen combines on one polypeptide chain a DNA recognition and a DNA helicase function.

These differences may reflect the special constraints imposed on a viral genome, as the *protein-around-DNA* motif requires less space on the small and parsimonious viral DNA; moreover, merging of the recognition and the helicase function may facilitate fast and frequent viral replication cycles. This does not imply that T antigen functions constitutively in an unregulated manner. In fact, its activity as a replication initiator is tightly controlled by phosphorylation, as described by I. Moarefi et al. (p. 157, this Vol.).

As in bacteria, promoter elements in the vicinity of the origin enhance the replication potential of the SV40 origin in vivo. Possibly, *trans*-acting factors keep the origin clear of inhibiting proteins such as nucleosomes. Another and not mutually exclusive possibility is that promoter-bound *trans*-activating proteins directly influence the establishment of a functional initiator complex (see M. DePamphilis et al., this Vol., p. 93, for a thorough discussion of this point).

Recently, a biochemical assay has become available for the study of *papilloma virus* DNA in vitro. The functions of two proteins, the viral E1 and E2 proteins, are required for the initiation of replication (see Fig. 2 on p. 96). Interestingly, one of these, the E2 protein, is also known as a transcription factor. For replication initiation, it combines with the E1 replication protein, a putative DNA helicase, to from a preinitiation complex at the papilloma origin of replication (Ustav et al. 1991; Yang et al. 1991).

4 Origins in Yeast Genomes

As we have seen, bacteria and viruses carry one origin of replication in their genomes. In contrast, all cellular eukaryotic genomes contain many replication starting points, distributed along chromosomal DNA at distances of about 50 to several hundred kbp. Replication forks are established at these sites and proceed in opposite directions until they meet the advancing forks of adjacent replication units (Huberman and Riggs 1968).

Unfortunately, no in vitro assay as simple as that used for the study of bacterial and viral DNA replication exists for the investigation of eukaryotic DNA replication. Therefore, most conclusions regarding the mechanism of replication initiation remain uncertain at present. However, important advances have been made, most notably throught studies using the budding yeast *Saccharomyces cerevisiae* as a model eukaryotic organism.

S. cerevisiae became the most successful system for these studies because plasmids with certain cloned genomic DNA sequences are able to replicate as independent units in the nucleus. The DNA sequences which convey this property to plasmids are known as "autonomously replicating sequences" or ARS elements. Evidence has shown that ARS elements are quite common in the yeast genome (on average, once every 30–40 kbp) and that a number of these ARS elements serve as replications origins for chromosome replication, but others do not. However, no replication starting point appears to exist in the yeast genome without an ARS element (for review see Gasser 1991).

5′———T/ATTTAT/CA/GTTTTT/A (T) – – – – ——————— 3′
3′———A/TAAATA/GT/CAAAAA/T (A) – – – – ——————— 5′

"consensus sequence"
(A element)

3′region
(functionally distinct B elements)
- binding sites for auxiliary proteins, including transcription factors
- low free energy requirements for DNA unwinding

Fig. 4. Functional domains of yeast ARS elements. The consensus sequence is common to all yeast origins but the structure of the 3' region varies from one ARS element to another. Compare the strict sequence requirements of the bacterial origin of replication (Fig. 1) with the relatively loose organization of yeast ARS elements. Detailed descriptions of individual ARS elements are presented in the contributions to this volume by I. Hofmann et al. (Fig. 3, p. 86 and by B. Stillman Fig. 4, p. 137)

Structural features of ARS elements are (1) the A element or ARS consensus, an AT-rich sequence of 11–12 bp, common to all ARS elements; and (2) an adjacent region on the 3' side of the T-rich strand of the consensus (Fig. 4). While any one of many single base substitutions in the consensus element eliminates ARS activity, the 3' region can be altered by small deletions or insertions without much effect on replication efficiency. In fact, the 3' region can be as small as 19 bp and larger than 50 bp depending on the plasmid sequence context. An extensive linker substitution study has clearly shown that the 3' region is composed of several distinct functional sequences, the B elements. B elements can be binding sites for transcription factors or sites with low free energy requirements for localized DNA unwinding (Marahrens and Stillman 1992). Indeed, it has been shown that the 3' region frequently assumes an unwound state when placed in highly negatively supercoiled plasmids (Umek and Kowalski 1988).

These findings suggest that yeast replication starting sites consist of a modular array of short functional sequences, including an essential A element and several activating B elements whose number, order, structure and molecular function may vary among individual ARS regions.

An important step towards a biochemical analysis of the yeast origin function has recently been made by the identification of a large protein complex that assembles in the presence of ATP on a functional ARS region in vitro (Bell and Stillman 1992; see also B. Stillman, p. 127, this Vol.).

5 Is the Yeast ARS Region a Paradigm for Cellular Eukaryotic Replication Starting Sites?

It will be instructive to compare the structure of the origins in the budding yeast, *S. cerevisiae*, with that of the origins in fission yeast, *S. pombe*. This is an interesting comparison as *S. pombe* is known to be genetically more related to higher eukaryotes than to *S. cerevisiae*. However, unlike mammalian cells in culture, *S. pombe* cells are able to replicate transfected plasmid DNAs as extrachromosomal elements, allowing the identification of possible ARS sequences in the genome of this organism.

A study designed to isolate *S. pombe* ARS elements used Sau3A restricted *S. pombe* DNA randomly cloned in plasmid vectors for transfection. The restriction fragments had an average size of 200–300 bp. Significantly, however, all autonomously replicating clones recovered had incorporated genomic fragments of > 1 kbp, which as one common feature had an AT content of 69–76% (compared to an average AT content of ca. 60% of the *S. pombe* genome). A computer search revealed that all fragments contained at least one element resembling the *S. cerevisiae* ARS consensus. However, an engineered deletion of these elements had no effect on extrachromosomal replication in *S. pombe* (but caused a loss of ARS activity when tested in *S. cerevisiae*). Thus, an ARS sequence active *S. pombe* clearly has less well-defined sequence requirements than an ARS element in *S. cerevisiae*, but it is not known what these requirements are, except, perhaps, an unusually high AT content (Maundrell et al. 1988).

In contrast to yeast, mammalian ARS elements are difficult to identify because transfected DNA is rapidly lost at mitosis or integrated into the cellular genome. Exceptions to this general rule have been reported but they were not generally accepted because of their poor reproducibility (for references, see Burhans et al. 1990).

A possible strategy to partially circumvent this problem is the use of the Epstein-Barr virus (EBV) as a carrier for potential replication starting points (Heinzel et al. 1991; Krysan and Calos 1991). EBV-DNA replicates as an episome in infected mammalian cells. It contains two regions, a family of DNA repeats and a 65-bp palindromic element, important for nuclear retention and extrachromosomal replication. Removal of the palindromic element from an EBV-derived plasmid destroys the origin function, but cloned mammalian DNA sequences restore the replicative capacity of the EBV plasmid.

Surprisingly, almost all mammalian sequences confer the ability to replicate as an extrachromosomal element on EBV plasmids, provided that the incorporated DNA is longer than 10 kbp. Deletion analysis did not reveal any individual essential sequence or structural element, and replication initiation occurred at numerous locations (Krysan and Calos 1991).

Apparently, the size of the cloned mammalian fragments is important in this assay, and not a particular sequence. This does not mean that any DNA can replace the EBV origin. Indeed, not all human DNA fragments stimulate EBV replication with equal efficiency, and bacterial DNA functions very poorly as a replication starting site in this system.

An interpretation of the data is that EBV-borne autonomous replication does not depend on one or more essential elements but on a special combination and/or arrangement of some loosely defined sites; and that the probability for the presence of this collection of necessary sites increases with the lengths of the DNA fragment under study.

In its chromosomal context, replication initiation appears to be more selective than in plasmids, and much evidence exists suggesting that the starting points for mammalian replication are not randomly distributed in the cellular genome. Indeed, preferred starting sites exist, as shown by a variety of methods used to map replication starts in the vicinity of genes or gene clusters (see, for example, Taljanidisz et al. 1989).

It remains an interesting problem to determine how these chromosomal starting sites are selected. It may be useful to recall that euchromatic genomic regions with actively transcribed genes are usually (but not always) replicated earlier than heterochromatin with silent genes. Thus, chromatin structure may influence the selection of replication starts, and transcription factors in the promoter/enhancer regions of actively transcribed genes may serve to facilitate the access of replication proteins to appropriate DNA sites.

So far, only a few mammalian replication starting sites have been identified on the DNA level. The most precise data are from investigations of the dihydrofolate reductase (DHFR) gene region. Curiously, however, the results, as described in the literature, are contradictory.

One laboratory detected newly established replication forks over a stretch of about 30 kbp, consistent with the possibility that initiation occurs more or less randomly within a relatively large section of a chromosome (for review see Hamlin et al. 1991). In contrast, a second laboratory, studying the same genetic region, was able to define a 450-bp piece of DNA as a replication starting site (Burhans et al. 1990; see Fig. b on p. 107). This DNA segment is certainly interesting as it contains a number of conspicuous DNA elements: AT-rich regions; bent DNA; DNA regions which assume single-stranded conformation when cloned into negatively supercoiled plasmids; regions which function as ARS elements in yeast cells; binding sites for known transcription factors; binding sites for specific factors that associate with DNA helicases and the like (as described in detail by M. DePamphilis et al. and P. Held et al. in their contributions to this volume, see p. 93 and p. 113).

Taken at face value, these findings support the notion of a modular structure of an eukaryotic replication starting site, consisting of a variety of multiple closely spaced DNA elements, including some which are known to function as transcription factor binding sites.

However, presently, it is difficult to decide whether the 450-bp DNA segment in the DHFR genetic region may or may not serve as an origin (in the strict sense of a bacterial origin as defined in Fig. 1), simply because we do not yet possess the adequate assay systems. As already mentioned, an ARS-type assay does not adequately reflect origin activity in mammalian cells. An exception may be the EBV system. However, we cannot entirely exclude the possibility that this system tells us much about the functioning of the EBV genome and little about a cellular replication starting site.

We clearly need a biochemically defined assay system for the replication of cellular DNA in vitro. One could argue that such systems already exist, namely extracts from *Xenopus* eggs (Blow and Laskey 1986) or *Drosophila* embryos (Crevel and Cotterill 1991). However, DNA molecules when added to extracts from these cells are first converted to chromatin-like structures, and then enclosed in "pseudo-nuclei" before DNA replication is initiated (Blow and Laskey 1986; Newport 1987). Moreover, a round of semiconservative replication only occurs when nuclear formation is complete. It will be difficult to make this system amenable to biochemical studies, at least when this is meant to imply the purification of initiator proteins, and to use these proteins for the activation of isolated replication starting sites in vitro.

Indeed, some scientists believe that an in vitro replication of cellular DNA, employing purified proteins and isolated "origin" sequences, will be impossible for *a priori* reasons. This opinion is based on the observation that, in intact nuclei, DNA replication occurs in clusters, each one containing many hundred replication forks (Cox and Laskey 1991), possibly suggesting that the replication machinery is associated with fixed sites on the nuclear matrix, and that the DNA is reeled through a stationary, structure-bound complex (reviewed by Cook 1991). It would indeed be very difficult or even impossible to reconstitute a structure of this complexity in the test tube.

However, the present discussion is reminiscent of a similar situation 20 years ago when some influential scientists maintained that true in vitro replication of the bacterial genome could not be accomplished. In fact, nobody has succeeded yet in replicating an entire bacterial genome in vitro. Nevertheless, we now have a relatively clear picture of the events at the bacterial replication origin and of the mechanisms regulating the process of replication initiation. This progress depended to a great extent on the power of reductionism. It was necessary to use as substrate a small plasmid with the cloned origin rather than the entire bacterial chromosome, and isolated enzymes rather than unfractionated protein extracts (Kornberg and Baker 1990).

In a similar vein, one may argue that the ongoing efforts to identify and clone bona fide genomic replication starting regions are the necessary first steps. The logical next step towards a biochemically defined replication system will be the identification and isolation of functional eukaryotic initiator proteins that bind to DNA and prepare it for the assembly of a replication apparatus.

Acknowledgement. We thank E. Fanning for her comments on the manuscript and members of our laboratory for discussions.

References

Andrews BJ (1992) Dialogue with the cell cycle. Nature 355:393–394

Bell SP & Stillman B (1992) ATP-dependent recognition of eukaryotic origins of DNA replication by a multiprotein complex. Nature 357:128–134

Bow JJ & Laskey RA (1986) Initiation of DNA replication in nuclei and purified DNA by a cell-free extract of *Xenopus* eggs. Cell 47:577–587

Borowiec JA & Hurwitz J (1988) Localized melting and structural changes in the SV40 origin of replication induced by T antigen. EMBO J 7:3149–3158

Bramhill D & Kornberg A (1988) A model for initiation at origins of DNA replication. Cell 54:915–918

Burhans WC, Vassilev LT, Caddle MS, Heintz NH & DePamphilis ML (1990) Identification of an origin of bidirectional DNA replication in mammalian chromosomes. Cell 62:955–965

Cook PR (1991) The nucleoskeleton and the topology of replication. Cell 66:627–635

Cox LS & Laskey RA (1991) DNA replication occurs at discrete sites in the pseudonuclei assembled from purified DNA. Cell 66:271–275

Crevel G & Cotterill S (1991) DNA replication in cell-free extracts from *Drosophila melanogaster*. EMBO J 10:4361–4369

Fanning E & Knippers R (1992) Structure and function of simian virus 40 large T antigen. Ann Rev Biochem 61:55–85
Gasser SM (1991) Replication origins, factors and attachment sites. Curr Opinion Cell Biol 3:407–413
Gille H, Egan JB, Roth A & Messer W (1990) The FIS protein binds and bends the origin of chromosomal DNA replication, ori C, of *Escherichia coli*. Nucl Acids Res 19:4167–4172
Gille H & Messer W (1991) Localized DNA melting and structural perturbations in the origin of replication, Ori C, of *Escherichia coli* in vitro and in vivo. EMBO J 10:1579–1584
Hamel PA, Gallie BL & Phillips RA (1992) The retinoblastoma protein and the cell cycle. Trends Genet 8:180–185
Hamlin JL, Vaughn JP, Dijkwel PA, Leu TH & Ma C (1991) Origins of replication: timing and chromosomal position. Curr Opinion Cell Biol 3:414–421
Heinzel SS, Krysan PJ, Tran CT & Calos MP (1991) Autonomous DNA replication in human cells is affected by the size and the source of the DNA. Mol Cell Biol 11:2263–2272
Herschman HR (1991) Primary response genes induced by growth factors and tumor promoters. Annu Rev Biochem 60:281–319
Huberman JA & Riggs AD (1968) On the mechanism of DNA replication in mammalian chromosomes. J Mol Biol 32:327–341
Kornberg A & Baker TA (1992) DNA replication. 2nd edn. WH Freeman, New York
Krysan PJ & Calos MP (1991) Replication initiates at multiple locations on an autonomously replicating plasmid in human cells. Mol Cell Biol 11:1464–1472
Marahrens Y & Stillman B (1992) A yeast chromosomal origin of DNA replication defined by multiple functional elements. Science 255:817–823
Maundrell K, Hutchinson A & Shall S (1988) Sequence analysis of ARS elements in fission yeast. EMBO J 7:2203–2209
Newport J (1987) Nuclear reconstitution in vitro: stages of assembly around protein-free DNA. Cell 48:205–217
Proud CG (1992) Protein phosphorylation in translational control. Curr Top Cell Regul 32:243–369
Skarstad K, Baker TA & Kornberg A (1990) Strand separation required for the initiation of replication at the chromosomal origin of *E. coli* is facilitated by a distant RNA-DNA hybrid. EMBO J 9:2341–2348
Taljanidisz J, Popowski J & Sarkar N (1989) Temporal order of gene replication in Chinese hamster ovary cells. Mol Cell Biol 9:2881–2889
Umek RM & Kowalski D (1988) The ease of DNA unwinding as a determinant of initiation of yeast replication origins. Cell 52:559–567
Von Meyenburg K & Hansen FG (1987) Regulation of chromosome replication. In: Neidhardt FC (ed) *Escherichia coli* and *Salmonella typhimurium*. Cellular and molecular biology. Am Soc Microbiol, Washington, DC, pp 1555–1577
Wessel R, Schweizer J & Stahl H (1992) Simian virus 40 T-antigen helicase is a hexamer which forms a binary complex during bidirectional unwinding from the viral origin of DNA replication. J Virol 66:804–815

Regulation of DNA Replication in Model Systems

Regulation of DNA Replication in *Escherichia coli*

E. Boye[1], A. Lyngstadaas[1], A. Løbner-Olesen[2], K. Skarstad[1], and S. Wold[1]

1 Introduction

The circular chromosome of *Escherichia coli* is replicated bidirectionally from a unique origin *oriC* (Master and Broda 1971; Bird et al. 1972). The molecular actors and interactions required for initiation and propagation of a replication fork have to a large extent been characterized and described by biochemical experiments combined with evidence and methods from bacterial genetics and physiology (for reviews, see von Meyenburg and Hansen 1987; Baker and Kornberg 1991; Bremer and Churchward 1991). An *oriC*-containing plasmid (minichromosome) may be initiated and completely replicated in vitro by purified proteins (Kaguni and Kornberg 1984; Funnell et al. 1986), so that the minimal requirements for minichromosome replication have been defined. In contrast to our detailed knowledge of the biochemical steps of DNA replication, information about regulation of the process remains scant. Regulation of DNA replication occurs largely at the level of initiation, but evidence for downstream control points have been presented (Atlung et al. 1987; Skarstad et al. 1989; Løbner-Olesen et al. 1989). This work gives an overview of the factors known or suggested to affect the regulation of DNA replication in *E. coli*.

2 Results and Discussion

2.1 The C Period

The pioneering work of Cooper and Helmstetter (1968) divided the *E. coli* cell cycle into a C period, when DNA replication occurs, and a D period, which is the time between termination of DNA replication and cell division. The duration of the C and D periods was shown to be constant and about 40 and 20 min, respectively. To allow the cells to grow with a doubling time down to 20 min, multiple rounds of DNA replication are going on simultaneously, so that fast-growing cells may contain several origins of replication. Numerous experiments have since confirmed the basic concepts of the Cooper-Helmstetter model. At slow growth, the C period is prolonged (Kubitschek and Newman 1978; Skarstad et al. 1983; 1985; Allman et al. 1991), but it is not

[1] Department of Biophysics, Institute for Cancer Research, Montebello, 0310 Oslo, Norway.
[2] Department of Molecular, Cellular, and Developmental Biology, University of Colorado, Boulder, CO 80309-0347, USA.

43. Colloquium Mosbach 1992
DNA Replication and the Cell Cycle

clear whether the C period varies continuously with growth rate or is constant above a certain growth rate. Even if the C period may vary continuously with growth rate it is fair to say that DNA replication is largely regulated at the level of initiation.

2.2 Initiation Mass

A combination of the Cooper-Helmstetter model with cell mass measurements of the closely related bacterium *Salmonella typhimurium* (Schaechter et al. 1958) led Donachie (1968) to deduce that the initiation mass of *E. coli* is constant. Initiation mass was defined as the cell mass at initiation divided by the number of origins to be initiated. Although there is general agreement that the initiation mass is almost constant, there are few descriptions in the literature of how the initiation mass varies with growth rate. The most extensive results published (Churchward et al. 1981) indicate that it increases by a factor of 2 when the doubling time decreases from 120 to 40 min. It may be argued that the mass at initiation is not the relevant parameter, but rather the amount of protein required for initiation (Bremer and Churchward 1991).

2.3 Transcription

Before initiation of replication can occur in vivo, de novo protein synthesis (Maaløe and Hanawalt 1961) and a transcriptional event (Lark 1972; Messer 1972) are necessary. Therefore, addition of inhibitors of protein synthesis or of transcription preclude further initiations. We do not know the identity of the protein(s) which has to be synthesized shortly before initiation. We also do not know the nature of the transcriptional event, but some evidence has emerged.

First, a transcriptional event in the vicinity of *oriC* stimulates initiation in vitro (Baker and Kornberg 1988; Skarstad et al. 1990), a phenomenom called transcriptional activation. This activation is independent of the direction of transcription and transcripts do not have to enter *oriC*. The gene immediately counterclockwise to *oriC* *(gid)* is transcribed away from *oriC*, and this transcript has been implicated in transcriptional activation of initiation (Asai et al. 1990; 1992; Ogawa and Okazaki 1991).

Second, the transition points from RNA to DNA synthesis within *oriC* represent the start sites for DNA replication. Transcripts from the *mioC* gene located clockwise for *oriC* have been reported to coincide with the RNA/DNA transition points within *oriC* (Kohara et al. 1985; Rokeach and Zyskind 1986), which has been taken to suggest that the transcripts serve as primers for leading strand synthesis. This predicts that regulation of initiation is linked to regulation of *mioC* transcription.

2.4 The Stringent Response

Shifts in growth rate or amino acid starvation induce the stringent response, which is mediated by guanosine tetraphosphate, ppGpp (for review, see Cashel and Rudd 1987). Stringent control of initiation of minichromosome (Lycett et al. 1980; Wein-

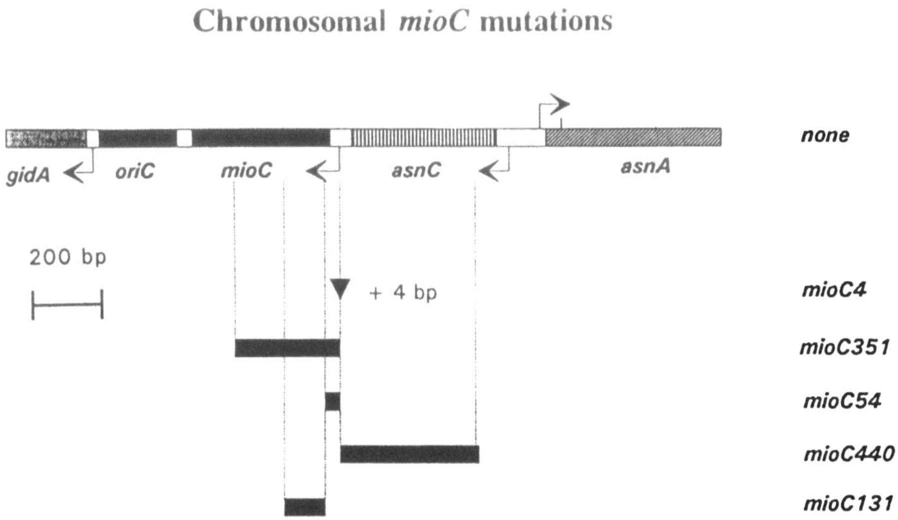

Fig. 1. Schematic map of the *oriC–mioC* region. *Arrows* mark the starting points and direction of transcription, the *black rectangles* in the *lower part* represent the extension of chromosomal deletions and the *triangle* the 4 bp insertion

berger and Helmstetter 1989) and chromosome replication (Levine et al. 1991) suggest the possibility that ppGpp regulates initiation and provides a necessary but still unidentified coupling mechanism between general cell growth and the rate of initiation.

2.5 The mioC Gene

By using *oriC*-dependent minichromosomes numerous workers have shown that the presence of the *mioC* gene greatly stimulates initiation of replication (see e.g. Stuitje and Meijer 1983; Lother et al. 1985; Tanaka and Hiraga 1985; Løbner-Olesen et al. 1987), presumably because it supplies a transcriptional activity close to *oriC* (see above). To study the effect of *mioC* transcription on chromosomal replication we have constructed strains with several different *mioC* mutations on the chromosome itself (Løbner-Olsen and Boye, submitted). The 4 bp insertion *mioC*4 (Fig. 1) and the larger deletions *mioC*54, *mioC*131, *mioC351*, and *mioC*440 all significantly reduce the transcriptional activity from the *mioC* promoter into *oriC* and are the same mutations that have been shown to dramatically decrease minichromosome replication (Løbner-Olesen et al. 1987). We have employed a very sensitive method (flow cytometry) to monitor the growth and DNA replication kinetics of the mutant cells and their wild type parent. The parent strain LJ24 (Rasmussen et al. 1991) and the mutants were grown under several different growth conditions to find a phenotype of the *mioC* mutant cells.

For a shift up in growth rate the cells were grown in minimal medium with glycerol as carbon source, giving a doubling time of about 100 min. At time 0 casamino

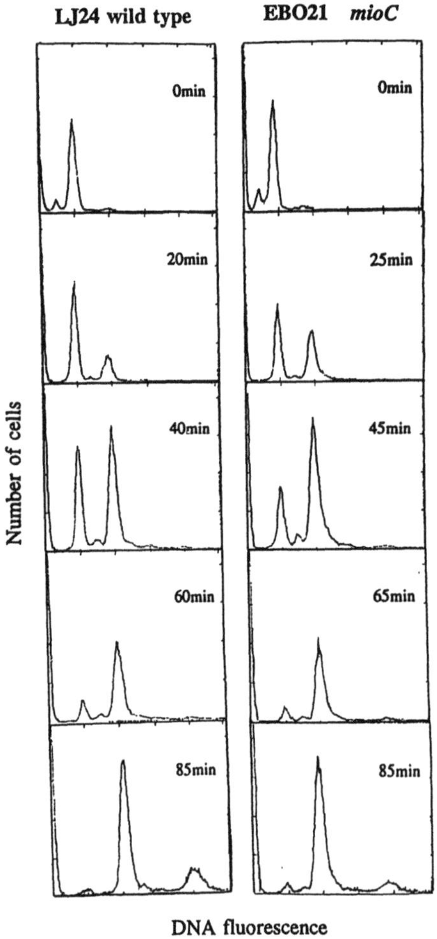

Fig. 2. DNA histograms of strains LJ24 and EBO21 *mioC4* after a shift up in growth rate. At the times indicated after the shift up, samples were removed and incubated in the presence of RIF and CPX before fixation, staining, and flow cytometry

acids and glucose was added, which gave a doubling time of 38 min. At different times after the shift, samples were removed and incubated in the presence of rifampicin (RIF), to stop initiation, and cephalexin (CPX), to stop cell division. This treatment allows the replication forks to proceed to the termini so that all cells contain fully replicated chromosomes. The number of chromosomes present reflects the number of *oriC* copies present at the time of drug addition (Boye and Løbner-Olesen 1991). The cells were fixed in ethanol, stained for DNA, and subjected to flow cytometry as described (Skarstad et al. 1985).

The wild type parent LJ24 (Fig. 2, left) and strain EBO21 *mioC*4 (Fig. 2, right) both contained mostly two chromosomes after RIF/CPX treatment. After the shift, both strains accumulated an increasing fraction of 4-chromosome cells and finally some 8-chromosome cells (Fig. 2, bottom panels). It may be concluded that the DNA replication kinetics are similar in the two strains after a shift up. All the other *mioC* mutants behaved in the same way (data not shown).

In addition to a shift up all strains were grown at different growth rates under steady state conditions, taken through a shift down experiment and they were subjected to amino acid starvation. Flow cytometry was employed to measure the number of origins per cell, the DNA contents, and the cell size. In no experiment were we able to demonstrate any difference between any of the *mioC* mutants and their wild type parent (Løbner-Olesen and Boye, submitted) and we conclude that the *mioC* gene has little, if any, effect on chromosomal replication under the conditions tested.

2.6 Minichromosomes as a Model System

Most of what we know about the biochemistry of *oriC* initiation stems from work with minichromosomes. It has been generally accepted that minichromosomes have the same requirements for replication as the chromosome, their replication is initiated in synchrony with that of the chromosome (Leonard and Helmstetter 1986), and they are replicated once per cell cycle (Koppes and von Meyenburg 1987; Jensen et al. 1990). However, the above results demonstrate that there is a difference in replication control for the chromosome and for a minichromosome, suggesting that minichromosomes may not be valid as models for regulatory aspects of chromosome replication.

2.7 The DnaA Protein

Although a number of proteins participate in initiating replication at *oriC* only the DnaA protein is specific for initiation. A complex of 20–30 molecules of DnaA bind to the four recognition sequences (DnaA boxes) within *oriC* and accomplish a separation of the two DNA strands in a specific region of *oriC* (Kornberg 1988), in a crucial step to establish semiconservative replication. The *dnaA* gene is autoregulated (Atlung et al. 1985; Braun et al. 1985) and the concentration of DnaA is rate limiting for initiation (Atlung et al. 1987; Pierucci et al. 1987; Xu and Bremer 1988; Skarstad et al. 1989). If the intracellular concentration of DnaA is artificially increased, initiation occurs earlier in the cell cycle and at a lower initiation mass (Løbner-Olesen et al. 1989). These results clearly show that the concentration or activity of DnaA protein plays an important role in regulating initiation. Models have been formulated based on the idea that DnaA is the major, and maybe the only, determinant of the timing of initiation (Mahaffey and Zyskind 1989; Hansen et al. 1991).

2.8 Synchrony of Initiation

When two or more copies of *oriC* are present in a cell they are initiated in synchrony (Skarstad et al. 1986). Flow cytometry experiments show that wild type cells contain 2^n fully replicated chromosomes (n = 0, 1, 2, 3, ..) after incubation in the presence of RIF. The asynchrony phenotype occurs when a large fraction of cells contain 3, 5, 6, or 7 chromosomes. Certain mutations in *dnaA* (Skarstad et al. 1988), *recA* (Skarstad

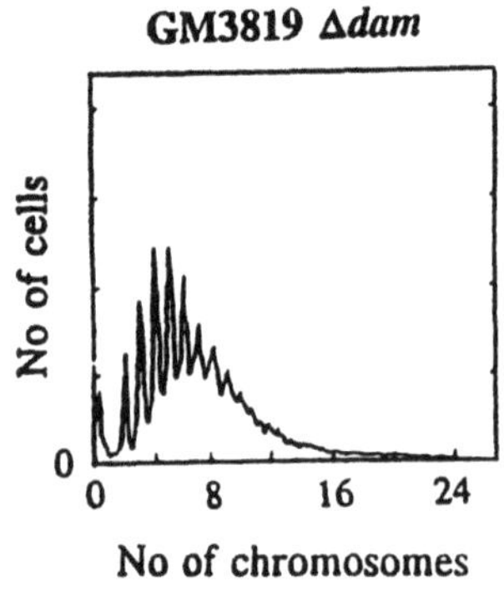

Fig. 3. DNA histogram of strain GM3819 *dam-16* grown in LB with 0.2% glucose before incubation in the presence of RIF and CPX

and Boye 1988), and in *dam, rpoC,* and *dnaC* (Boye et al. 1988) exhibit the asynchrony phenotype.

2.9 Adenine Methylation

After passage of the replication fork, adenines in the sequence 5'-GATC-3' are methylated by the *dam* gene product, leaving the newly replicated DNA hemimethylated for a few minutes. Mutants lacking Dam methyltransferase have the asynchrony phenotype (Boye et al. 1988; Boye and Løbner-Olesen 1990). After run-out of DNA replication in the presence of RIF the DNA histogram of strain GM3819 *dam-16* displays distinct peaks representing cells with 2, 3, 4, 5, 6, 7, and 8 fully replicated chromosomes (Fig. 3). The underlying broad distribution of DNA which is not resolved into peaks probably reflect cells that are unable to complete all rounds of replication. The reason for the asynchrony in *dam* mutant cells is not known, but the observed binding of newly replicated and therefore hemimethylated *oriC* to the membrane (Ogden et al. 1988; Landoulsi et al. 1990) has been suggested to be a mechanism to secure that each origin is replicated once and only once (Boye and Løbner-Olesen 1990; Boye 1991).

2.10 Other Proteins Interacting with oriC

Recently, several proteins have been shown to bind to *oriC*, but their functions in initiation are not yet clear. The IciA protein binds the region where the initial separation of the DNA strands occur, and it has been shown to block initiation in vitro (Hwang and Kornberg 1990). We have studied IciA-less cells and IciA overproducers but we have been unable to find a phenotype that matches the dramatic in vitro effects of the protein (Boye, Hwang, and Kornberg, unpublished).

The proteins Fis and IHF both participate in site-specific recombination, and both of them have specific binding sites within *oriC* (Polaczek 1990; Gille et al. 1991; Filutowicz et al. 1992). IHF can substitute for the otherwise required histone-like protein HU in vitro (Skarstad et al. 1990), and mutants lacking IHF (Filutowicz and Roll

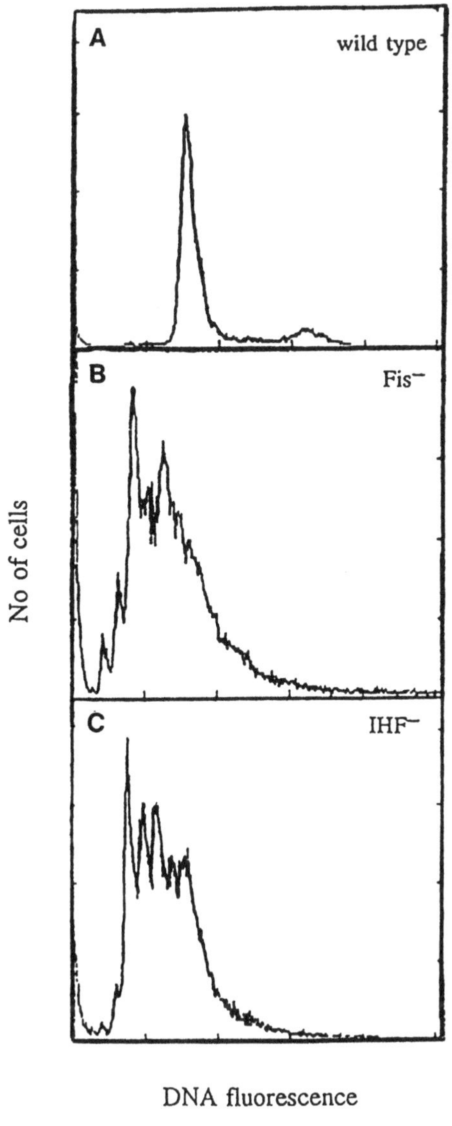

Fig. 4. DNA histograms of strains CSH26 (**A**) and the otherwise isogenic strains lacking Fis (*B*) and IHF (**C**) proteins. The cells were grown in LB with 0.2% glucose with doubling times of 20 min (**A**), 26 min (**B**), and 25 min (**C**), before incubation in the presence of RIF and CPX

1990) or Fis (Gille et al. 1991; Filutowicz et al. 1992) have been shown to have aberrant DNA replication control.

In contrast to the wild type cell CSH26, which mainly contains eight fully replicated chromosomes after run-out in the presence of RIF and CPX (Fig. 4A), both the otherwise isogenic *fis*::*Km* mutant (Fig. 4B) and the *himA*::*Tc himD*::*Cm* double mutant lacking both subunits of IHF (Fig. 4C), display the asynchrony phenotype. One interpretation of these results is that the binding of Fis and IHF to *oriC* is necessary to achieve synchronous initiations. Alternatively, the two proteins affect the expression of other genes or the function of other protein which in turn affect initiation kinetics.

A protein residing in the outer membrane specifically binds *oriC* when hemimethylated, i.e. shortly after initiation of the origin (Landoulsi et al. 1990). This hitherto unidentified protein has been proposed to sequester *oriC* in the membrane and prevent multiple initiations within the same cell cycle. The protein has not been purified and its binding site to hemimethylated *oriC* is not known. For the same reason as a *dam* mutant cell, namely lack of *oriC* sequestration in the membrane, cells without this *oriC*-binding protein are predicted to have the asynchrony phenotype.

Recently, a novel *oriC*-binding protein has been identified. The Rob protein (right-hand side of *oriC* binding protein) specifically protects a region at the right-hand border of *oriC* (Skarstad, Hwang, Thöny, and Kornberg, J. Biol. Chem., in press). The protein has been purified, its gene cloned and sequenced, but its possible function in initiation is not known.

2.11 What is Asynchrony?

The asynchrony phenotype may arise in different ways, and we shall consider two alternatives that are principally different. First, initiations may be truly asynchronous, in which case all origins are initiated at different points in time. Second, most of the origins are initiated at the correct time, but one or two origins lag behind and are initiated later. A principal difference between the two alternatives is that in the first case the cells have lost the ability to initiate at the correct time (defective timing), while in the last case the cells know when they shall initiate and most initiations are synchronous, but the ability to coordinate multiple successful initiations have been lost (defective synchrony).

In an attempt to discriminate between these two alternatives a *dam* deletion strain (ALO452 *dam-16*) and its isogenic wild type (LJ24, Rasmussen et al. 1991) were grown at a low growth rate, so that the cells contained either one or two copies of *oriC*. In this experiment, minimal medium with 0.5% proline as the carbon source gave a doubling time of 8 h (LJ24) and 6 h (ALO452). The DNA histograms of exponentially growing cells showed that most cells contained one fully replicated chromosome (the left-hand peak in Fig. 5A, B), some contained two fully replicated chromosomes (the right-hand peak in Fig. 5A, B), while a fraction of the cells had an intermediate DNA content and were in the process of replicating their single chromosome. Isocontour plots of DNA content versus scattered light (cell size) visualize how the cells start, when newborn, with one chromosome and initially grow in size without any DNA replication (Fig. 5C, D). After awhile DNA replication is initiated, resulting in an increase in both cell size and DNA content until the cells have doubled their DNA content. The cells grow in size for awhile before cell division, and the divided cells reenter the cycle at the starting point at the lower tip of the graphs in panels C and D. The cell size distribution was considerably broader for the *dam* mutant than for the wild type parent (not shown).

In the case of a defective timing mechanism cells would initiate their one origin at all times in the cell cycle (and at all cell masses), which would give a broad size distribution of cells moving from the one-chromosome position to the two-chromosome position. This is not observed. The two strains appear to initiate their single chromo-

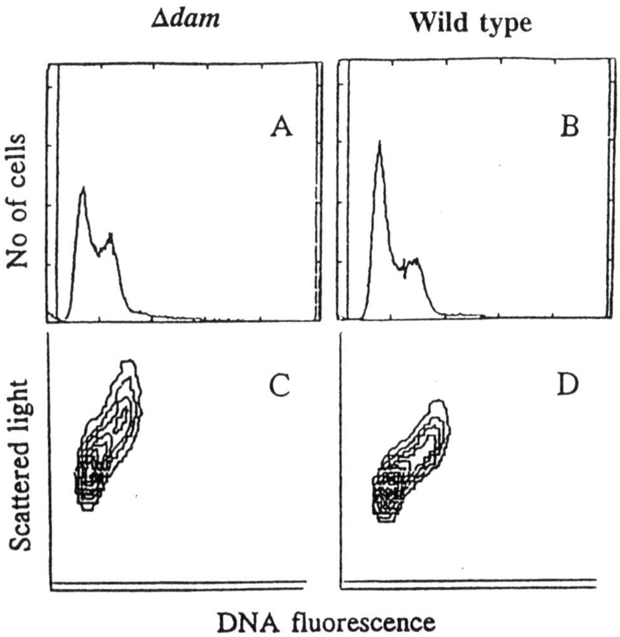

Fig. 5. DNA and DNA/light scatter histograms of strain LJ24 and ALO452 *dam-16*. The cells were growing in minimal medium with proline as carbon source at doubling times of 8 h (LJ24) and 6 h (ALO452) before preparation for flow cytometry. Note that the DNA fluorescence scale is the same in **A** and **B**, and in **C** and **D**, but the *upper* and *lower panels* cannot be compared directly

some at a relatively well defined point in the cell cycle, although the mass at initiation is somewhat larger for the *dam* mutant cell. Similar experiments with the asynchronous *dnaA46* mutant gave DNA histograms for the wild type and mutant strains that were almost indistinguishable (data not shown). It is known that *dam* and *dnaA* mutant strains have gross deficiencies in their initiation mechanism when several initiations have to be coordinated. Nonetheless, the above data suggest, and is consistent with, the hypothesis that the timing of single initiations occur normally.

The inter-initiation time, i.e. the time between two successive replications of a specific chromosomal region, has been shown by density shift analyses to be constant for rapidly growing wild type cells but highly variable for a *dnaA* (Tippe-Schindler et al. 1979) and a *dam* mutant (Bakker and Smith 1989). A prediction of the above hypothesis is that inter-initiation time is constant for both mutants when growing slowly.

Acknowledgements. We are grateful to Marit Osland for her excellent technical assistance and to Walter Messer for the *fis* and *him* mutant strains. This work was supported by the Public Health Service grant GM39582 from the N.I.H., U.S.A., by the Norwegian Cancer Society, by the Danish Center for Microbiology, and by the NATO Collaborative grant no 890467.

References

Allman R, Schjerven T & Boye E (1991) Cell cycle parameters of *Escherichia coli* K-12. J Bacteriol 173:7970–7974

Asai T, Takanami M & Imai M (1990) The AT richness and *gid* transcription determine the left border of the replication origin of the *E. coli* chromosome. EMBO J 9:4065–4072

Asai T, Chen C-P, Nagata T, Takanami M & Imai M (1992) Transcription in vivo within the replication origin of the *Escherichia coli* chromosome: a mechanism for activating initiation of replication. Mol Gen Genet 231:169–178

Atlung T, Clausen ES & Hansen FG (1985) Autoregulation of the *dnaA* gene of *Escherichia coli* K-12. Mol Ren Genet 200:442–450

Atlung T, Løbner-Olesen A & Hansen FG (1987) Overproduction of DnaA protein stimulates initiation of chromosome and minichromosome replication in *Escherichia coli*. Mol Gen Genet 106:51–59

Baker TA & Kornberg A (1988) Transcriptional activation of initiation of replication from the *E. coli* chromosomal origin: an RNA-DNA hybrid near *oriC*. Cell 55:113–123

Bakker A & Smith DW (1989) Methylation of GATC sites is required for precise timing between rounds of DNA replication in *Escherichia coli*. J Bacteriol 171:5738–5742

Bird R, Louarn J, Martuscelli J & Caro L (1972) Origin and sequence of chromosome replication in *Escherichia coli*. J Mol Biol 70:549–566

Boye E, Løbner-Olesen A & Skarstad K (1988) Timing of chromosomal replication in *Escherichia coli*. Biochim. Biophys Acta 951:359–364

Boye E & Løbner-Olesen A (1990) The role of *dam* methyltransferase in the control of DNA replication in *E. coli*. Cell 62:981–989

Boye E & Løbner-Olesen A (1991) Bacterial growth studied by flow cytometry. Res Microbiol 142:131–135

Boye E (1991) A turnstile for initiation of DNA replication. Trends Cell Biol 1:107–109

Braun RE, O'Day K & Wright A (1985) Autoregulation of the DNA replication gene *dnaA* in *E. coli* K-12. Cell 40:159–169

Bremer H & Churchward G (1991) Cyclic chromosome replication in *E. coli*. Microbiol Rev 55:459–475

Campbell JL & Kleckner N (1990) *E. coli oriC* and the *dnaA* promoter are sequestered from *dam* methyltransferase following the passage of the replication fork. Cell 62:967–979

Cashel M & Rudd KE (1987) The stringent response. In: Neidhardt FC, Ingraham JL, Low KB, Magasanik B, Schaechter M & Umbarger HE (eds) *Escherichia coli* and *Salmonella typhimurium. Cellular and molecular biology*. American Society for Microbiology, Washington, DC, pp 1410–1438

Churchward G, Estiva E & Bremer H (1981) Growth rate-dependent control of chromosome replication initiation in *Escherichia coli*. J Bacteriol 145:1232–1238

Cooper S & Helmstetter CE (1968) Chromosome replication and the division cycle of *Escherichia coli* B/r. J Mol Biol 31:519–540

Donachie WD (1968) Relationship between cell size and time of initiation of DNA replication. Nature 219:1077–1079

Filutowicz M & Roll J (1990) The requirement of IHF protein for extrachromosomal replication of the *Escherichia coli oriC* in mutant deficient in DNA polymerase I activity. New Biol 2:818–827

Filutowicz M, Ross W, Wild J & Gourse RL (1992) Involvement of Fis protein in replication of the *Escherichia coli* chromosome. J Bacteriol 174:398–407

Funnell BE, Baker TA & Kornberg A (1986) Complete enzymatic replication of plasmids containing the origin of the *Escherichia coli* chromosome. J Biol Chem 261:5616–5624

Gille H, Egan JB, Roth A & Messer W (1991) The Fis protein binds and bends the origin of chromosomal DNA replication, *oriC*, of *Escherichia coli*. Nucl Acids Res 19:4167–4172

Hansen FG, Christensen BB & Atlung T (1991) The initiator titration model: computer simulation of chromosome and minichromosome control. Res Microbiol 142:161–167

Jensen MR, Løbner-Olesen A & Rasmussen K (1990) *Escherichia coli* minichromosomes: absence of copy number control and random segregation. J Mol Biol 215:257–265

Kaguni JM & Kornberg A (1984) Replication initiated at the origin (*oriC*) of the *E. coli* chromosome reconstituted with purified enzymes. Cell 38:183–190

Kohara Y, Tohdoh N, Jiang X & Okazaki T (1985) The distribution and properties of RNA primed initiation sites of DNA synthesis at the replication origin of *Escherichia coli* chromosome. Nucl Acids Res 13:6847–6866

Koppes LJH & von Meyenburg K (1987) Nonrandom minichromosome replication in *Escherichia coli* K-12. J Bacteriol 169:430–433

Kornberg A (1988) DNA replication. J Biol Chem 263:1–4

Kornberg A & Baker TA (1991) DNA replication. WH Freeman & Co, New York

Kubitschek HE & Newman CN (1978) Chromosome replication during the division cycle in slowly growing, steady-state cultures of three *Escherichia coli* B/r strains. J Bacteriol 136:179–190

Leonhard AC & Helmstetter CE (1986) Cell cycle-specific replication of *Escherichia coli* minichromosomes. Proc Natl Acad Sci USA 83:5101–5105

Levine A, Vannier F, Dehbi M, Henckes G & Seror SJ (1991) The stringent response blocks DNA replication outside the *ori* region in *Bacillus subtilis* and at the origin in *Escherichia coli*. J Mol Biol 219:605–613

Lycett GW, Orr E & Pritchard RH (1980) Chloramphenicol releases a block in initiation of chromosome replication in a *dnaA* strain of *Escherichia coli* K-12. Mol Gen Genet 178:329–336

Løbner-Olesen A, Skarstad K, Hansen FG, von Meyenburg K & Boye E (1989) The DnaA protein determines the initiation mass of *Escherichia coli* K-12. Cell 57:881–889

Maaløe O & Hanawalt PC (1961) Thymine deficiency and the normal DNA replication cycle. J Mol Biol 3:144–155

Mahaffey JM & Zyskind JW (1989) A model for the initiation of replication in *Escherichia coli*. J Theor Biol 140:453–477

Masters M & Broda P (1971) Evidence for the bidirectional replication of the *Escherichia coli* chromosome. Nature New Biol 232:137–140

Ogawa T & Okazaki T (1991) Concurrent transcription from the *gid* and *mioC* promoters activate replication of an *Escherichia coli* minichromosome. Mol Gen Genet 230:193–200

Ogden GB, Pratt MJ & Schaechter M (1988) The replicative origin of the *E. coli* chromosome binds to cell membranes only when hemimethylated. Cell 54:127–135

Pierucci O, Helmstetter CE, Rickert M, Weinberger M & Leonard AC (1987) Overexpression of the *dnaA* gene in *Escherichia coli* B/r: chromosome and minichromosome replication in the presence of rifampin. J Bacteriol 169:1871–1877

Polaczek P (1990) Bending of the origin of replication of *Escherichia coli* by binding of IHF at a specific site. New Biol 2:265–271

Rokeach LA & Zyskind JW (1986) RNA terminating within of *E. coli* origin of replication: stringent regulation and control by DnaA protein. Cell 46:763–771

Schaechter M, Maaløe O & Kjeldgaard NO (1958) Dependency on medium and temperature of cell size and chemical composition during balanced growth of *Salmonella typhimurium*. J Gen Microbiol 19:592–606

Skarstad K, Steen HB & Boye E (1983) Cell cycle parameters of slowly growing *Escherichia coli* B/r studied by flow cytometry. J Bacteriol 154:656–662

Skarstad K, Steen HB & Boye E (1985) *Escherichia coli* DNA distributions measured by flow cytometry and compared with theoretical computer simulations. J Bacteriol 163:661–668

Skarstad K, Boye E & Steen HB (1986) Timing of initiation of chromosome replication in individual *Escherichia coli* cells. EMBO J 5:1711–1717

Skarstad K, von Meyenburg K, Hansen FG & Boye E (1988) Coordination of chromosome replication initiation in *Escherichia coli*: effects of different *dnaA* alleles. J Bacteriol 170:2549–2554

Skarstad K, Løbner-Olesen A, Atlung T, von Meyenburg K & Boye E (1989) Initiation of DNA replication in *Escherichia coli* after overproduction of the DnaA protein. Mol Gen Genet 218:50–56

Skarstad K, Baker TA & Kornberg A (1990) Strand separation required for initiation of replication at the chromosomal origin of *E. coli* is facilitated by a distant RNA–DNA hybrid. EMBO J 9:2341–2348

Tippe-Schindler R, Zahn G & Messer W (1979) Control of the initiation of DNA replication in *Escherichia coli*. I. Negative control of initiation. Mol Gen Genet 168:185–195

von Meyenburg K & Hansen FG (1987) Regulation of chromosome replication. In: Neidhardt FC, Ingraham JL, Low KB, Magasanik B, Schaechter M & Umbarger HE (eds) *Escherichia coli* and *Salmonella typhimurium. Cellular and molecular biology*. American Society for Microbiology, Washington, DC, pp 1555–1577

Weinberger M & Helmstetter CE (1989) Inhibition of protein synthesis transiently stimulates initiation of minichromosome replication in *Escherichia coli*. J Bacteriol 171:3591–3596

Xu YC & Bremer H (1988) Chromosome replication in *Escherichia coli* induced by oversupply of DnaA. Mol Gen Genet 211:138–142

Protein-Primed Replication of Bacteriophage ø29 DNA

M. Salas[1], J. Méndez[1], J. A. Esteban[1], A. Bernad[1], M. S., Soengas[1], J. M. Lázaro[1], M.A. Blasco[1] and L. Blanco[1]

1 Introduction

The replication of bacteriophage ø29 DNA starts at both ends of the molecule and takes place by a protein-priming mechanism (reviewed in Salas 1991) in which the ø29 DNA polymerase catalyzes the covalent linkage of dAMP to the OH group of serine 232 in the terminal protein. The ø29 DNA polymerase also catalyzes elongation, that proceeds by a strand-displacement mechanism (Blanco et al. 1989). In addition to initiation and DNA polymerization activities, the ø29 DNA polymerase has pyrophosphorolytic (Blasco et al. 1991) and 3'-5'exonuclease activities (Blanco and Salas 1985a), the latter being involved in proofreading (Garmendia et al. 1992). All these characteristics, together with the existence of amino acid sequence similarities among ø29 DNA polymerase and other prokaryotic and eukaryotic DNA polymerases (Bernad et al. 1987, 1989; Blanco et al. 1991), make of the ø29 enzyme a good candidate for structure-function studies. Three amino acid segments (named Exo I, Exo II and Exo III), predicted to contain the critical amino acid residues forming the 3'-5'exonuclease active site, were identified in the N-terminal portion of ø29 DNA polymerase and other prokaryotic and eukaryotic DNA polymerases (Bernad et al. 1989). In addition, six conserved segments, predicted to form the polymerization active site, are located in the C-terminal portion of DNA polymerases (Bernad et al. 1987; Blanco et al. 1991). Site-directed mutagenesis studies in the conserved regions of ø29 DNA polymerase have demonstrated the predicted location of enzymatic activities in two structurally separated domains: the synthetic activities (protein-primed initiation and DNA polymerization) are located in the C-terminal portion whereas the 3'-5'exonuclease activity resides in the N-terminal third of the polypeptide (Bernad et al. 1989, 1990a, b; Blanco et al. 1991; Blasco et al. 1992; Soengas et al. 1992).

[1] Centro de Biología Molecular (CSIC-UAM), Universidad Autónoma, Canto Blanco, 28049 Madrid, Spain.

43. Colloquium Mosbach 1992
DNA Replication and the Cell Cycle

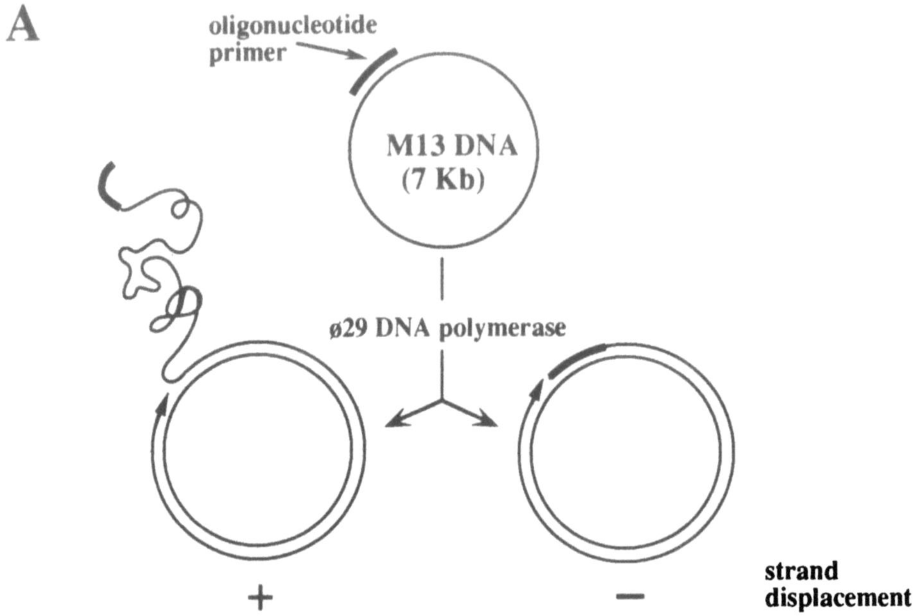
A
oligonucleotide primer
M13 DNA (7 Kb)
ø29 DNA polymerase
+
−
strand displacement

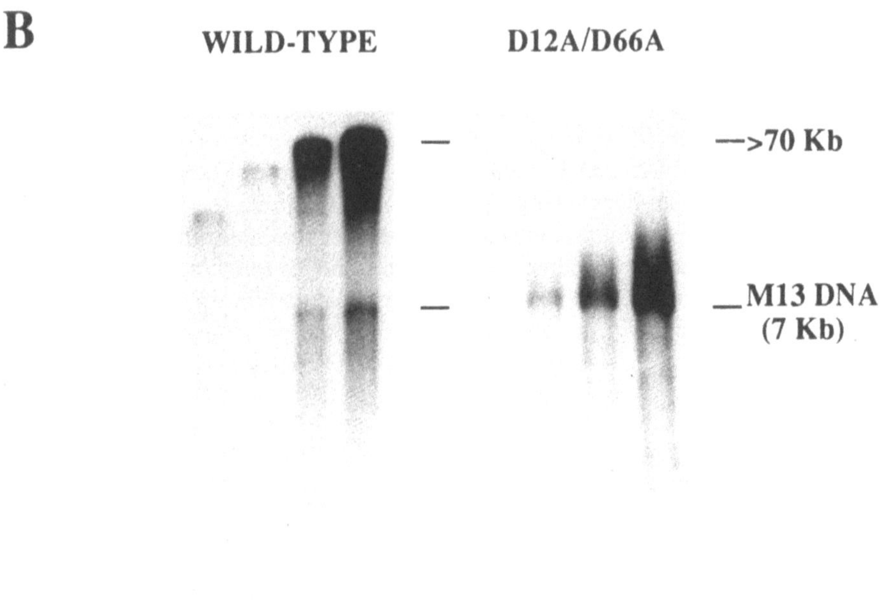
B
WILD-TYPE
D12A/D66A
—>70 Kb
_M13 DNA (7 Kb)
5 10 20 40
5 10 20 40
time, min

2 Results and Discussion

2.1. Overlapping Structural Domains for the 3'-5'Exonuclease and Strand-Displacement Activities

ø29 DNA polymerase has the intrinsic ability to carry out strand-displacement coupled to DNA polymerization (Blanco et al. 1989). Interestingly, all the ø29 DNA polymerase mutants in the Exo I, Exo II and Exo III motifs, in addition to inactivating the 3'-5'exonuclease activity (Bernad et al. 1989; Soengas et al. 1992), were strongly diminished in their ability to carry out strand-displacement, a property essentially required to replicate ø29 DNA. On the contrary, none of these mutants were affected in their protein-priming or DNA polymerization activities. Furthermore, none of the mutations in the C-terminal portion of ø29 DNA polymerase specifically affected strand-displacement synthesis (Bernad et al. 1990b; Blanco et al. 1991, Blasco et al. 1991). Figure 1 shows the strand-displacement synthesis (> 70 kb) carried out by the wild-type ø29 DNA polymerase on singly-primed M13 DNA, in comparison with an exonuclease-deficient mutant (D12A/D66A; Bernad et al. 1989). In the latter case, accumulation of synthesized DNA in the position corresponding to full length M13 DNA, was found. Taking into account these results, we propose that the strand-displacement activity of ø29 DNA polymerase resides in the N-terminal domain, probably overlapping with the 3'-5'exonuclease active site.

2.2. Protein-Primed Replication of Single-Stranded Oligonucleotides

Using an in vitro system that replicates ø29 DNA-TP, with ø29 DNA polymerase and terminal protein (TP) as the only proteins (Blanco and Salas 1985b), we asked whether the recognition of the replication origin by these proteins occurs before unwinding of the double helix or, on the contrary, a single-stranded DNA is exposed first and then the proteins bind to the template strand. As shown in Fig. 2, a 29mer oligonucleotide with the sequence corresponding to the template strand of the ø29 DNA right replication origin was an active template for both initiation (TP-dAMP) and complete replication [TP-$(dNMP)_{29}$]. Taking into account that an exonuclease-deficient ø29 DNA polymerase was used to prevent 3'-5'exonucleolytic degradation of the template oligonucleotide, the lower efficiency of elongation obtained when a double-stranded oligonucleotide, corresponding to the right replication origin, was used [Fig. 2; ø29 oriR (29mer)], is due to a defect in the initiation of strand-displace-

Fig. 1. Defective strand-displacement synthesis catalyzed by an exonuclease-deficient ø29 DNA polymerase. **A** Scheme of the strand-displacement assay on singly primed M13 DNA, carried out as described (Blanco et al. 1989). **B** Analysis by alkaline agarose gel electrophoresis and autoradiography of the products synthesized either by wild-type or exonuclease-deficient (D12A/D66A; Bernad et al. 1989) ø29 DNA polymerase, using the assay described in A

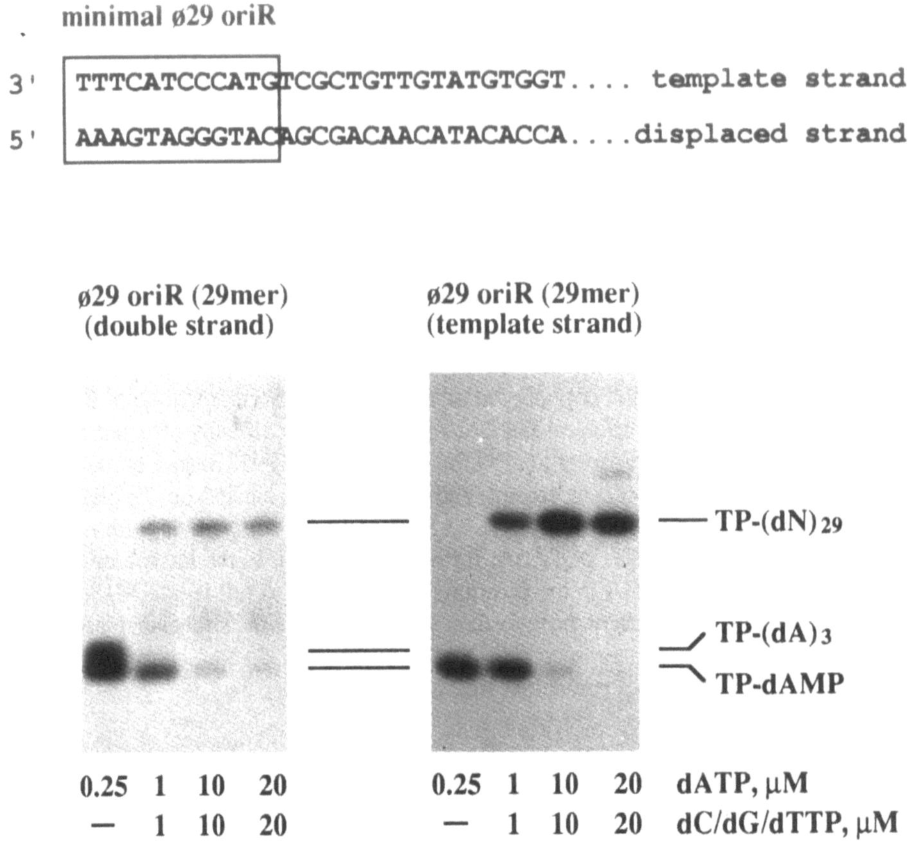

Fig. 2. TP-primed replication of single-stranded oligonucleotides. *Top* Sequences of both template and displaced strands corresponding to the 29 terminal nucleotides of the right ø29 DNA replication origin. The minimal sequences required as functional origin in double-stranded conformation are boxed (Gutiérrez et al. 1988). *Bottom* Analysis by SDS-polyacrylamide gel electrophoresis and autoradiography of the formation of the initiation complex (TP-dAMP) and fully replicated product [TP-$(dNMP)_{29}$], using either the template strand or the corresponding double-stranded form of the oligonucleotide indicated at the top of the figure, assayed as described (Méndez et al. 1992). The type and concentration of dNTP substrates used is indicated

ment synthesis (specific of exonuclease-deficient ø29 DNA polymerase mutants) that partially blocks elongation at position about TP-$(dNMP)_6$ (Soengas et al. 1992).

2.3. A Sliding-Back Mechanism for Protein-Primed DNA Replication

A mutational analysis of the ø29 DNA right replication origin was carried out using oligonucleotides with single changes introduced in the first, second and third positions from the 3'end (Méndez et al 1992). These oligonucleotides were assayed for

Table 1. Nucleotide selection and efficiency of the protein-primed initiation of mutated single-stranded ø29 DNA templates[a]

		Substrate usage (%)				
Template	3' Sequence	dATP	dCTP	dGTP	dTTP	Activity (%)
oriR(12)t	TTTCATCCCATG	100	1.3	4.3	9.7	100
T1A	ATTCATCCCATG	100	1.3	5.0	10	110
T1C	CTTCATCCCATG	100	2.0	3.0	6.3	96
T1G	GTTCATCCCATG	100	2.0	5.5	13	54
T2A	TATCATCCCATG	27	1.3	5.8	100	96
T2C	TCTCATCCCATG	6.7	0.8	100	4.1	140
T2G	TGTCATCCCATG	13	100	2.1	5.8	69
T3G	TTGCATCCCATG	100	17	5.8	6.5	73

[a] The initiation assay was carried out independently with each of the four (α-^{32}P)dNTPs, in the presence of 125 ng of TP, 125 ng of an exonuclease-deficient ø29 DNA polymerase, and 200 ng (51 pmol) of each template oligonucleotide, as described (Méndez et al. 1992). As natural template, a 12mer oligonucleotide with the sequence of the template strand of the ø29 DNA right origin [oriR(12)t] was used. Point substitutions in the oriR(12)t sequence are indicated using a 3-symbol code: the first letter is the nucleotide to be changed; the number indicates its position from the 3'end; the second letter is the nucleotide introduced (e.g. T1A is the oligonucleotide in which the first T has been changed to A). The amount of TP-dNMP formed is expressed as a percentage of dNTP substrate usage, relative to the dNTP giving the maximal initiation reaction (preferred substrate), taken as 100%. Activity reflects the TP-dNMP formation corresponding to the preferred substrate in each case, relative to that obtained with the natural sequence (oriR(12)t). 100% means the formation of 2.6 fmol of TP-dAMP.

TP-primed initiation with each of the four (α-^{32}P)dNTPs as substrate. The results obtained indicated that the initiation reaction with these single-stranded DNA templates is directed by the second nucleotide from the 3'end (see Table 1). This unexpected result was also obtained using as template double-stranded DNA molecules containing 77 bp of the ø29 left origin of DNA replication, with a T- > G transversion in the second nucleotide at the 3'end (not shown).

The presence of a 3'terminal sequence of at least two T's in the template was required to efficiently elongate the initiation complex (Méndez et al. 1992). Alteration of this repetition strongly diminished the replication capability. Furthermore, although initiation is directed by the second nucleotide of the template, all the nucleotides of the template, even the 3'terminal one, are replicated, as determined by truncated elongation assays (Méndez et al. 1992).

Therefore, we propose a sliding-back mechanism for the transition from initiation to elongation that could account for the necessity of a terminal repetition for efficient elongation and also for the recovery of the information corresponding to the 3'end of the template. As shown in Fig. 3A, during initiation, the second nucleotide of the template (T) directs the linkage of dAMP to the primer protein (TP). During transi-

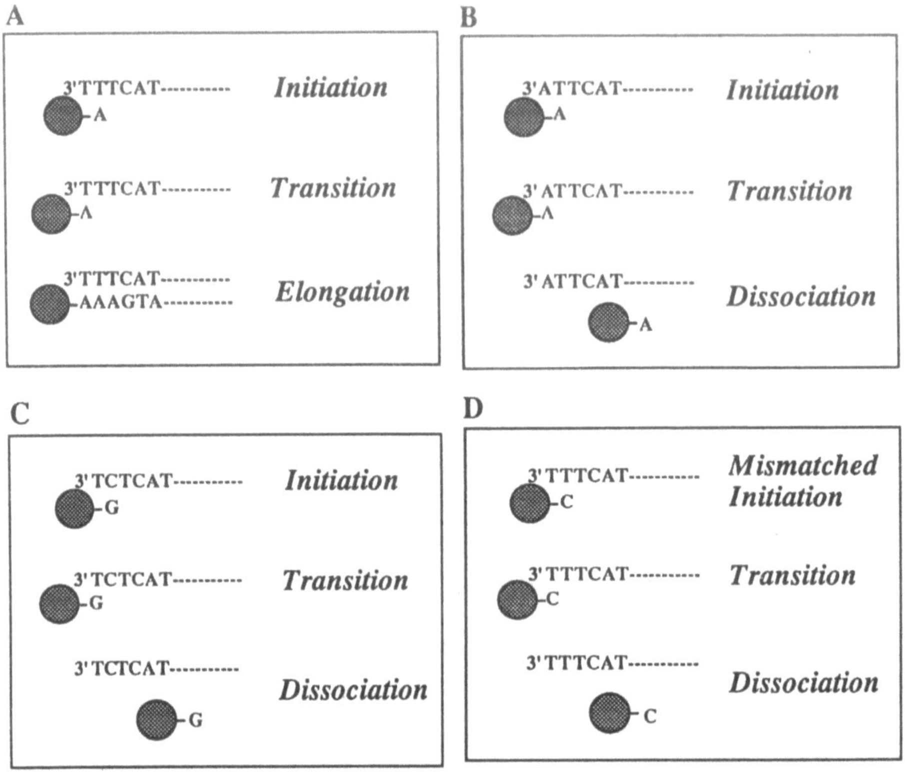

Fig. 3. Sliding-back mechanism for the transition from initiation to elongation in ø29 DNA replication. **A** Protein-primed initiation occurs opposite the second 3'-nucleotide of the template (initiation). The initiation complex slides back, setting the initiating dAMP base-paired to the 3' terminal nucleotide of the template (transition). The translocated initiation complex is elongated to fully replicate the template (elongation). **B, C** Normal initiation but aborted transition occurring with template oligonucleotides mutated in the first or the second 3'-nucleotide, respectively (Méndez et al. 1992). **D** Non-exonucleolytic proofreading: mismatched initiation on natural template would be discriminated at the transition step

tion, this TP-dAMP initiation complex slides backwards, locating dAMP in front of the first nucleotide of the template (T). After this transition step, that probably leads to the dissociation of the DNA polymerase/TP heterodimer, the following nucleotide (A) is incorporated, using again the second T of the template as a director. This mechanism could explain the low replication activity obtained with the oligonucleotides with changes in the first and second terminal T's (Méndez et al. 1992). In both cases, as schematized in Figs. 3B and 3C, respectively, initiation occurs with the dNTP complementary to the second base at the 3'end of the DNA; after the transition step, the first nucleotide incorporated is located in front of the first nucleotide of the template, producing an incorrect base-pairing; this mismatch would produce dissociation of initiation complexes, that would not be elongated. Furthermore, this sliding-back mechanism could contribute to increase the fidelity of the protein-primed initiation reaction. Thus, if a mismatched initiation occurs on the natural template sequence (see Fig. 3D), transition to elongation would not be efficient, and the incorrect initiation complex would dissociate from the DNA.

It is worth noting that genomes that contain terminal proteins have some sequence repetition at the DNA ends (reviewed in Salas 1991), suggesting that the sliding-back mechanism proposed to initiate linear ø29 DNA replication could be extrapolable to other systems that use proteins as primers.

Acknowledgements. This investigation has been aided by research grant 5R01 GM27242–12 from the National Institutes of Health, by grant nº PB90–0091 from Dirección General de Investigación Científica y Técnica, by grant BIOT CT 91–0268 from the European Economic Community, and by an institutional grant from Fundacíon Ramón Areces. J. M., J. A. E. and M. A. B. were predoctoral fellows from Ministerio de Educacíon y Ciencia, A. B. was a postdoctoral fellow from the Spanish Research Council, and M. S. S. was a predoctoral fellow from Universidad Autónoma de Madrid.

References

Bernad A, Blanco L & Salas M (1990a) Site-directed mutagenesis of the "YCDTDS" amino acid motif of the ø29 DNA polymerase. Gene 94:45–51

Bernad A, Lázaro JM, Salas M & Blanco L (1990b) The highly conserved amino acid sequence Tyr-Gly-Asp-Thr-Asp-Ser in α-like DNA polymerases is required by phage ø29 DNA polymerase for protein-primed initiation and polymerization. Proc Natl Acad Sci USA 87:4610–4614

Bernad A, Zaballos A, Salas M & Blanco L (1987) Structural and functional relationships between prokaryotic and eukaryotic DNA polymerases. EMBO J 6:4219–4225

Bernad A, Blanco L, Lázaro JM, Martin G & Salas M (1989) A conserved 3'-5'exonuclease active site in prokaryotic and eukaryotic DNA polymerases. Cell 59:219–228

Blanco L & Salas M (1985a) Characterization of a 3'-5'exonuclease activity in the phage ø29 DNA polymerase. Nucl Acids Res 13:1239–1249

Blanco L & Salas M (1985b) Replication of ø29 DNA with purified terminal protein and DNA polymerase: synthesis of full length DNA. Proc Natl Acad Sci USA 82:6404–6408

Blanco L, Bernad A, Lázaro JM, Martin G, Garmendia C & Salas M (1989) Highly efficient DNA synthesis by the phage ø29 DNA polymerase. Symmetrical mode of DNA replication. J Biol Chem 264:8935–8940

Blanco L, Bernad A, Blasco MA & Salas M (1991) A general structure for DNA-dependent DNA polymerases. Gene 100:27–38

Blasco MA, Bernad A, Blanco L & Salas M (1991) Characterization and mapping of the pyrophosphorolytic activity of the phage ø29 DNA polymerase. J Biol Chem 266:7904–7909

Blasco MA, Lázaro JM, Bernad A, Blanco L, Salas M (1992) DNA polymerase active site. Mutants in conserved residues Typ^{254} and Typ^{390} are affected in dNTP binding. J Biol Chem 267: in press

Garmendia C, Bernad A, Esteban JA, Blanco L & Salas M (1992) The bacteriophage ø29 DNA polymerase, a proofreading enzyme. J Biol Chem 267:2594–2599

Gutiérrez J, Garmendia C & Salas M (1988) Characterization of the origins of replication of bacteriophage ø29 DNA. Nucl Acids Res 16:5895–5914

Méndez J, Blanco L, Esteban JA, Bernad A & Salas M (1992) Initiation of ø29 DNA replication occurs at the second 3'nucleotide of the linear template: A sliding-back mechanism for protein-primed DNA replication. Proc Natl Aced Sci USA 89: in press

Salas M (1991) Protein-priming of DNA replication. Annu Rev Biochem 60:39–71

Soengas MS, Esteban JA, Lázaro JM, Bernad A, Blasco MA, Salas M & Blanco L (1992) Site-directed mutagenesis at the Exo III motif of ø29 DNA polymerase; overlapping structural domains for the 3'-5' exonuclease and strand-displacement activities. The EMBO J 11: in press

Initiation of Bacteriophage λ DNA Replication in Vitro in a System Reconstituted with Purified Proteins

R. McMacken[1]

1 Introduction

Extensive genetic and biochemical studies during the past three decades have established bacteriophage λ as a prototype for elucidation of the fundamental molecular mechanisms responsible for initiation and regulation of chromosomal DNA replication (see Keppel et al. 1988) for a recent review of λ DNA replication).

The 48.5 kilobase pair bacteriophage λ chromosome encodes only two proteins, the λ *O* and *P* gene products (Fig. 1) that are required for replication of the viral genome. However, λ, like most temperate coliphages, relies heavily on its *Escherichia coli* host to supply most of the proteins needed for this complex process. The required host proteins can be grouped into three categories. The first category consists of *E. coli* replication proteins that participate in propagation of a replication fork along the bacterial chromosome (i.e., DnaB helicase, primase, single-stranded DNA-binding protein, and DNA polymerase III holoenzyme). The second group, surprisingly, includes three of the approximately 20 known *E. coli* heat shock proteins (i.e., DnaJ, DnaK, and GrpE proteins). These three heat shock proteins are not absolutely required for replication of the host chromosome. Finally, λ DNA replication requires the function of both DNA gyrase, which produces negative superhelical template DNA and acts to relieve torsional strain created by replication fork movement, and RNA polymerase, which provides transcriptional activation of the template DNA (see Sect 2.6). Genetic analysis of λ DNA replication, reviewed by Furth and Wickner (1983), suggests that the primary function of the λ O and P replication proteins is to recruit host proteins and direct them to act on the λ genome rather than on the bacterial chromosome.

The linear vegetative λ chromosome is rapidly circularized and converted to a supercoiled form following its injection into a sensitive cell. λ replicates primarily in a circle-to-circle mode via theta-structure intermediates during the first 15 min. of the latent period (Tomizawa and Ogawa 1968). Thereafter, λ replication is predominantly accomplished via a rolling-circle replication process, but no new initiations take place during this mode of replication (Enquist and Skalka 1973). In vivo, replication of the phage chromosome proceeds either bidirectionally (70%) or unidirectionally right or left (30%) from *ori*λ (Inman and Schnos 1970). The bacteriophage λ replication origin (*ori*λ) is located in the center of the *O* gene, situated approximately 20% from the right end of the chromosome (Fig. 1). The genetically defined origin contains four 19-

[1] Department of Biochemistry, Johns Hopkins University School of Hygiene and Public Health, 615 North Wolfe Street, Baltimore, MD 21205, USA.

43. Colloquium Mosbach 1992
DNA Replication and the Cell Cycle

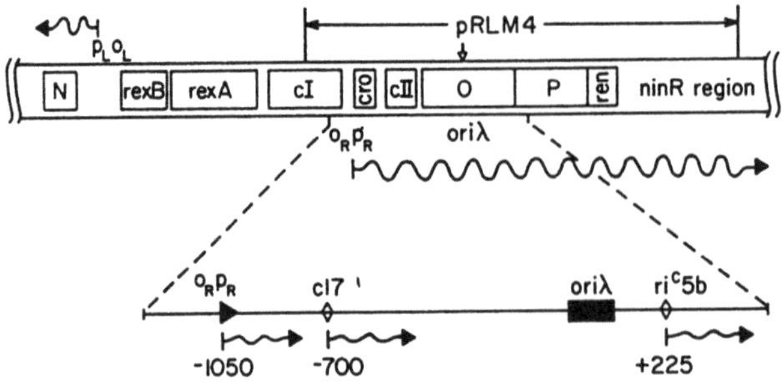

Fig. 1. Genetic and transcriptional map of the region surrounding the λ replication origin with locations of promoters that can serve to transcriptionally activate λ DNA replication. *Above* The genetic map of the λ chromosome between nucleotide positions 35 000 and 42 000 (Sanger et al. 1982) is depicted. The position of the λ replication origin (*ori*λ) is denoted with a *downward arrow*. The *wavy lines* indicate the regions transcribed from the primary promoters (p_R, p_L) used during vegetative growth of the virus. The portion of the λ chromosome contained in the *ori*λ plasmid (pRLM4), used as the template for in vitro replication, is indicated. *Below* An expanded map of the region around the λ replication origin. The positions of promoters in wild-type or mutant λ that can direct transcriptional activation of λ DNA replication are shown. The *wavy lines* depict the direction of transcription that is initiated at each promoter. The *numbers below each promoter* indicate the approximate number of base pairs from the starting point of transcription to the center of the four tandem λ O protein recognition sites located in *ori*λ

base pair repeating sequences (iterons) which have been shown to be recognition sites for the λ O initiator protein (Tsurimoto and Matsubara 1981), and an adjacent 40-base pair segment that is A+T-rich.

We and others have established crude soluble systems that support the specific replication of supercoiled plasmid templates (e.g., pRLM4; Fig. 1) that contain *ori*λ (Anderl and Klein 1982; Tsurimoto and Matsubara 1982; Wold et al. 1982). In these systems, λ O and P initiators were added to a soluble extract (Fuller et al. 1981) of *E. coli* cells that contains all of the proteins required for replication of the bacterial chromosome. Characterization of the properties of *ori*λ plasmid replication in this system indicated that initiation and propagation of λ DNA replication occurred in a completely physiological manner (Wold et al. 1982). Encouraged by these results, we set out to establish a defined system, composed entirely of purified proteins, that would support the physiological replication of *ori*λ plasmids. This chapter reviews the establishment of such a system and our initial characterization of the molecular events connected with the initiation and regulation of λ DNA replication in vitro.

2 Results and Discussion

2.1. Establishment and Characterization of a Nine-Protein System for λ Replication

It was apparent from studies of other multiprotein DNA replication systems that a thorough analysis of the biochemical mechanisms involved in λ DNA replication

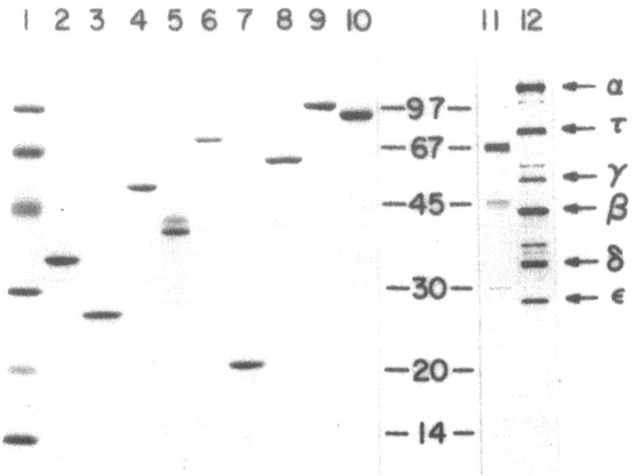

Fig. 2. Analysis of purified λ and *E. coli* replication proteins by SDS-PAGE. Samples of each protein (5–10 μg for *lanes 2–10* and 0.56 μg for *lane 12*) required for initiation of λ DNA replication in vitro were electrophoresed through polyacrylamide and visualized by staining with Coomassie Blue (*lanes 1–10*) or silver (*lanes 11 and 12*). The samples for *lanes 1–12* were: MW markers, λ O, λ P, DnaB, DnaJ, DnaK, SSB, DnaG primase, DNA gyrase A subunit, DNA gyrase B subunit, MW markers, and DNA polymerase III holoenzyme, respectively. Polypeptides that have been shown to be subunits of the DNA polymerase III holoenzyme are identified

Table 1. Requirements for the reconstitution of *ori*λ plasmid DNA replication with purified proteins

Component omitted or added	DNA synthesis (pmol)
None	311
– λ O protein	8
– λ P protein	13
– DnaB protein	6
– DnaJ protein	30
– DnaK protein	7
– SSB	10
– DnaG primase	19
– DNA polymerase III holoenzyme	18
– DNA gyrase[a]	50
– ATP	16

[a] Coumermycin was present at 50 μM.

The standard λ replication reaction was carried out as previously described (Mensa-Wilmot et al. 1989b), except that the individual components were omitted or added as indicated.

would require the reconstitution of a defined in vitro system composed of a set of highly purified proteins. Those λ and *E. coli* proteins identified by genetic and biochemical studies as being probable participants in λ DNA replication were purified to near homogeneity (Fig. 2; Mensa-Wilmot et al. 1989b). As indicated in Table 1, a combination of nine phage- and bacterial-encoded proteins was found to be sufficient to support replication of supercoiled *ori*λ plasmid DNA (Mensa-Wilmot et al. 1989b). Omission of any of these proteins from the system led to a striking reduction in the level of λ DNA synthesis. However, if the *ori*λ plasmid template DNA is amply supercoiled, DNA gyrase is not absolutely required for initiation of λ DNA replication. DNA synthesis is initiated on each supercoiled molecule in the absence of DNA gyrase, but ceases after approximately 100 positive supercoils accumulate in the unreplicated portion of the template (Alfano and McMacken 1988; Mensa-Wilmot et al. 1989b).

The 9-protein system is highly specific for supercoiled DNA templates bearing *ori*λ. Plasmid templates containing replication origins from related lambdoid phage 82, from *E. coli*, or from the replicative form of phage M13 were not replicated to any significant extent (Table 2). *Ori*λ plasmid template DNA underwent a single round of circle-to-circle replication in the reconstituted multiprotein system (Mensa-Wilmot et al. 1989b). Since the *E. coli* enzymes that process RNA primers and join Okazaki fragments (i.e., DNA polymerase I, RNase H, and DNA ligase) were omitted from the standard in vitro system, the daughter molecules contained one or more interruptions in the newly synthesized strand. Electron microscope (Fig. 3) and gel electrophoretic analysis of the pathway of *ori*λ plasmid DNA replication indicated that initiation occurred at or near *ori*λ and proceeded through theta structures (Fig. 3 A, B) to a late Cairns form (Fig. 3 C). The immediate product of *ori*λ plasmid replication was a multiply intertwined, catenated dimer (Fig. 3 D) in which each daughter molecule was catenated 20–25 times with its sister molecule (Mensa-Wilmot et al. 1989b). DNA gyrase slowly removed these linkages in the standard system.

We found that this 9-protein reconstituted system had three primary deficiencies when compared to the characteristics of λ DNA replication in vivo. First, replication in vitro is almost never bidirectional, being approximately 50% unidirectional leftward and 50% unidirectional rightward (Mensa-Wilmot et al. 1989b; Learn and McMacken, unpubl. data). Second, the GrpE heat shock protein, absolutely required for

Table 2. Template specificity of DNA replication in the reconstituted multiprotein system

Template DNA	Origin type	DNA synthesis (pmol)
None	---	0
pRLM4	λ	480
pRLM5	phage 82	2
M13*oriC*26	*E. coli*	0
M13mp8 RF	M13 RF	0

Reactions were carried out as previously described (Mensa-Wilmot et al. 1989b), except that the added template DNA (215 ng) was varied as indicated

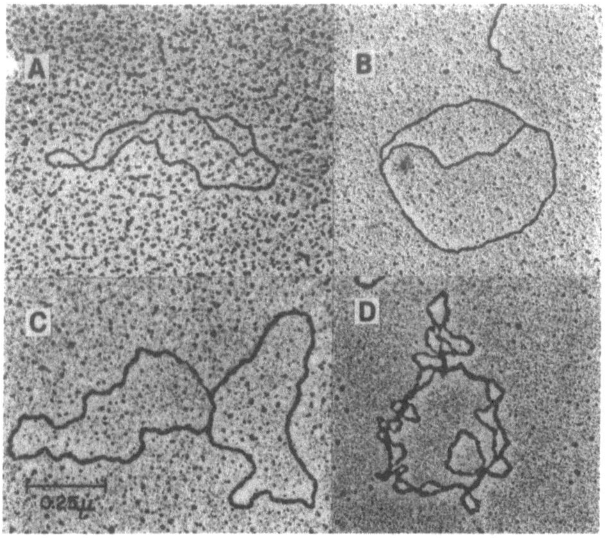

Fig. 3. Electron micrographs of *ori*λ plasmid replication intermediates formed in the reconstituted multiprotein replication system. **A** An early Θ-structure; **B** an intermediate Θ-structure; **C** a typical late Cairns intermediate; and **D** a multiply intertwined catenated dimer. (see text for details)

λ DNA replication in vivo, is not essential in vitro. Addition of purified GrpE to the 9-protein system, however, does stimulate *ori*λ DNA replication somewhat and permits the level of required DnaK to be reduced five to ten fold (Alfano and McMacken 1989a; Zylicz et al. 1989). Third, RNA polymerase transcription is unnecessary for initiation of *ori*λ plasmid DNA replication in the reconstituted system. Yet, one of the hallmarks of λ DNA replication in vivo is its absolute dependence on transcription (Dove et al. 1969; Dove et al. 1971). Restoration of the coupling between transcription and λ DNA replication in the purified protein system required addition of *E. coli* HU protein to the purified protein system (Mensa-Wilmot et al. 1989; see Sect. 2.6).

2.2. Ordered Assembly of Nucleoprotein Structures at Ori λ

We discovered that the λ O and P proteins, in conjunction with the *E. coli* DnaJ and DnaK heat shock proteins and ATP, were capable of mediating the sequence-independent transfer of a molecule of DnaB protein onto any SSB-coated single-stranded DNA to form a preprimosome. Recognition of the DnaB preprimosome by primase enables the synthesis of RNA primers, which can be extended by DNA polymerase III holoenzyme to synthesize a complementary DNA strand. We have termed this reaction the λ single-strand replication reaction (LeBowitz and McMacken 1984; LeBowitz et al. 1985). These studies suggested that several important steps in the initiation of DNA replication at duplex *ori*λ sequences occur prior to priming and DNA synthesis. We collaborated with Dodson and Echols to use electron microscopy to demonstrate that multiple distinguishable nucleoprotein structures are formed at *ori*λ prior to priming of DNA replication. In the first step, λ O protein binds to the four iterons at *ori*λ and self-associates to form a nucleosomelike structure, termed an O-some (Dodson et al. 1985). Determination of

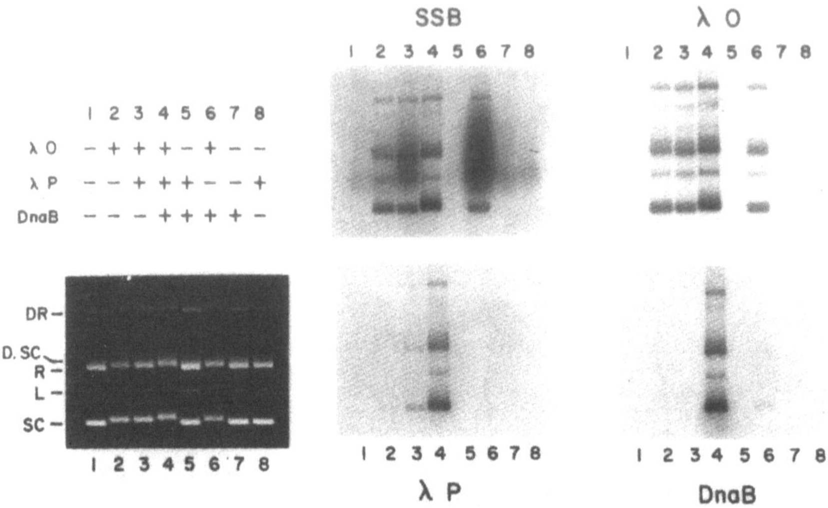

Fig. 4. Assembly of an *ori*λ: O-P-DnaB complex. All samples contained negatively supercoiled pRLM4 DNA and SSB. *O*, *P*, and *DnaB* replication proteins were added as indicated. Nucleoprotein complexes were formed, cross-linked with glutaraldehyde, isolated by gel filtration, resolved by electrophoresis in agarose, and characterized by immunoblotting as previously described (Alfano and McMacken 1989b). The *left panel* is a photograph of a gel that had been stained with ethidium bromide. All other panels are immunoblots that have been developed with antibody preparations that are specific for the indicated replication protein. The banding positions of the various topological forms of the template are indicated in the DNA panel. *SC* Negatively supercoiled monomeric DNA; *L* linear DNA; *R* nicked circular monomeric DNA; *D. SC* supercoiled dimeric DNA; *DR* nicked circular dimeric DNA

the stoichiometry of O protein binding to the phage replication origin indicates that the O-some contains four dimers of O (Um and McMacken, unpubl. data). Addition of both λ P protein and *E. coli* DnaB protein to the O-some results in the formation of a larger, more asymmetric, second-stage structure (Dodson et al. 1985). P binds avidly to DnaB in solution to form a P · DnaB complex (Wickner 1979; Mallory et al. 1990), which, in turn associates with the O-some. Both immunoblotting (Fig. 4) (Alfano and McMacken 1989b) and immunoelectron microscopy (Dodson et al. 1989) analysis of the second-stage nucleoprotein structure indicated that it contained all three proteins, O, P, and DnaB, required for its formation.

The DnaJ heat shock protein specifically recognizes the *ori*λ · O · P · DnaB nucleoprotein structure and binds to this prepriming intermediate to form a third-stage *ori*λ · O · P · DnaB · DnaJ complex (Fig. 5; Alfano and McMacken 1989b). DnaJ also binds to the O-some, but with substantially reduced affinity. DnaK protein, the hsp70 homologue of *E. coli*, binds to each of the first three prepriming intermediates (Fig. 5; Alfano and McMacken 1989b; Dodson et al. 1989). We suggest, however, that DnaK binds functionally only to the third-stage structure containing DnaJ, i.e., the functional role of DnaJ is to target the action of DnaK to a specific nucleoprotein substrate. The complete preinitiation intermediate (fourth-stage nucleoprotein structure)

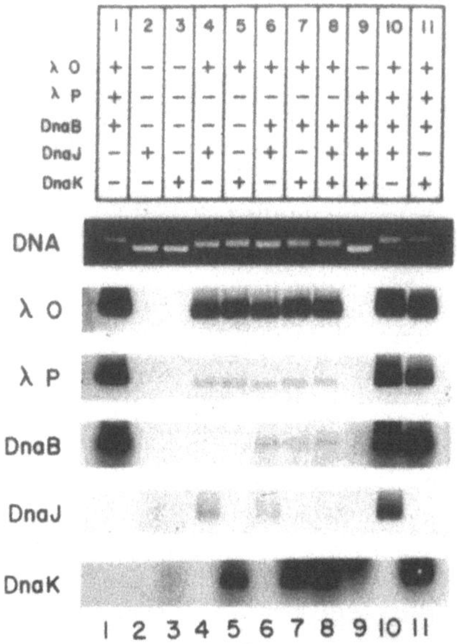

Fig. 5. Assembly of nucleoprotein structures that contain DnaJ or DnaK heat shock proteins. Formation of nucleoprotein complexes and their analysis by immunoblotting were carried out as described in the legend to Fig. 4. All samples contained negatively supercoiled pRLM4 DNA, SSB and ATP (4 mM). *O*, *P*, *DnaB*, *DnaJ*, and *DnaK* were added as indicated. For simplicity, only the regions of the gel or immunoblots containing negatively supercoiled monomer DNA are shown, since binding of specific proteins to this form of the plasmid template was representative of their binding to other topological forms. (Alfano and McMacken 1988)

contains O, P, DnaB, DnaJ, and DnaK, as well as some SSB (Alfano and McMacken 1988, 1989b).

In the absence of ATP, each of the four prepriming nucleoprotein structures is relatively stable. Each can be isolated by gel filtration and the first three can be shown to be physiologically active intermediates in the initiation of λ DNA replication in vitro (Alfano and McMacken 1989b; Zylicz et al. 1989). The complete fourth-stage preinitiation complex, however, following isolation by gel filtration, needs to be supplemented with additional DnaK to convert it into an active replication intermediate (Alfano and McMacken 1989b). Addition of ATP to the fourth-stage complete preinitiation complex results in a dramatic rearrangement of the nucleoprotein structure and is accompanied by a localized unwinding of the supercoiled *ori*λ template (Dodson et al. 1986). This unwinding, first observed by electron microscopy, produces a fifth prepriming nucleoprotein structure, termed the unwound complex, in which 600–1300 base pairs of duplex DNA rightward from *ori*λ is unwound and coated with SSB (Dodson et al. 1986). The unwound complex can also be readily observed by immunoblotting (Fig. 6, lanes 5 and 6; Alfano and McMacken 1988, 1989b), when the six prepriming proteins (i.e., O, P, DnaB, DnaJ, DnaK, and SSB) are incubated with supercoiled *ori*λ plasmid DNA in the presence of a hydrolyzable ribonucleoside triphosphate. We infer that the unwinding is mediated by DnaB protein, since this key *E. coli* replication protein has intrinsic DNA helicase activity (LeBowitz and McMacken 1986). In the absence of DNA gyrase, positive superhelical strain accumulates in the duplex portion of the template during DNA unwinding. The available evidence suggests that this torsional strain blocks complete unwinding of the template by DnaB. The unwound complex contains DnaB (Alfano and McMacken 1988, 1989b), which in turn serves as the recognition signal for primase

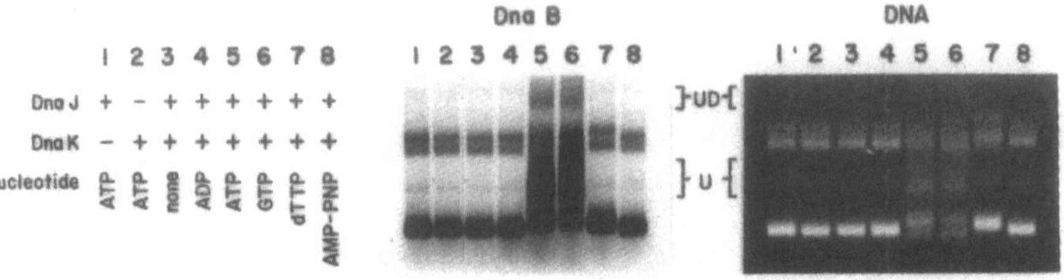

Fig. 6. Nucleotide and protein requirements for the assembly of prepriming and unwound complexes at *oriλ*. All samples contained negatively supercoiled pRLM4 DNA and proteins SSB, O, P and DnaB. *DnaJ* and *DnaK* proteins and *nucleotides* were added as indicated. Each nucleotide was present at a final concentration of 4 mM. The single immunoblot presented was incubated with antibody specific for DnaB. The *panel labeled DNA* is a photograph of a gel stained with ethidium bromide. The *bracket labeled U* indicates the position of the unwound complex formed on monomeric supercoiled plasmid DNA. The *bracket labeled UD* indicates the position of a similar complex formed on a supercoiled *oriλ* plasmid dimer

to synthesize RNA primers for DNA chain elongation (LeBowitz et al. 1985; Alfano and McMacken 1988, 1989b; Mensa-Wilmot et al. 1989b).

2.3 DnaJ- and DnaK-Mediated Disassembly of Prepriming Nucleoprotein Structures

Formation of a P · DnaB complex in solution results in a strong inhibition of each of the intrinsic enzymatic activities of DnaB (Mallory et al. 1990), including its DNA helicase activity (Stephens and McMacken, unpubl. data). Thus, we inferred that the P · DnaB interaction needs to be disrupted by the cooperative action of the DnaJ and DnaK heat shock proteins to permit initiation of unwinding of the *oriλ* plasmid template. We made use of the discovery that all prepriming nucleoprotein structures except the unwound complex could be formed on linearized *oriλ* templates (Alfano and McMacken 1988). Formation of the unwound complex is prevented, because initiation of unwinding by DnaB at *oriλ* requires a negatively supercoiled template. Thus, with a linear *oriλ* DNA template we were able to distinguish protein rearrangements caused by DNA unwinding from those caused by protein turnover mediated by the action of the DnaJ and DnaK heat shock proteins (Alfano and McMacken 1989a). In the absence of ATP, all six prepriming proteins assemble into the complete preinitiation complex, as assayed by immunoblotting (Fig. 7, lane 1). Addition of ATP to the incubation mixture results in the partial disassembly of this fourth-stage prepriming nucleoprotein structure (Fig. 7, lane 2). Nearly all of the bound P, DnaB and DnaJ proteins are released from the template by the action of the *E. coli* heat shock proteins, and smaller amounts of O, DnaK, and SSB are released as well (Alfano and McMacken 1989a; Dodson et al. 1989). In the absence of GrpE protein three fold higher levels of DnaK protein were required to achieve maximal protein turnover (Fig. 7, lane 3). We conclude that the fourth-stage preinitiation complex is partially disassembled via an ATP-dependent reaction mediated by the DnaJ and DnaK heat

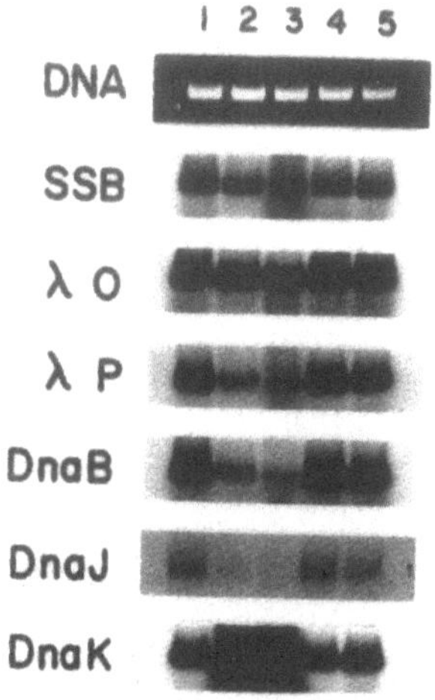

Fig. 7. Effect of ATP and DnaK concentration on the disassembly of nucleoprotein structures formed at *oriλ* on linear templates. Reaction mixtures containing *O, P, DnaB, DnaJ, DnaK,* and *SSB* were assembled as described for immunoblotting of nucleoprotein complexes (Alfano and McMacken 1989b) except for the following changes. All mixtures contained pRLM4 DNA that had been linearized by digestion with *Bam*HI. The reaction mixture used for the sample applied to *lane 3* contained a three fold increase in the standard level of DnaK protein. The nucleotide cofactor present in the reaction mixtures was varied: *lane 1* none; *lanes 2, 3* ATP (4 mM); *lane 4* AMP (4 mM); *lane 5* ATPγS (1 mM). Nucleoprotein complexes were formed, isolated, and analyzed by immunoblotting as previously described (Alfano and McMacken 1989b). The *panel labeled DNA* is a photograph of a replicate agarose gel that had been stained with ethidium bromide. All other panels are nitrocellulose immunoblots of a replicate agarose gel. The target antigen that was stained in each immunoblot is indicated at the *left*

shock proteins. We infer that this disassembly reaction, when carried out with a supercoiled *oriλ* plasmid, permits DnaB to initiate unwinding of the duplex template.

2.4 Scheme for the Pathway of Initiation of λ DNA Replication

We present in Fig. 8 a hypothetical scheme for the prepriming pathway portion of the initiation of λ DNA replication. The O-some (Fig. 8 a) is produced by the binding of a single dimer of λ O to each of the four iterons (black boxes) present in *oriλ* and the subsequent self-assembly of the O dimers. If the template is negatively supercoiled, formation of the O-some may cause the A + T-rich region to unwind (Schnos et al. 1988). However, based on the sensitivity of this DNA segment to various chemical probes (Sampath, Learn and McMacken, unpubl. data), it is apparent that the A + T-rich segment of *oriλ* does not exist in a stable single-stranded configuration in the O-some. Rather, it is present in an undefined non-B-DNA configuration in which only a limited number of the nucleotide residues can be modified by single-stranded DNA probes.

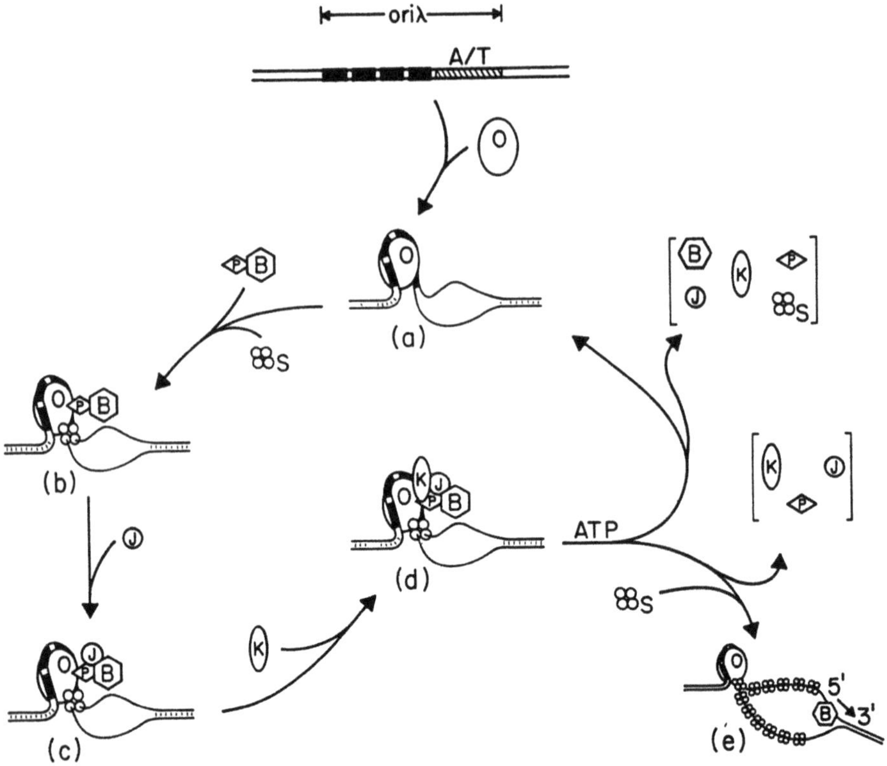

Fig. 8. Model for initiation of bacteriophage λ DNA replication (see text for details). Proteins are symbolized by single shapes and letters for simplicity. *O* λ O protein; *P* λ P protein; *S* SSB; *B* DnaB; *J* DnaJ; *K* DnaK. The *filled boxes* represent the four iterons (O protein recognition sequences) contained within *ori*λ. As discussed in the text, DnaK is capable of adding to structures **a, b** and **c**. For simplicity, we have depicted DnaK as adding to the *ori*λ: O-P-DnaB-DnaJ structure, which is the largest prepriming structure that we have isolated in a functional form

The primary functional role of the λ P protein is to recruit the host DnaB helicase to *ori*λ. P binds tightly to DnaB in solution to form a P · DnaB complex, which subsequently associates with the O-some to form the second-stage *ori*λ · O · P · DnaB nucleoprotein prepriming complex (Fig. 8 b). It is interesting that neither P alone nor DnaB alone binds efficiently to the O-some (Alfano and McMacken 1989b). In the next step, DnaJ binds to the second-stage prepriming complex to form an *ori*λ · O · P · DnaB · DnaJ nucleoprotein structure (Fig. 8 c). This is followed by the binding of DnaK to yield the complete preinitiation complex (Fig. 8 d). DnaB is believed to remain tightly tethered to λ P protein in this structure. SSB is not required for the assembly of any of the first four prepriming complexes, but does apparently interact directly with bound O protein in these nucleoprotein structures (Alfano and McMacken 1988).

Partial disassembly of the fourth-stage complex ensues when ATP is added to the reaction mixture. On approximately one-half of the template molecules, a DnaB helicase molecule present in the preinitiation complex successfully initiates unwinding of the template and binding of SSB to the displaced single strands stabilizes the nascent unwound complex (Fig. 8 e). We presume that P, DnaJ, and much of the bound DnaK protein need to be released from the preinitiation complex to enable DnaB to regain its helicase activity. Priming and DNA synthesis proceed spontaneously on the unwound complex when the in vitro reaction contains primase, DNA polymerase III holoenzyme, rNTPs, and dNTPs. On the remaining 50% of the template molecules, the heat shock protein-mediated disassembly reaction releases DnaB into solution together with other proteins from the preinitiation complex. It is not certain if the O-some is also disassembled during this reaction, but O is found bound to *ori*λ following the disassembly reaction (Alfano and McMacken 1989a; Dodson et al. 1989). Such unreplicated template molecules can be efficiently recycled through the prepriming initiation pathway (Alfano and McMacken 1989a, b).

2.5 Functional Studies of the E. coli DnaJ, DnaK, and GrpE Heat Shock Proteins

More recently, we have been examining how the DnaK hsp70 protein interacts with substrate proteins and have also been exploring how the DnaJ and GrpE heat shock proteins modulate the activity of DnaK (Jordan, Mehl and McMacken, unpubl. data) in order to better understand how these proteins facilitate λ DNA replication. These preliminary results are briefly summarized below. Based on the finding of Flynn et al. (1989) that proteins evolutionarily related to DnaK bind small peptides, we have searched and found a similar peptide-binding activity associated with DnaK. The binding of peptides to DnaK produces a ten fold stimulation in the weak intrinsic ATPase activity of this bacterial heat shock protein. The peptide-mediated stimulation of the ATPase activity of DnaK is length-dependent. Peptides shorter than five residues do not stimulate DnaK's ATPase, whereas peptides longer than eight or nine residues stimulate maximally. Equilibrium binding experiments indicate that a DnaK monomer contains a single binding site for each peptide and ATP. Analysis of the peptide-DnaK interaction by NMR showed that the peptide binds to DnaK in an extended conformation, with the primary interaction occurring with the peptide backbone and not its amino acid side chains (Landry et al. 1992).

We have found that peptides which interact tightly with DnaK serve as potent inhibitors of the reconstituted 9-protein λ DNA replication system. This suggests that the peptide-binding site on DnaK is used by this heat shock protein during its function in the initiation of λ DNA replication. We discovered that DnaJ protein also stimulates the ATPase activity of DnaK, apparently by binding to the same site on DnaK as does peptide. DnaJ, however, acts at concentrations 100-fold lower than those required for peptide-mediated enhancement of DnaK's ATPase. We infer that DnaJ contains an amino acid sequence that functions as a particularly potent activator of DnaK ATPase function. Thus, DnaJ present in the third-stage preinitiation complex may serve to target DnaK action to this particular nucleoprotein structure. In related studies, we have found that GrpE binds as a dimer to DnaK and increases the on and off rates for ATP and ADP binding to DnaK. The precise molecular role of ATP hydrolysis by DnaK in the protein disassembly process at *ori*λ is uncertain at present. It is known, however, that just the binding of ATP to DnaK promotes the rapid release of bound peptide (Jordan and McMacken, unpubl. data).

2.6 Transcriptional Activation of λ DNA Replication in Vitro

Studies of λ DNA replication in vivo demonstrated that the λ cI repressor could directly block λ DNA replication even when all required phage- and host-encoded replication proteins were present in the cell (Thomas and Bertani 1964). Dove and colleagues demonstrated that this effect of repressor resulted from an inhibition of transcription from the phage p_r promoter (Fig. 1) and that transcription directed across or near *ori*λ was required for the initiation of λ DNA replication (Dove et al. 1969, 1971). Later, it was demonstrated that transcription which emanated from the *ri*[c]5b promoter (Fig. 1), located more than 100 base pairs downstream from *ori*λ, and which was directed away from the phage origin, could still provide the requisite transcriptional activation (Furth et al. 1982). Although physiological in most respects, the reconstituted multiprotein λ DNA replication system (Mensa-Wilmot et al. 1989b) did not provide the transcriptional control of λ DNA replication that had been observed in vivo (Dove et al. 1971) or in the crude soluble in vitro system (Wold et al. 1982). We therefore searched for a factor or factors in crude extracts of *E. coli* that were capable of restoring transcriptional control of λ DNA replication to the purified-protein system. Mensa-Wilmot et al. (1989a) purified a small, basic protein which acted as a strong inhibitor of *ori*λ plasmid DNA replication when it was added to the 9-protein replication system (Fig. 9). This protein was identified as *E. coli* HU protein, a major component of the bacterial nucleoid (Mensa-Wilmot et al. 1989a). Addition to the λ replication system of sufficient HU protein (150 ng) to coat two-thirds of the *ori*λ plasmid template caused a greater than 90% inhibition of λ DNA replication (Fig. 9).

The HU-mediated inhibition could be largely counteracted if *E. coli* RNA polymerase and the four rNTPs were also present in the reaction mixture (Fig. 10). Addition of rifampicin or omission of either RNA polymerase or rNTPs prevented transcriptional activation of λ DNA replication in the reconstituted system containing inhibitory amounts of HU (Mensa-Wilmot et al. 1989a). RNA polymerase, itself, also

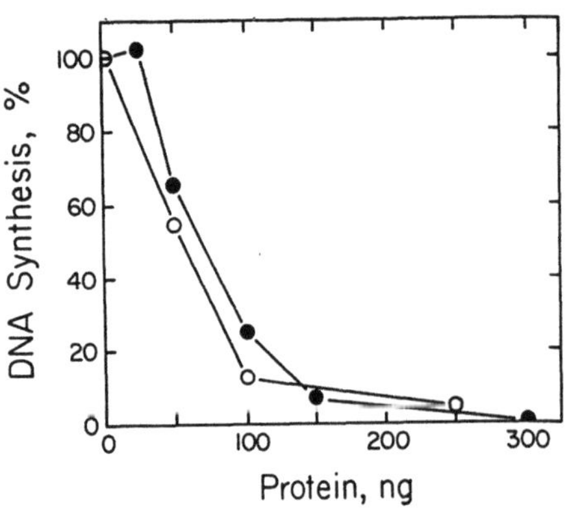

Fig. 9. HU protein inhibits *ori*λ plasmid DNA replication in vitro. Standard *ori*λ plasmid replication mixtures (Mensa-Wilmot et al. 1989b) were assembled on ice and supplemented with the indicated amounts of HU protein (*filled circles*) or purified *ori*λ replication inhibitor (*open circles*). DNA synthesis was measured after a 30-min incubation at 30 °C. 100% DNA synthesis represents incorporation of 350 pmol of labeled deoxynucleotide into acid-insoluble material

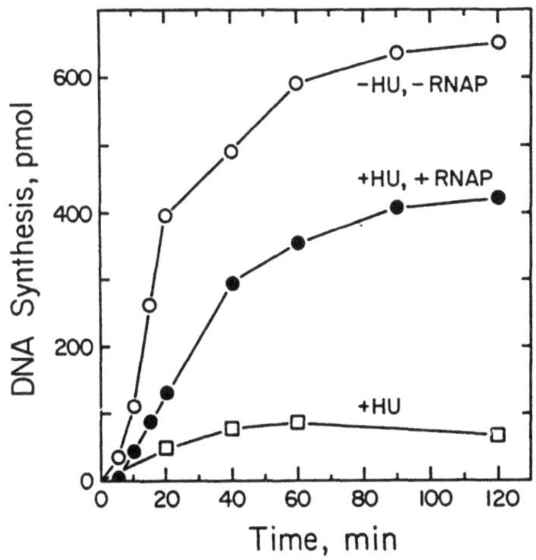

Fig. 10. Time course of *oriλ* DNA replication in the presence and absence of HU and RNA polymerase. Standard *oriλ* replication reaction mixtures were assembled at 0 °C and supplemented with rNTPs and, where indicated, with HU protein (100 ng). After a 10-min incubation on ice, RNA polymerase (*RNAP*; 1.6 μg) was added, as indicated (*filled circles*), and all reaction mixtures were incubated at 30 °C. DNA synthesis was measured at the indicated times after the start of incubation at 30 °C

inhibits λ DNA replication in vitro, especially when rNTPs are not present (Mensa-Wilmot et al. 1989a). This inhibitory effect of RNA polymerase binding may account for the incomplete recovery of λ DNA replication following transcriptional activation (Fig. 10). Our current efforts in this area are aimed at understanding the molecular basis for HU-mediated inhibition of λ DNA replication. We had previously demonstrated that initiation of λ DNA replication requires a negatively supercoiled template (Wold et al. 1982) and that the superhelical tension must be maintained until the DnaB helicase had been inserted between the two strands of the *oriλ* template (Alfano and McMacken 1988). We have prepared *oriλ* plasmid templates with defined levels of superhelicity and demonstrated that the template superhelical density must be more negative than –0.045 to obtain initiation of λ DNA replication. Our preliminary results suggest that HU inhibits λ DNA replication by removing superhelical tension from the *oriλ* template as a consequence of its capacity to bind and wrap the template in left-handed toroidal supercoils. For example, we find that the degree of inhibition of λ DNA replication obtained at a defined HU concentration is inversely related to the negative superhelical density of the *oriλ* template. In related studies, we have found that HU binds preferentially to negatively supercoiled DNA and that HU exists in a different quarternary structure on supercoiled DNA than it does on relaxed DNA (Huang and McMacken, unpubl. data).

Estimates of the average superhelical density of DNA in *E. coli* range from –0.025 to –0.035 (Bliska and Cozzarelli 1987). Thus, there apparently is not sufficient superhelical tension at equilibrium to permit initiation of λ DNA replication in vivo. Recently, it has been found that transcription can generate transient supercoils (Liu and Wang 1987; Tsao et al. 1989). It is possible, therefore, that such transient supercoils provide the additional superhelical tension needed for initiation of λ DNA replication and that this consequence of transcription is the fundamental reason that transcriptional activation is absolutely required for replication of the phage chromosome in vivo. We are currently seeking direct evidence for this possibility.

References

Alfano C & McMacken R (1988) The role of template superhelicity in the initiation of bacteriophage λ DNA replication. Nucl Acids Res 16:9611–9630
Alfano C & McMacken R (1989a) Heat shock protein-mediated disassembly of nucleoprotein structures is required for the initiation of bacteriophage λ DNA replication. J Biol Chem 264:10709–10718
Alfano C & McMacken R (1989b) Ordered assembly of nucleoprotein structures at the bacteriophage λ replication origin during the initiation of DNA replication. J Biol Chem 264:10699–10708
Anderl A & Klein A (1982) Replication of λ dv DNA in vitro. Nucl Acids Res 10:1733–1740
Bliska JB & Cozzarelli NR (1987) Use of site-specific recombination as a probe of DNA structure and metabolism in vivo. J Mol Biol 194:205–218
Dodson M, Echols H, Wickner S, Alfano C, Mensa-Wilmot K, Gomes B, LeBowitz J, Roberts JD & McMacken R (1986) Specialized nucleoprotein structures at the origin of replication of bacteriophage λ : localized unwinding of duplex DNA by a six-protein reaction. Proc Natl Acad Sci USA 83:7638–7642
Dodson M, McMacken R & Echols H (1989) Specialized nucleoprotein structures at the origin of replication of bacteriophage λ: protein association and disassociation reactions responsible for localized initiation of replication. J Biol Chem 264:10719–10725
Dodson M, Roberts J, McMacken R & Echols H (1985) Specialized nucleoprotein structures at the origin of replication of bacteriophage λ: complexes with λ O protein and with λ O, λ P, and *Escherichia coli* DnaB proteins. Proc Natl Acad Sci USA 82:4678–4682
Dove WF, Hargrove E, Ohashi M, Haugli F & Guha A (1969) Replicator activation in λ Jpn J Genet (Suppl) 44:11–22
Dove WF, Inokuchi H & Stevens WF (1971) Replication control in phage λ. In: Hershey AD (ed) The bacteriophage lambda. Cold Spring Harbor Laboratory, Cold Spring Harbor, New York, pp 747–771
Enquist LW & Skalka A (1973) Replication of bacteriophage λ DNA dependent on the function of host and viral genes. I. Interaction of red, gam and rec. J Mol Biol 75:185–212
Flynn GC, Chappell TG & Rothman JE (1989) Peptide binding and release by proteins implicated as catalysts of protein assembly. Science 245:385–390
Fuller RS, Kaguni JM & Kornberg A (1981) Enzymatic replication of the origin of the *Escherichia coli* chromosome. Proc Natl Acad Sci USA 78:7370–7374
Furth ME, Dove WF & Meyer BJ (1982) Specificity determinants for bacteriophage λ DNA replication. III. Activation of replication in λ*ri*[c] mutants by transcription outside of *ori*. J Mol Biol 154:65–83
Inman RB & Schnos M (1970) Position of branch points in replicating λ DNA. J Mol Biol 51:61–73
Landry SJ, Jordan R, McMacken R & Gierasch LM (1992) Different conformations for the same polypeptide bound to chaperones DnaK and GroEL. Nature 355:455–457
LeBowitz JH & McMacken R (1984) The bacteriophage λ O and P protein initiators promote the replication of single-stranded DNA. Nucl Acids Res 12:3069–3088
LeBowitz JH & McMacken R (1986) The *Escherichia coli* dnaB replication protein is a DNA helicase. J Biol Chem 261:4738–4748
LeBowitz JH, Zylicz M, Georgopoulos C & McMacken R (1985) Initiation of DNA replication on single-stranded DNA templates catalyzed by purified replication proteins of bacteriophage λ and *Escherichia coli*. Proc Natl Acad Sci USA 82:3988–3992
Liu LF & Wang JC (1987) Supercoiling of the DNA template during transcription. Proc Natl Acad Sci USA 84:7024–7027
Mallory JB, Alfano C & McMacken R (1990) Host-virus interactions in the initiation of bacteriophage λ DNA replication. Recruitment of *Escherichia coli* DnaB helicase by λ P replication protein. J Biol Chem 265:13297–13307

Mensa-Wilmot K, Carroll K & McMacken R (1989a) Transcriptional activation of bacteriophage λ DNA replication in vitro. Regulatory role of histonelike protein HU of *Escherichia coli*. EMBO J 8:2393–2402

Mensa-Wilmot K, Seaby R, Alfano C, Wold MS, Gomes B & McMacken R (1989b) Reconstitution of a nine-protein system that initiates bacteriophage λ DNA replication. J Biol Chem 264:2853–2861

Sanger F, Coulson AR, Hong GF, Hill DF & Petersen GB (1982) Nucleotide sequence of bacteriophage λ DNA. J Mol Biol 162:729–773

Schnos M, Zahn K, Inman RB & Blattner FR (1988) Initiation protein induced helix destabilization at the λ origin: a prepriming step in DNA replication. Cell 52:385–395

Thomas R & Bertani L (1964) On the control of the replication of temperate bacteriophages superinfecting immune hosts. Virology 24:241–253

Tomizawa J & Ogawa T (1968) Replication of phage λ DNA. Cold Spring Harbor Symp Quant Biol 33:533–551

Tsao Y-P, Wu H-Y & Liu LF (1989) Transcription-driven supercoiling of DNA: direct biochemical evidence from in vitro studies. Cell 56:111–118

Tsurimoto T & Matsubara K (1981) Purified bacteriophage λ O protein binds to four repeating sequences at the λ replication origin. Nucl Acids Res 9:1789–1799

Tsurimoto T & Matsubara K (1982) Replication of λ dv plasmid in vitro promoted by purified λ O and P proteins. Proc Natl Acad Sci USA 79:7639–7643

Wickner SH (1979) DNA replication proteins of *Escherichia coli* and phage λ. Cold Spring Harbor Symp Quant Biol 431:303–310

Wold MS, Mallory JB, Roberts JD, LeBowitz JH & McMacken R (1982) Initiation of bacteriophage λ DNA replication in vitro with purified λ replication proteins. Proc Natl Acad Sci USA 79:6176–6180

Zylicz M, Ang D, Liberek K & Georgopoulos C (1989) Initiation of λ DNA replication with purified host- and bacteriophage-encoded proteins: the role of the dnaK, dnaJ and grpE heat shock proteins. EMBO J 8:1601–1608

Mechanisms Controlling Hepadnaviral Nucleocapsid Assembly and Replication

H. Schaller[1], Ch. Kuhn[1] and R. Bartenschlager[1,2]

1 Introduction

Hepadnaviruses, with the human hepatitis B virus (HBV) as type member, comprise a family of small enveloped animal viruses which are characterized by containing a circular DNA genome of about 3 kb which is not covalently closed and only partially double-stranded, and to which a protein is covalently attached at the 5'-end of the coding strand. This unconventional genome structure reflects an equally unconventional mode of genome replication involving protein-primed reverse transcription of an RNA pregenome. In contrast to the basically related, much better understood retroreplication of retroviruses, hepadnaviral reverse transcription takes place already before release of the mature virus particle from the host cell. Consequently, hepadnaviruses carry a DNA genome in the extracellular state (Fig. 1) and are therefore differentiated as pararetroviruses from the RNA containing classical retroviruses.

As confirmed for retroelements from a variety of organisms, eukaryotic retroreplication is confined to cytoplasmic core particles whose formation requires selective and ordered interactions between its individual components: the structural protein (the *gag* gene product, in HBV the C-gene product), the genomic RNA, and products of the *pol* gene (in HBV P-gene) possessing the enzymatic activities required for reverse transcription. This chapter will summarize recent data on structure and function of the hepadnaviral P-protein, and in particular discuss its central role in the early steps in the assembly of replication-competent core particles.

2 Results and Discussion

2.1 Hepadnaviral Genome Organization

With the smallest DNA genome known for animal viruses, hepadnaviruses have developed highly specialized mechanisms for optimal use of their restricted genome space (Fig. 2). These include extensively overlapping reading frames and overlapping transcription units with heterogeneous transcription initiation sites allowing the production of two core gene products (particulate, cytoplasmic core protein and secreted

[1] ZMBH, University of Heidelberg, Im Neuenheimer Feld 282, D-6900 Heidelberg, FRG.
[2] Present address: F. Hoffmann-La Roche Ltd., Pharmaceutical Research-New Technologies, CH-4002 Basel, Switzerland.

43. Colloquium Mosbach 1992
DNA Replication and the Cell Cycle

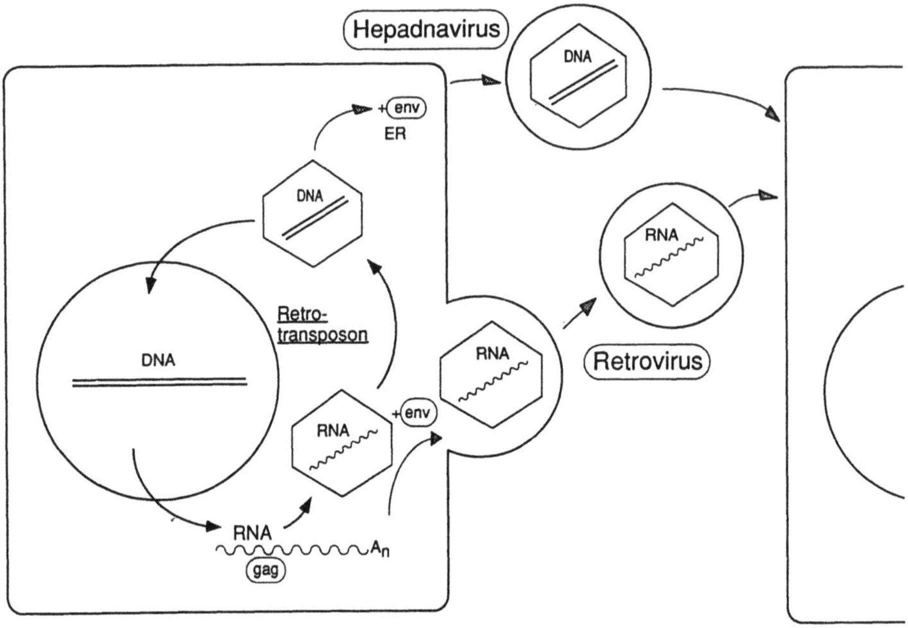

Fig. 1. Basic replication cycle of eukaryotic retroelements. From a nuclear DNA copy an RNA (pre)genome is transcribed and exported into the cytoplasm. After translation into *gag* and *pol* gene products the genomic RNA assembles with these proteins to form core particles capable of reverse transcription. Core particles are then either directly reimported into the nucleus, as in the case of retrotransposons, or exported as enveloped virus particles capable of secondary infection

HBeAg) and three variant preS/S proteins. As an exception, the P-gene product is not translated from an mRNA of its own, but instead by internal initiation from the genomic RNA. Internal translation initiation also appears to give rise to variants of the X-gene product (R. Schneider, pers. comm.).

2.2 P-Protein Domain Structure and Replication Functions

From its size and from sequence similarities to the retroviral *pol* gene product (Toh et al. 1983), the P-gene has been the obvious candidate for encoding the virion-associated DNA polymerase/reverse transcriptase activity detected quite early in HBV research (Kaplan et al. 1973). However, direct biochemical studies of the hepadnaviral enzyme were not possible because of its low abundance in both the HBV-infected cell or the virus particle, and furthermore by the failure to solubilize the enzyme from virions (Radziwill et al. 1988). Thus, analysis of P-protein structure and function relied exclusively on indirect detection methods, in particular on the endogenous polymerase reaction which utilizes the ability of the particle-associated P-protein to incorporate radioactively labelled deoxynucleotides into the incomplete viral genome with which it is associated. Using this assay for a mutational analysis of transiently ex-

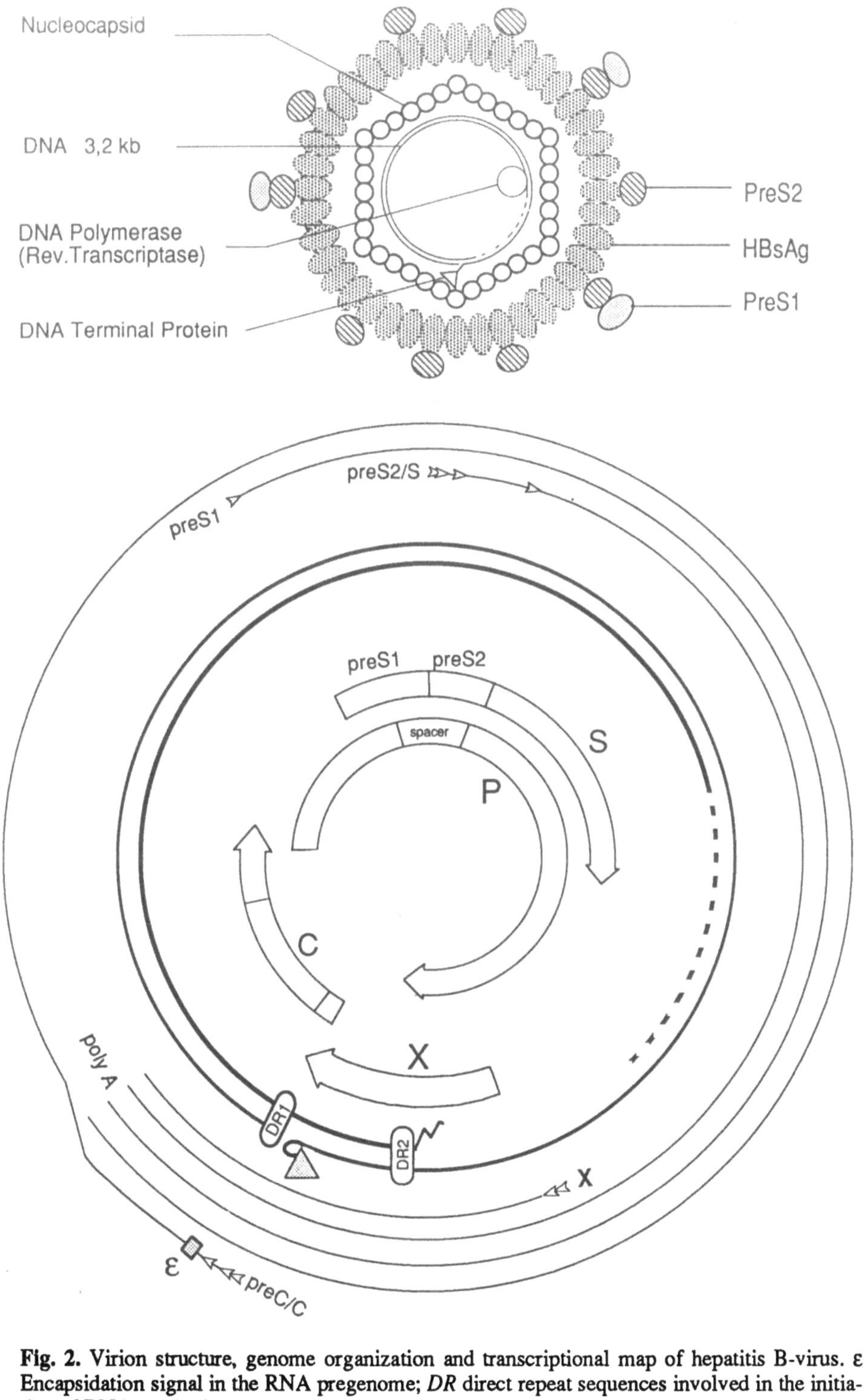

Fig. 2. Virion structure, genome organization and transcriptional map of hepatitis B-virus. ε Encapsidation signal in the RNA pregenome; *DR* direct repeat sequences involved in the initiation of DNA synthesis

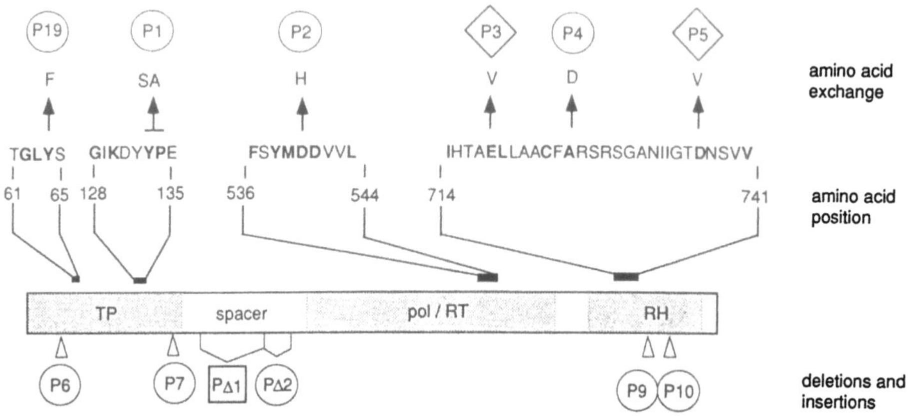

Fig. 3. Domain structure and mutational analysis of the HBV P-protein. *Shaded segments* indicate amino acid (aa) sequences well conserved among hepadnaviruses (and to retroviruses in the pol and RH regions). Mutational changes in conserved amino acids are indicated *above*; in-phase insertions of four aa (Δ), and a 90 aa deletion *(PΔ1)* are presented *above* and *below* the P-domains, respectively. Effects on DNA synthesis; (◊), first DNA strand only; (□, O), 40% or < 2% of wild-type DNA synthesis, respectively

pressed HBV genomes, we found the hepadnaviral P-protein to be organized in several functional domains (Bartenschlager and Schaller 1988; Radziwill et al. 1990). As outlined in Fig. 3 these are arranged in the order: terminal protein (TP, the putative primer of minus-strand DNA synthesis (Gerlich and Robinson 1980); a non-essential spacer region reverse transcriptase/DNA polymerase (Pol/RT), and RNase H (RH). While point mutations or in-phase insertions in each of the functional domains can abolish viral DNA synthesis, the spacer region, as exemplified by mutant P Δ1, can be largely deleted without affecting known enzymatic activities. Compared to retroviruses, the hepadnaviral P-protein lacks the protease and integrase functions which are apparently not required for the hepadnaviral life cycle.

By mutational analysis it was also found that the hepadnaviral pregenomic RNA, although having a genetic organization resembling that of other retroelements with overlapping core *(gag)* and P pol *(pol)* open reading frames (Fig. 3), expresses the P-protein unlinked to the core gene (Chang et al. 1989; Schlicht et al. 1989), most likely by translation initiation from within the RNA pregenome. The interpretation that this mechanism resembles internal ribosome entry as in picornaviruses (Chang et al. 1990) has been questioned and a modified version of leaky scanning has been proposed (Jean-Jean et al. 1989; Lin and Lo 1992).

2.3 Direct Detection of the HBV DNA Polymerase

Despite the general progress in understanding the basic mechanisms of P-gene expression and function, the physical structure of the P-protein remained essentially unknown due to the lack of a detection method that was sensitive enough to cope with its low abundance. To overcome this problem we recently developed a novel detection method which uses ^{32}P-labelling in vitro of the HBV P-protein at newly introduced target sites for protein kinase A (PKA). By phosphorylating the sequence RRXSX (Edelman et al. 1987) near either terminus of P-proteins expressed from recombinant vaccinia viruses, we initially showed that it was possible to detect these proteins in amounts as low as 5 pg, which provided an increase of sensitivity over standard techniques by two orders of magnitude (Bartenschlager et al. 1992). The biological activity of such a modified P-protein was not significantly affected, as shown subsequently by transient expression of an HBV genome carrying an appropriately modified P-gene: core particles produced from this construct displayed somewhat reduced endogenous polymerase activity, but were otherwise indistinguishable from nucleocapsids produced from a wild-type construct (Bartenschlager et al. 1992). After in vitro phosphorylation, the nucleocapsid-associated P-protein was then characterized to possess the size of 90 kDa as expected for the full length protein with a fraction of the protein being covalently linked to the viral DNA. These results indicate that initiation of reverse transcription and covalent linkage of the polymerase to its DNA product do not detectably alter the protein's primary structure which is in agreement with an earlier report detecting the P-protein via its linkage to radiolabelled viral DNA (Bartenschlager and Schaller 1988). They suggest that the hepadnaviral P-protein, in contrast to the retroviral *pol* protein, may perform its replication functions as a multidomain polypeptide without requiring prior proteolytic processing.

2.4 The RNA Encapsidation Signal

Packaging of the RNA pregenome into nucleocapsids is a highly specific process in which subgenomic RNAs, serving as mRNAs for the preS/S gene products, and also the preC mRNA which differs from the pregenome only by an extension of about 30 nucleotides at its 5'-end (Fig. 2), are excluded from encapsidation (Enders et al. 1985; Junker-Niepmann et al. 1990; Nassal et al. 1990). As a first step towards analyzing the underlying mechanism, we have recently identified an 85-nucleotide-long RNA sequence near the 5'-end of the HBV RNA pregenome (ε, Fig. 2; Bartenschlager et al. 1990; Junker-Niepmann et al. 1990), which is essential for pregenome encapsidation and also sufficient to mediate packaging of non-viral RNA sequences. This nucleotide sequence is remarkably well conserved between mammalian hepadnaviruses (Fig. 4), and this degree of sequence stability is unusually high for a region not encoding overlapping reading frames. The encapsidation sequence is further characterized by containing a number of inverted repeat sequences with the potential to form secondary and possibly tertiary structures which could be envisaged to be recognized in the packaging reaction. One such stem-loop structure, which is further characterized by possessing a conserved CUGU motif in its central loop and a second pyrimidine-rich bulge in its stem (Fig. 4), is found in the same position in all hepadnaviruses, including the evolutionary distant duck hepatitis B-virus which shares only little homology with the primary HBV DNA sequence in the precore region (Junker-Niepmann et al. 1990).

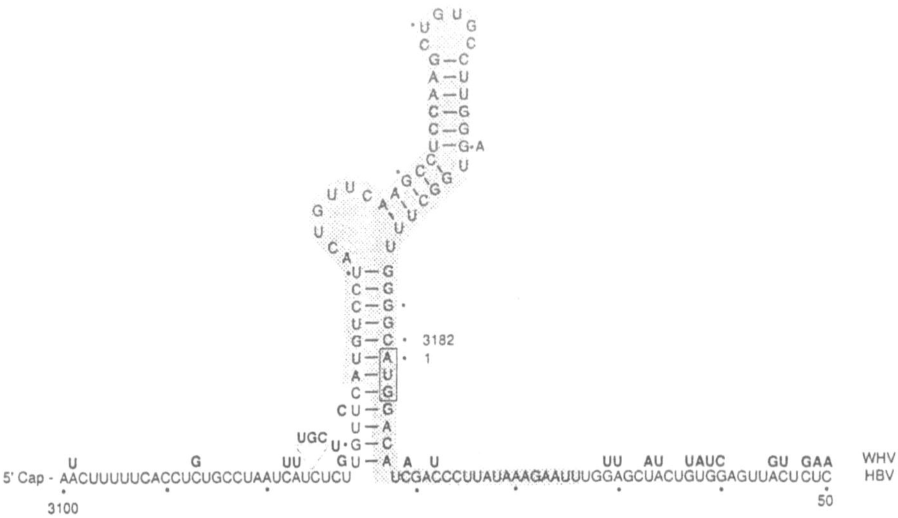

Fig. 4. Secondary structure prediction and sequence conservation of the hepatitis B-virus encapsidation signal ε. The ε sequence *(shaded area)*, as defined by deletion analysis and by sequence conversation, is presented in the context of the 5'-terminus of the *HBV* RNA pregenome and folded into one of several possible secondary structures. Base changes in the related woodchuck hepatitis B virus *(WHV)* are indicated in *boldface* above the HBV sequence

2.5 The P-Gene Product is Essential for Genomic RNA Encapsidation

Having defined the RNA encapsidation signal, we asked whether interaction with the core protein was sufficient for selective pregenome encapsidation, or whether any other viral protein was required. Since the P-protein is also an essential constituent of the replication-competent core particle, it appeared to be a logical candidate for participating in the encapsidation reaction. To clarify this point, several stop and frameshift mutations scattering throughout the HBV P-gene were transiently expressed in hepatoma cells and encapsidated RNAs were quantitated by nuclease protection or primer extension analysis (Bartenschlager et al. 1990). The data obtained from these experiments, and analogous results from the DHBV system (Hirsch et al. 1990), convincingly demonstrated that a P-gene product carrying all functional domains was required for efficient incorporation of the RNA pregenome into nucleocapsids. Interestingly, point mutations in the P-gene inactivating one of its enzymatic activities had no influence on its packaging function, indicating that the P-protein participates as a structural component in the encapsidation reaction. Moreover, the importance of the hepadnaviral P-protein for efficient RNA packaging was confirmed by the finding that the P-gene product was also required for the encapsidation of non-viral RNAs carrying a functional HBV ε-sequence at its 5'-end (Bartenschlager et al. 1990). Finally, as a surprising feature of the encapsidation reaction, it was found that the P-protein was encapsidated in *cis*, i.e. together with the genomic RNA molecules from which it was translated (Bartenschlager et al. 1990; Hirsch et al. 1990). This finding suggested that the P-protein might be a limiting factor in the encapsidation reaction.

2.6 Nucleocapsid Assembly Is Initiated By P-Protein Binding to the ε-Sequence in the Viral RNA Pregenome

As mentioned above, hepadnaviruses as well as other pararetroviruses, such as cauliflower mosaic virus, express their *pol* gene unfused to the preceding core gene (Chang et al. 1989; Schlicht et al. 1989; Schulze et al 1990). This raises the question of how a freely diffusible, reverse transcriptase/DNA polymerase is efficiently incorporated into the nucleocapsid and how promiscuous reverse transcription of cellular RNAs is circumvented. A first hint to an answer came from the aforementioned observation that the P-gene product was essential for packaging RNA molecules that carried a functional encapsidation signal. This result suggested that a specific interaction of the P-protein with the ε-RNA sequence was a prerequisite for the formation of replication-competent nucleocapsids. While this interaction seemed to be essential for the encapsidation of the RNA genome, it was not known whether packaging of the P-protein relied on the same mechanism. Alternatively, one could assume that the P-protein could bind to core protein subunits prior to, or during nucleocapsid formation, resulting in a mechanism of P-protein packaging that is independent of the viral RNA.

To decide between these alternatives, nucleocapsids produced from mutated HBV genomes by transient expression were analyzed for encapsidation of phosphorylatable P-proteins (Bartenschlager and Schaller 1992). The results of this analysis demonstrated that P-gene mutants defective in RNA packaging function were also defective for P-protein encapsidation. Furthermore, P-protein packaging was also abolished when the RNA encapsidation signal was inactivated, but could be restored by *trans*-complementation with recombinant RNA molecules carrying a functional ε-sequence. Taken together, these results demonstrate that packaging of the P-protein depends on the concomitant encapsidation of the viral RNA genome and suggest that direct binding of the hepadnaviral P-protein to the ε-RNA sequence ensures efficient packaging of both components into the nucleocapsid.

2.7 A Model for the Control of Hepadnaviral Assembly

Based on the findings described above, we suggest a model for the early steps of hepadnaviral nucleocapsid assembly which is presented in Fig. 5 in comparison with the analogous events in retroviral assembly and replication. Accordingly, hepadnaviral nucleocapsid formation is initiated by direct binding of the P-protein to the ε-sequence in the RNA pregenome. Presumably, this RNA/P protein complex is then recognized by one or several core protein molecules, forming an initial assembly complex, which upon the addition of further core polypeptides is converted to the replication-competent nucleocapsid. The finding that the P-protein acts preferentially in *cis* indicates that the initial P-protein/ε-RNA interaction is essentially irreversible due to a very tight P-protein/RNA binding and/or rapid addition of core protein subunits.

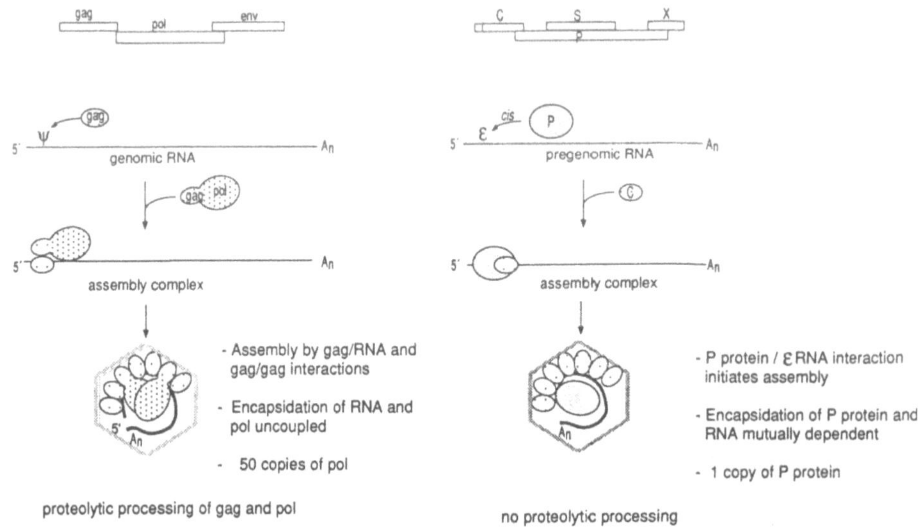

Fig. 5. Schematic comparison of the early steps leading to the formation of replication-competent nucleocapsids of retroviruses and hepadnaviruses. The genomic organization is given schematically at the *top of each panel*. The hepadnaviral core gene start codon, contained in the encapsidation signal, is indicated as a *filled circle*. The retroviral encapsidation signal is depicted as Ψ. For further details, see text

This mechanism of core particle assembly differs strikingly from cognate events in retroviruses. Here, packaging of genomic RNA is mediated by the core protein produced from the *gag* gene (Gorelick et al. 1988; Luban and Goff 1991) and is independent from the *pol* protein (Oertle and Spahr 1990). Encapsidation of the latter is accomplished by synthesis of *gag/pol* fusions followed by coassembly into the capsid structure via the *gag* moiety. This mechanism of nucleocapsid formation accounts for the fact that packaging of the retroviral *pol* protein can occur in the absence of genomic RNA (Levin et al. 1974) and explains why retroviruses contain about 20–70 *pol* protein molecules per particle (Panet et al. 1975; Krakower et al. 1977). In the case of hepadnaviruses, this number is much lower. In initial quantitations we calculated that there is probably only one or two P-protein molecules in the replication-competent HBV nucleocapsid (Bartenschlager and Schaller 1992). Thus, direct binding of the P-protein, as monomer or dimer, to the ε-sequence not only ensures encapsidation of the viral polymerase, but also provides a mechanism to regulate the number of encapsidated molecules.

References

Bartenschlager R & Schaller H (1988) The aminoterminal domain of the hepadnaviral P gene encodes the terminal protein (genome-linked protein) believed to prime reverse transcription. EMBO J 7:4185–4192

Bartenschlager R & Schaller H (1992) Hepadnaviral assembly is initiated by polymerase binding to the encapsidation signal in the viral RNA genome. EMBO J 10 (in press)

Bartenschlager R, Junker-Niepmann M & Schaller H (1990) The P gene product of hepatitis B virus is required as a structural component for genomic RNA encapsidation. J Virol 64:5324–5332

Bartenschlager R, Kuhn C & Schaller H (1992) Expression of the P-protein of the human hepatitis B virus in a vaccinia virus system and detection of the nucleocapsid-associated P-gene product by radiolabelling at newly introduced phosphorylation sites. Nucl Acids Res 20:195–202

Blum H, Zhen-Sheng Z, Galun E, von Weizsäcker F, Garner B, Liang TJ & Wands JR (1992) Hepatitis B virus X protein is not central to the viral life cycle in vitro. J Virol 66:1223–1227

Bosch V, Bartenschlager R, Radziwill G & Schaller H (1988) The duck hepatitis B virus P-gene codes for protein strongly associated with the 5'-end of the viral DNA minus strand. Virology 166:475–485

Chang L-J, Ganem D & Varmus H (1990) Mechanisms of translation of the hepadnaviral polymerase (P) gene. Proc Natl Acad Sci USA 87:5158–5162

Chang L-J, Pryciak P, Ganem D & Varmus H (1989) Biosynthesis of reverse transcriptase of hepatitis B viruses involves de novo translational initiation not ribosomal frameshifting. Nature (London) 337:364–367

Edelman AM, Blumenthal D & Krebs E (1987) Protein serine/threonine kinases. Annu Rev Biochem 56:567–613

Enders G, Ganem D & Varmus H (1985) Mapping the major transcripts of ground squirrel hepatitis B virus: the presumtive template for reverse transcription is terminally redundant. Cell 42:297–304

Gerlich W & Robinson W (1980) Hepatitis B virus contains protein attached to the 5'-terminus of its complete DNA strand. Cell 21:801–809

Gorelick R, Henderson L, Hanser J & Reins A (1988) Point mutant of Moloney murine leukemia virus that fail to package viral RNA: evidence for specific RNA recognition by "zinc finger like" protein sequence. Proc Natl Acad Sci USA 85:8420–8424

Hirsch R, Lavine J, Chang LJ, Varmus HE & Ganem D (1990) Polymerase gene products of hepatitis B virus are required for genomic RNA packaging as well as for reverse transcription. Nature (London) 344:522–525

Jean-Jean O, Weimer T, deRecondo M, Will H & Rossignol J (1989) Internal entry of ribosomes and ribosomal scanning involved in hepatitis B virus gene expression. J Virol 63:5451–5454

Junker-Niepmann M, Bartenschlager R & Schaller H (1990) A short *cis* acting sequence is required for hepatitis B virus pregenome encapsidation and sufficient for packaging of foreign RNA. EMBO J 9:3389–3396

Kaplan P, Greenman R, Gerin J, Purcell R & Robinson W (1973) DNA polymerase associated with human hepatitis B virus antigen. J Virol 12:995–1005

Krakower J, Barbacid M & Aaronson S (1977) Radioimmunassay for mammalian type C viral reverse transcriptase. J Virol 22:331–339

Levin J, Grimley P, Ramseur J & Berezesky I (1974) Deficiency of 60 to 70S RNA in murine leukemia virus particles assembled in cells treated with actinomycin D. J Virol 14:152–161

Luban J & Goff S (1991) Binding of human immunodeficiency virus type 1 (HIV-1) RNA to recombinant HIV-1 *gag* polyproptein. J Virol 65:3203–3212

Nassal M, Junker-Niepmann M & Schaller H (1990) Translational inactivation of RNA function: discrimination against a subset of genomic transcripts during hepatitis B virus nucleocapsid assembly. Cell 63:1357–1363

Oertle S & Spahr P-F (1990) Role of the *gag* polyprotein precursor in packaging and maturation of Rous sarcoma virus genomic RNA. J Virol 64:5757–5763

Panet A, Baltimore D & Hanafusa H (1975) Quantitation of avian RNA tumor virus reverse transcriptase by radioimmunoassay. J Virol 16:146–152

Radziwill G, Tucker W & Schaller H (1990) Mutational analysis of hepatitis B virus P gene product: domain structure and RNaseH activity. J Virol 64:613–620
Radziwill G, Zentgraf H, Schaller H & Bosch V (1988) The duck hepatitis B virus DNA polymerase is tightly associated with the viral core structure and unable to switch to an exogenous template. Virology 163:123–132
Schlicht H-J, Radziwill G & Schaller H (1989) Synthesis and encapsidation of duck hepatitis B virus reverse transcriptase do not require formation of core-polymerase fusion proteins. Cell 56:85–92
Schulze M, Hohn T & Jiricny J (1990) The reverse transcriptase gene of cauliflower mosaic virus is translated separately from the capsid gene. EMBO J 9:1177–1185
Toh H, Hayashida H & Miyata T (1983) Sequence homology between retroviral reverse transcriptase and putative polymerase of hepatitis B virus and cauliflower mosaic virus. Nature (London) 305:827–829

Basic Elements of Eukaryotic Replication Process

Mammalian DNA Helicases, DNA Polymerases and DNA Polymerase Auxiliary Proteins

U. Hübscher[1], G. Cullmann[1], P. Thömmes[1], B. Strack[1], E. Ferrari[1], B. Senn[1], A. Georgaki[1], T. Weiser[1], M. W. Berchtold[1], and V. N. Podust[1]

1 Introduction

DNA replication must guarantee the faithful duplication of the genetic information in advance of cell division. Research progress in the last few years suggested that mechanisms of DNA replication might be similar in most organisms investigated so far (Kornberg and Baker 1992). DNA synthesis at the advancing replication fork requires the coordinate collaboration of different types of enzymes and other proteins which appear to be in a higher order structure to ensure the precise and rapid duplication of the more than 10^9 base pairs in higher eukaryotic cells.

The different enzymes and other proteins might be involved in DNA replication in the following way (reviewed e.g. in Stillman 1989; Thömmes and Hübscher 1990b; So and Downey 1992):

1. Sequence specific DNA binding proteins bind and possibly alter the structure of specific DNA sequences, called origins of replication.
2. The thus recognized and possibly structurally altered double-stranded DNA is separated by enzymes called DNA helicases.
3. Chain initiation is subsequently made by an enzyme called primase. It synthesizes short RNA transcripts of about 10 nucleotides.
4. The single-stranded DNA is very labile and has therefore immediately to be protected by a single-strand DNA binding protein (SSB).
5. DNA polymerases add deoxyribonucleoside 5'-monophosphates to the 3'OH group of the RNA initiator chain by hydrolyzing deoxyribonucleoside 5'-triphosphates. The accurate synthesis is guided by the correct base-pairing and by correction mechanisms constantly occurring while DNA is replicated. Therefore, the catalytic subunit of many replicative DNA polymerases possesses a second enzymatic activity, namely a 3'-5' exonuclease. These enzymes can act as proofreaders. When DNA polymerases incorporate a wrong deoxyribonucleoside 5'-monophosphate correct base pairing does not occur. DNA polymerase can not continue DNA replication, since the 3'-5' exonuclease first has to exonucleolytically cleave the incorrect nucleotide, thus allowing replication to continue. DNA helicases operate in front of DNA polymerases to denature the DNA.

[1] Department of Pharmacology and Biochemistry, University of Zürich-Irchel, Winterthurerstrasse 190, CH-8057 Zürich, Switzerland.

43. Colloquium Mosbach 1992
DNA Replication and the Cell Cycle

6. Since all DNA polymerases synthesize in the 5'-3' direction there has to be discontinuous replication at the lagging strand. Thus, small pieces of DNA, called Okazaki fragments, are made frequently. This also means that an Okazaki fragment reaches the one that has been synthesized previously. Then RNase H and 5'-3' exonuclease remove the primer of the previous Okazaki fragment and the gap is filled by DNA polymerase and the two Okazaki fragments are finally sealed by DNA ligase.
7. All replicative DNA polymerases require in addition auxiliary proteins for efficient priming, efficient primer recognition, processivity, accuracy, coordination at the replication fork and recycling at the lagging strand. Thus different subassemblies of multipolypeptide DNA polymerases are required for leading and lagging strand replication, respectively.
8. Release of torsional stress and separation of two daughter molecules are finally achieved by DNA topoisomerases.

During the last few years it became evident that two or likely three different DNA polymerases are involved in nuclear DNA replication (reviewed in Hübscher and Thömmes 1992). The current model of eukaryotic DNA replication, however, involves the two DNA polymerases α and δ as the leading and lagging strand enzymes, respectively (Tsurimoto et al. 1990). The gene for a third DNA polymerase termed ε (Burgers et al. 1990) has been found to be essential for viability in yeast suggesting a role of this enzyme in DNA replication (Morrison et al. 1990; Araki et al. 1992).

In this communication we summarize our progress with:
1. calf thymus DNA helicases,
2. calf thymus DNA polymerases δ and ε, both of which can form stable holoenzyme complexes in the presence of the three auxiliary proteins, namely replication factor A (RF–A), replication factor (RF–C) and proliferating cell nuclear antigen (PCNA), and
3. the molecular cloning of the cDNA for the catalytic subunit mouse DNA polymerase δ.

2 Results and Discussion

2.1 Calf Thymus DNA helicases

DNA in its double-stranded form is energetically favoured and therefore extremely stable. However, the DNA is not static in its structure but has a continuous dynamic. For many DNA metabolic events such as DNA replication, DNA repair, homologous recombination, site-specific recombination and transcription of the double-stranded structure has to be brought into a single-stranded form transiently. The "motto" for this transient single-strandedness clearly has to be "only when absolutely required and as short as possible", in order not to expose the relatively labile single-stranded DNA to the environment. Nature can perform this process with a remarkable set of enzymes called DNA helicases.

In higher eukaryotic cells an increasing number of DNA helicases have been found in the last two years (reviewed in Thömmes and Hübscher 1992). In our laboratory initial approaches included the isolation of calf thymus proteins with DNA dependent ATPase activities, one of which was the first mammalian DNA helicase identified (Hübscher and Stalder 1985). However, there were several cases where DNA dependent ATPases did not show any DNA helicase activities or DNA helicases had only a weak associated ATPase activity (P. Thömmes, B. Strack and U. Hübscher, unpubl. observ.). Improvements to identify DNA helicases were achieved by using the strand displacement assay first described for the characterization of *Escherichia coli* DNA helicases (Matson et al. 1983). For the successful isolation of several DNA helicases from calf thymus the extract had to be prefractionated, thus decreasing the presence of nucleases, which interfere with the sensitive displacement assay. Essential progress was eventually made when established and refined isolation procedure for the three replicative DNA polymerases α, δ and ε became available (Weiser et al. 1991; see also Sect. 2.2). This led to the simultaneous isolation of four different DNA helicases termed A, B, C and D (Thömmes et al. 1992). These four enzymes could be distinguished by the following criteria (Table 1): (1) putative molecular weights after sodium dodecyl sulfate polyacrylamide gel electrophoresis, (2) glycerol gradient sedimentation behaviour under low and high salt conditions, (3) salt requirements, (4) binding to double-stranded and single-stranded DNA, (5) nucleoside- and deoxynucleoside 5'-triphosphate requirements and (6) by the direction of unwinding.

Next we looked into the interaction of these helicases with RF-A, a well-characterized protein. RF-A (Fairman and Stillman 1988), also called replication protein A, (Wold and Kelly 1988) or single-stranded binding protein (Wobbe et al. 1987), is a heterotrimeric protein involved in cellular DNA replication and repair and has been identified in human cells and also in the yeast *Saccharomyces cerevisiae* (Brill and Stillman 1989). This protein has the same polypeptide composition (70, 34 and 11 kDa) in calf thymus (Fig. 1; see also Thömmes et al. 1992). RF-A stimulates most efficiently DNA helicase A (Table 1), and enables this enzyme to unwind long single-stranded DNA up to 800 bases (Fig. 1). The effect of RF-A appears to be specific since the corresponding single-stranded DNA binding proteins from *Escherichia coli* and bacteriophage T4 had no or even a negative effect on DNA helicase A, and human RF-A, as an example of a heterologous RF-A, had no stimulatory effect, suggesting a species-specific interaction (Thömmes et al. 1992).

What can we speculate about the physiological role of calf thymus DNA helicase A? The circumstantial evidence summarized in Table 2 might indicate that DNA helicase A plays a similar role as a cellular DNA helicase at the replication fork as it is attributed to the SV40 T antigen helicase (reviewed by Fanning 1992). Both enzymes have a 3'- > 5' direction of unwinding indicating that they might bind preferentially to the leading strand. T antigen was found to interact with DNA polymerase α (Dornreiter et al. 1990) while DNA helicase A, under many chromatographic separation conditions, does not separate from DNA polymerase α (B. Strack and U. Hübscher, unpublished observations). Both enzymes have a low K_m for nucleoside 5'-triphosphates and a strong preference for ATP and dATP as their hydrolyzable energy source. Furthermore, RF-A stimulates both DNA helicases trough a species specific

Table 1. Properties of calf thymus DNA helicases A, B, C and D[a]

Property	Helicase A	Helicase B	Helicase C	Helicase D
Tendency to copurify with	pol α	pol δ/ε	RF-A/PCNA	RF-A/PCNA
Polypeptides copurifying and cosedimenting with DNA helicase activity (Dalton)	47 000	100 000	45 000	100 000, 40 000
Direction of movement	3'→ 5'	5'→ 3'	5'→ 3'	5'→ 3'
S value				
Low salt (50 mM KCl)	9.2	5.8	9.3	> 14
High salt (500 mM KCl)	6.7	5.8	6.2	6.2
Relative activity on the short substrate in the presence of additional[a]: 100 mM NaCl (%)	44	65	287	37
100 mM KCl (%)	60	56	335	42
DNA binding	ss	ds/ss	ss	ss
Preferred nucleotide	A = dA > C > dC	A = dA > all other	A = dA > > all other	A = dA
Stimulation by homologous RF-A				
Short substrate (-fold)	15	3	3	8–10
Long substrate (-fold)	8	2	1–2	1–2

[a] For details, see Thömmes et al. (1992).

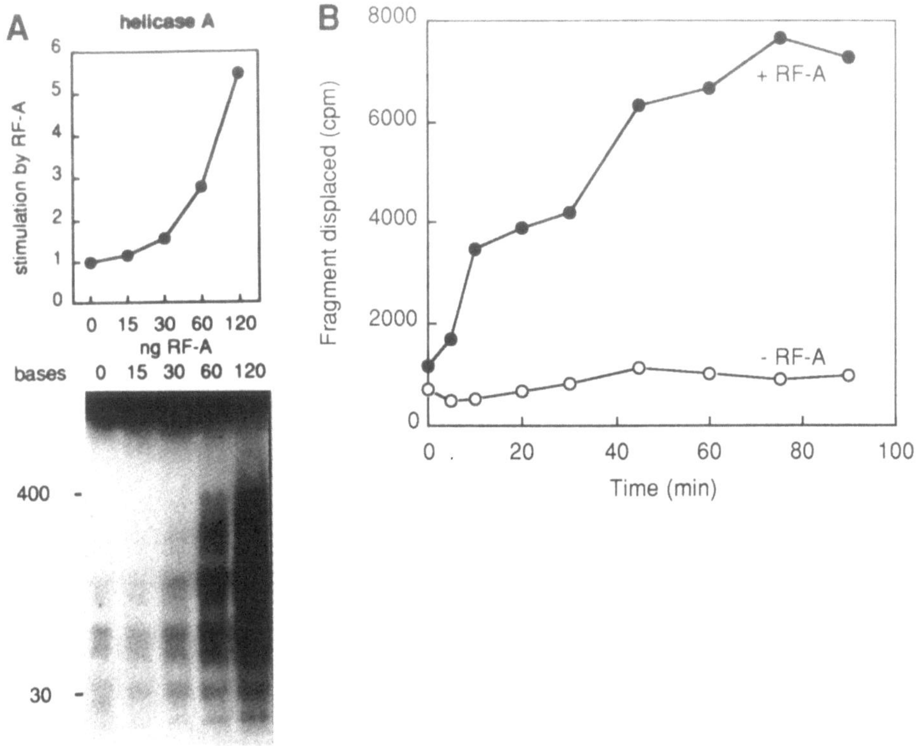

Fig. 1. Calf thymus RF-A enables DNA helicase A to unwind long single-stranded DNA fragments. Preparation of DNA helicase substrate containing a double stranded region of 30 to > 800 base pairs and DNA helicase assays were performed as described (Thömmes et al. 1992). **A** Increasing amounts of RF-A were titrated into a standard helicase assay containing 500 ng helicase A. The autoradiogram shown in the *lower panel* was quantitated to yield the graphic in the *upper panel.* **B** Kinetics by using 500 ng DNA helicase A in the presence or absence of 120 ng RF-A.

interaction with T antigen helicase in the initiation of SV40 replication (Matsumoto et al. 1990) and with calf thymus DNA helicase A in its ability to unwind long DNA fragments (Thömmes ét al. 1992 and Fig. 2). Finally, it was found that T antigen helicase acts as a double hexamer at the replication fork (Wessel et al. 1992). The fact that DNA helicase A has a tendency to aggregate under low salt concentration points to another similarity. It has been proposed that replicative DNA helicases have an oligomeric form (Lohman 1992).

DNA helicases are now identified in increasing numbers in the calf (reviewed in Thömmes and Hübscher 1992) similar to the situation in *Escherichia coli* (Matson 1991). Biochemical properties of eukaryotic DNA helicases are only of indirect evidence for their suggested *in vivo* roles. Model replication forks to study their interaction with DNA polymerases α, δ and ε and their auxiliary proteins and identification of their corresponding genes will help to clarify their roles in DNA metabolic events.

Table 2. Circumstantial evidence that calf thymus DNA helicase A might be a cellular counterpart of SV40 T antigen helicase

Property	T antigen DNA helicase[a]	Calf thymus DNA helicase A[a]
Direction of unwinding	3'→ 5'	3'→ 5'
Interaction with DNA polymerase α/primase	Yes	Strong tendency to copurify[b]
Nucleotide requirement	A > dA > Py (dT = U)	A = dA > Py (C > dC)
K_m for nucleoside 5'-triphosphate	Low	Low
Interaction with RF–A	Species-specific	Species-specific
Oligomerization	Yes	Tendency to aggregate

[a] For details, see Fanning (1992) and Thömmes et al. (1992), respectively.
[b] B. Strack and U. Hübscher (unpubl. observ.).

2.2 Calf Thymus DNA Polymerases δ, ε and Corresponding Auxiliary Proteins

At least five different DNA polymerases have been identified in eukaryotic cells (reviewed in Wang 1991; Hübscher and Thömmes 1992). The subunit composition of the yeast enzymes is nearly identical to the one in higher eukaryotes which has led to a new nomenclature for eukaryotic DNA polymerases (Burgers et al. 1990). According to this new classification, all eukaryotic DNA polymerases are denoted by greek letters (α, β, γ, δ and ε, respectively). The discovery that the three DNA polymerases α, δ and ε in yeast are essential for viability (Wang 1991) has raised new questions about the functional tasks of these enzymes.

Our observation that a DNA polymerase from calf thymus which contained a 3' → 5' exonuclease activity was able to synthesize efficiently on the homopolymer poly(dA)/oligo(dT) containing long single-stranded regions suggested that this enzyme might be DNA polymerase δ (Focher et al. 1988). However, this enzyme was processive in the absence of PCNA, a protein that had been identified as an auxiliary protein for calf thymus DNA polymerase δ (Tan et al. 1986). Subsequent work indicated that this form of DNA polymerase δ on this homopolymer template was under no circumstances dependent on PCNA and it was further characterized and termed PCNA independent DNA polymerase δ (Focher et al. 1989; Focher et al. 1990). This enzyme is now named DNA polymerase ε according to the new nomenclature (Burgers et al. 1990). Soon thereafter we were able to simultaneously isolate DNA polymerases α, δ and ε from calf thymus (Weiser et al. 1991). On a hydroxylapatite column DNA polymerase δ could completely be separated from DNA polymerase ε. The two exonuclease-containing DNA polymerases δ and ε could be biochemically distinguished from each other by several criteria: (1) DNA polymerase ε was independent of PCNA for processivity on poly(dA)/oligo(dT), (2) DNA polymerase ε preferentially utilized RNA primers, while DNA polymerase δ preferred DNA primers, (3) template utilization studies in the absence of PCNA indicated that DNA

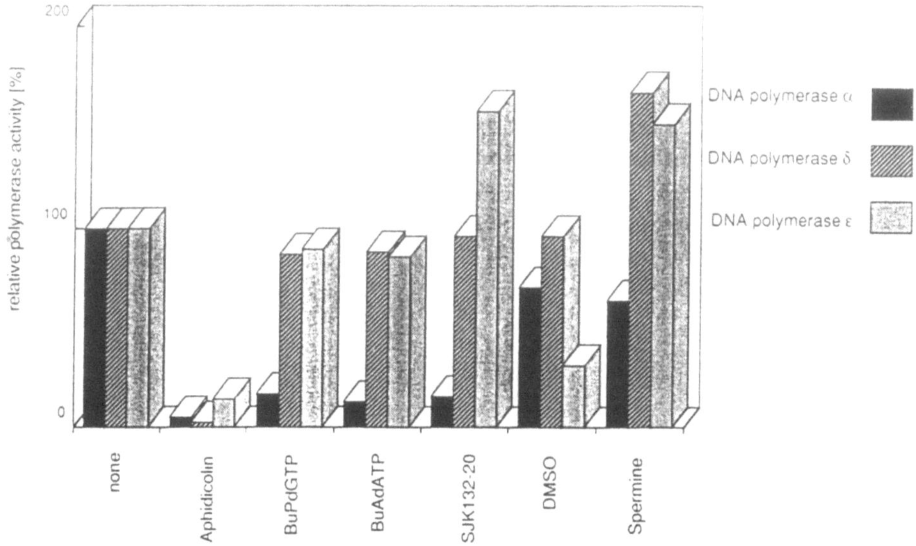

Fig. 2. Effect of inhibitors, antibodies and other substances on calf thymus DNA polymerases α, δ and ε. The final concentration of the tested substances were: aphidicolin (100 μg/ml), BuPdGTP (5 μM), BuAdATP (5 μM), SJK 132–20 antibody (0.5 μg/ per assay), dimethylsulfoxide (DMSO, 10% v/v), spermine (150 μM). Details of the assay systems are described in Weiser et al. (1991).

polymerase ε could to some but a low extent be active on natural DNA templates such as primed M13 or parvoviral DNA, (4) the elution mode from hydroxylapatite was different allowing complete separation of the two enzymes and (5) DNA polymerase ε was less sensitive to the chain terminator dideoxyTTP but was more inhibited by dimethyl sulfoxide. Figure 2 summarizes the effect of inhibitors, antibodies and other substances on calf thymus DNA polymerases α, δ and ε.

However, both DNA polymerases δ and ε had poor replication activity on naturally occurring DNA template/DNA primers such as singly DNA-primed M13 DNA or parvoviral DNA (Weiser et al. 1991). Our initial effort to isolate in the presence of ATP an efficient replication activity on singly DNA-primed M13 DNA resulted in DNA polymerase α type enzymes which were more complex than the four subunit DNA polymerase α/primase (Hübscher et al. 1987; Ottiger et al. 1987) but did not appear to have activities of DNA polymerases δ or ε (Senn 1992).

Detailed analysis of the SV40 *in vitro* replication system indicated that three proteins, namely RF-A, PCNA and RF-C had an effect on the synthetic activities of the DNA polymerases α and δ (Tsurimoto and Stillman 1989a). The analogous three proteins were identified in *Saccharomyces cerevisiae* (Yoder and Burgers 1991) and were shown to interact with DNA polymerase δ and under certain conditions (e.g. high salt) also with DNA polymerase ε (Burgers 1991). For efficient and fast replication all three proteins were required. Therefore, we used a complementation assay that enabled DNA polymerase δ and DNA polymerase ε to replicate a singly DNA-primed M13 DNA in the presence of stoichiometric amounts of SSB from *Escherichia coli* and PCNA. The complementation factor that was thus isolated (Fig. 3;

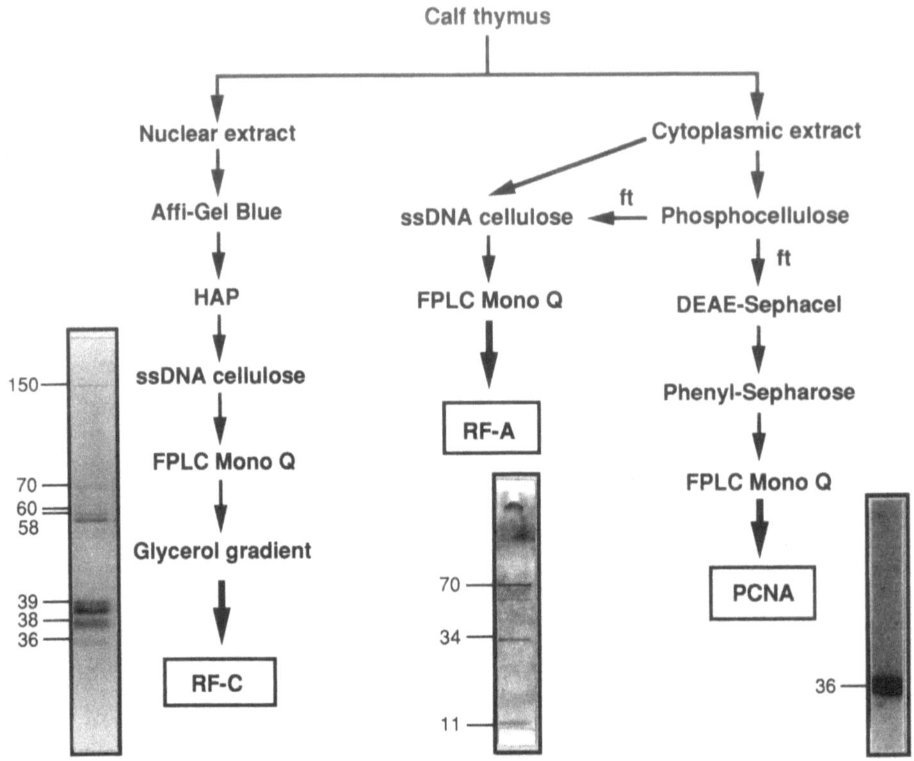

Fig. 3. Isolation scheme and polypeptide compositions of calf thymus RF–C, RF-A and PCNA. For details see, Thömmes et al. (1992) for isolation of RF-A, Podust et al. (1992) for isolation of RF–C and Prelich et al. (1987) for isolation of PCNA.

Podust et al. 1992) is very likely RF–C since the complex has a similar polypeptide composition as RF–C's from yeast (Yoder and Burgers 1991; Fien and Stillman 1992) and human (Tsurimoto and Stillman 1989b; Lee et al. 1991). In addition biochemical and physicochemical data are similar and the complex contains an ATPase that is stimulated by DNA and PCNA (Podust et al. 1992). RF–C is able to bind to primer/template junctions where single-stranded DNA is covered by an SSB (either *Escherichia coli* SSB, bacteriophage T4 gene 32 protein or RF-A). Furthermore, in the presence of ATP and PCNA calf thymus RF–C could form a stable and isolatable initiation complex, that could be replicated upon addition of DNA polymerase δ or ε (Podust et al. 1992). These data are similar to those described for RF–C's from yeast (Burgers 1991) and from human cells (Lee et al. 1991). As summarized in Table 3, it appears that RF–C is a multisubunit complex occurring in all eukaryotic cells and, as suggested by Tsurimoto and Stillman (1990), has functional counterparts in bacteriophage T4 (gene 44/62 complex) and in *Escherichia coli* (various subassemblies of the γ complex).

Table 3. A multisubunit DNA polymerase auxiliary protein from prokaryotes and eukaryotes[a]

Source	Molecular weight (kDa)	Name	Enzymatic activity	Authors
Bacteriophage T4	36/21	Gene 44/62-complex	ATPase stimulated by gene 45	Alberts (1987)
Escherichia coli	48/34/32/15/13	γ-Complex	ATPase stimulated by β	Maki and Kornberg (1988)
	48/34	$\gamma\delta$-Subcomplex	ATPase stimulated by β	Studwell and O'Donnell (1990)
	48/32	$\gamma\delta'$-Subcomplex	ATPase	O'Donnell and Studwell (1990)
Yeast	130/86/41/40/37/27	RF–C	ATPase stimulated by PCNA	Yoder and Burgers (1991)
	110/42/41/36	RF–C	ATPase stimulated by PCNA	Fien and Stillman (1992)
Calf thymus	155/70/60/58/39/38/36	RF–C	ATPase stimulated by PCNA	Podust et al. (1992)
Human	140/41/37	RF–C	ATPase stimulated by PCNA	Tsurimoto and Stillman (1989b)
	145/40/38/37/36.5	A1	ATPase stimulated by PCNA	Lee et al. (1991)

[a] The replication activity can be assayed on a SSB-coated primed single-stranded circular DNA in the presence of a "clamp" factor (gene 45 of bacteriophage T4, β-subunit of *Escherichia coli* DNA polymerase III holoenzyme or PCNA) and a DNA polymerase (gene 43 of bacteriophage T4, DNA polymerase III core or DNA polymerase δ or ε).

2.3 Formation of Stable Calf Thymus DNA Polymerase δ and ε Holoenzyme Complexes on SSB Coated Singly-DNA Primed M13 DNA

PCNA, RF–C and SSB were all required for efficient elongation of singly-DNA primed M13 DNA by DNA polymerase δ and ε (Fig. 4). In contrast to DNA polymerase δ, DNA polymerase ε could act at a residual activity on singly-DNA primed M13. We tested the ability of DNA polymerase δ and ε to form isolatable, active replication complexes. When SSB-coated singly-DNA primed M13 DNA was preincubated with RF–C, PCNA, ATP and DNA polymerase δ (or ε), complexes isolated via Bio-Gel A-5m could replicate DNA upon addition of the four dNTP's (Fig. 5). However, a further increase in DNA synthesis was observed when DNA polymerase δ (Fig. 5A) and DNA polymerase ε (Fig. 5C) were added to the filtrated complexes. This indicated that both DNA polymerases δ and ε were able to form stable holoenzyme complexes; it also suggested that their binding was partially reversible and holoenzyme formation thus appears to be a dynamic process. When holoenzyme complex formation was carried out in the absence of ATP only simultaneously addition of DNA polymerase δ (or ε), PCNA and ATP restored DNA synthesis (Fig. 5B

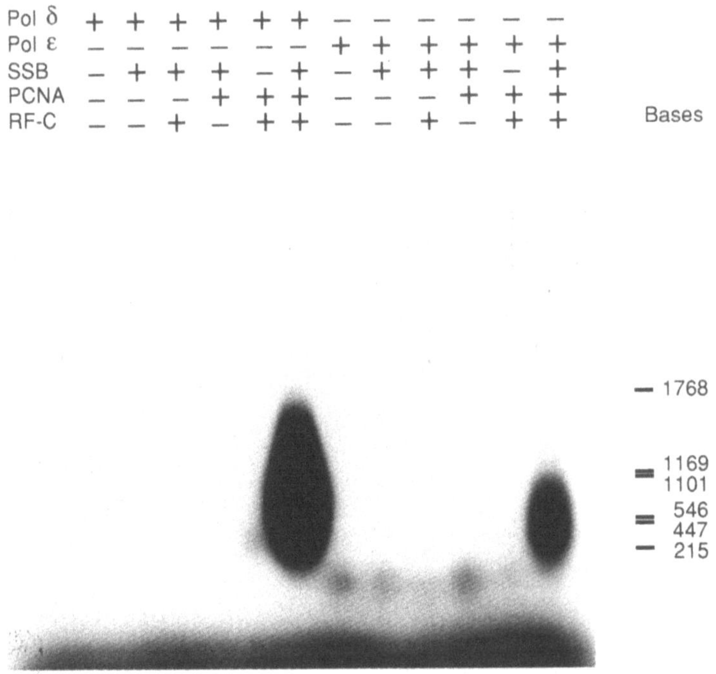

Fig. 4. Replication of singly-DNA primed M13 DNA is dependent on SSB, PCNA and RF–C. Replication of singly DNA-primed M13 DNA was carried out in a final volume of 25 μl containing: 40 mM Tris-HCl (pH 7.5), 1 mM dithiothreitol, 0.2 mg/ml bovine serum albumin, 10 mM $MgCl_2$, dATP, dGTP, dCTP each at 50 μM, 15 μM of [^{3}H]dTTP (1500 cpm/pmol), 100 ng of singly DNA-primed M13 DNA and the following amounts of proteins and ATP if included: pol δ (0.25 units), pol ε (0.35 units), *Escherichia coli* SSB (350 ng), RF–C (2.5 ng), PCNA (100 ng) and ATP (1 mM). For product analysis the reaction was stopped, prepared for electrophoresis and separated on a 1% alkaline agarose gel.

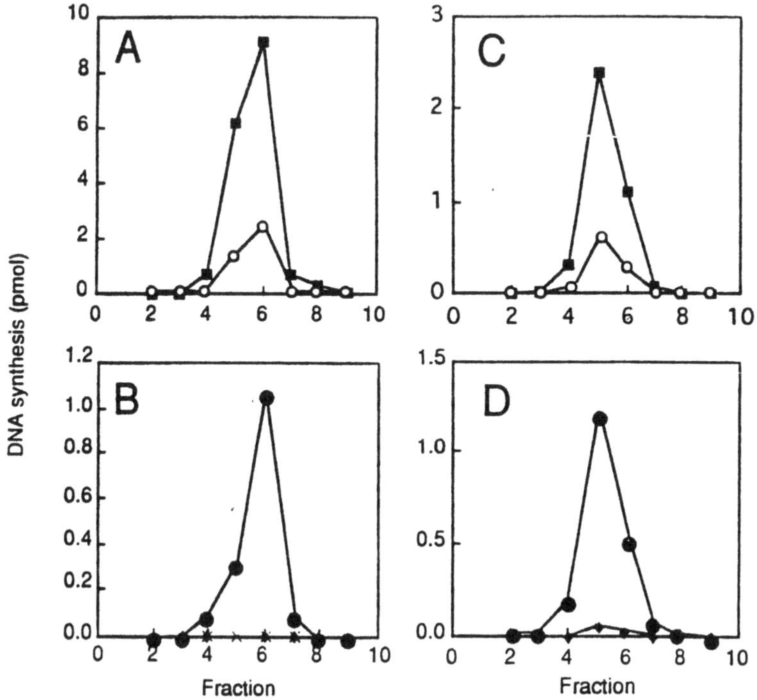

Fig. 5. Holoenzyme complex formation of calf thymus DNA polymerases δ and ε. Holoenzyme complex formation on singly-DNA primed M13 DNA was carried out, the products fractionated on a Bio-Gel A-5m column and the fractions supplemented with a dNTP mixture as described (Podust et al. 1992). **A** Holoenzyme complex formation was performed in the presence of ATP and the fractions from the column supplemented with dNTPs only (○) or with 0.2 units of pol δ (■). **B** Holoenzyme complex formation was performed in the absence of ATP and the fractions from the column were supplemented with dNTPs plus first with pol δ (0.25 units) and ATP (1 mM) (X), second with PCNA (100 ng) and ATP (1 mM) (◆) and third with pol δ (0.25 units), PCNA (100 ng) and ATP (1 mM) (●). **C** As **A**, but pol δ was replaced by 0.35 units of pol ε. **D** As **B**, but pol δ was replaced by 0.35 units pol ε.

and D, respectively). These data indicated that DNA polymerase ε as well as DNA polymerase ε could be retained by SSB-coated single DNA-primed M13 DNA only via the specific holoenzyme complex.

What are the functional tasks for DNA polymerase δ and ε holoenzyme at the replication fork? It was speculated (Nethanel and Kaufmann 1990) that two DNA polymerases (α/primase and another BuPdGTP resistant DNA polymerase) are required for replication of the lagging strand. In a region called "Okazaki zone" several Okazaki fragments could be initiated by the DNA polymerase α/primase complex. After synthesis of RNA and a short piece of DNA, the α/primase enzyme would leave to initiate another Okazaki fragment, thus allowing RNase H to remove the primer and to leave a DNA primer for further elongation. Indeed studies of the initiation of SV40 replication in vitro indicated that the processivity of polymerase α/primase in the presence of RF-A and T antigen DNA helicase is about 35 nucleotides only (Murakami et al. 1992). These data suggested that polymerase α/primase might leave the template after synthesis of only a very short DNA fragment in order to resume RNA synthesis of the next Okazaki fragment. DNA polymerase ε holoenzyme (rather than δ) could be a candidate for the completion of Okazaki fragments since we found that the replication speed on SSB-coated singly DNA-primed M13 DNA under standardized conditions was ten times lower for DNA polymerase ε than for DNA polymerase δ (Podust et al. 1992). A DNA polymerase at the lagging strand would need a fast recycling capacity rather than a high replication speed. At the moment we therefore favour the model by Burgers (1991) that proposes DNA polymerase δ as the leading and DNA polymerase α/primase and DNA polymerase ε as the lagging strand replicases.

2.4 Molecular Cloning of the cDNA for the Catalytic Subunit of Mouse DNA Polymerase δ

DNA polymerase α and δ belong, like most replicative DNA polymerases, to the group of α-type DNA polymerases (Blanco et al. 1991). These enzymes share six regions of similarity in the same linear arrangement. They are designated I to VI according to their extent of similarity. Region I is the most similar while region VI has the lowest degree of similarity. Finally, yeast DNA polymerase ε has a much weaker similarity to the α-type DNA polymerases (Morrison et al. 1990).

The sequence of the yeast *Saccharomyces cerevisiae* DNA polymerase δ (Boulet et al. 1989) allowed us to develop a PCR strategy for the cloning of mouse DNA polymerase δ. An amino acid comparison shows homology boxes between yeast DNA polymerase δ and herpesviral DNA polymerases (e.g. Herpes simplex virus type 1, Cytomegalovirus, Epstein Barr virus and Varizella zoster virus). The latter are called δ-like boxes (Boulet et al. 1989) and are not present in other prokaryotic and eukaryotic DNA polymerases. Therefore, it was possible to define a subgroup among the α-type DNA polymerases that consists of the *Saccharomyces cerevisiae* DNA polymerase δ and the herpesviral DNA polymerases.

We have cloned DNA polymerase δ from a mouse cDNA library by performing PCR with one primer corresponding to one of the δ-like homology boxes (see Fig. 6, pol δ-like) and the other to the DNA polymerase homology box I (see Fig. 6, pol-like), which is the most conserved region in α-type replicative DNA polymerases. The PCR product had a size of 1300 base pairs as expected from this region of the yeast enzyme. This PCR product was digested with Sau 3A and subcloned. One of the subclones (Sau 3A4, 170 base pairs) showed, throughout its entire length, 75% identity of the deduced amino acid sequence to yeast DNA polymerase δ, indicating that we had indeed cloned DNA polymerase δ. The homologous region covered the DNA polymerase homology boxes III and VI. Next, this subclone was used to screen a mouse NIH 3T3 cDNA library. One clone was obtained which partially overlapped with Sau 3A4 and continued downstream of box V towards the COOH terminus (Fig. 6). In addition the 1300 base pair PCR product was sequenced directly and the 3' end of the cDNA was PCR amplified by using an oligo (dT) primer corresponding to the poly(A) tail of the cDNA and a gene-specific primer. The resulting PCR product was subsequently sequenced.

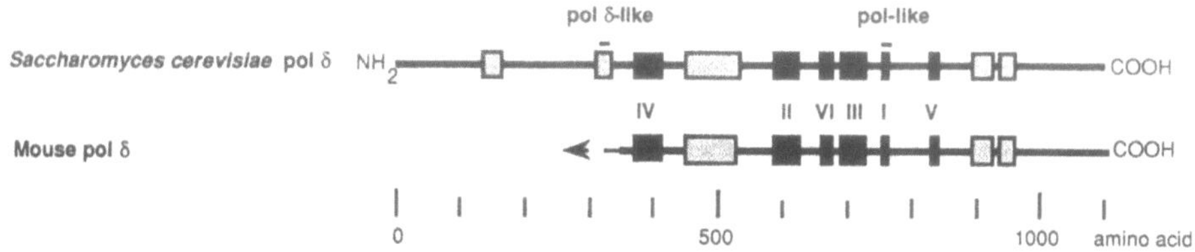

Comparison of the 2369 base pair region to other pol δ:

Mouse pol δ compared to:	Identity	Similarity	Authors
Calf pol δ	92.7%	94.7%	Zhang et al. (1991)
Human pol δ	91.6%	94.2%	Chung et al. (1991) Yang et al. (1992)
Schizosaccharomyces pombe pol δ	59.2%	75.3%	Pignède et al. (1991)
Saccharomyces cerevisiae pol δ	54.4%	72.9%	Boulet et al. (1989)
Plasmodium falciparum pol δ	48.6%	68.2%	Ridley et al. (1991)

Fig. 6. Molecular cloning of the cDNA of the catalytic subunit of mouse DNA polymerase δ. The *thick bar* represents the 2369 base pairs sequenced. *Black boxes* (*I–VI*) are α-type conserved regions and the *stippled boxes* are δ-like conserved regions found in DNA polymerase δ and herpes viral DNA polymerases. *Pol* DNA polymerase.

So far we have obtained the cDNA sequence of 2369 base pairs corresponding to about 70% of the DNA polymerase δ gene. This cDNA coding region is highly homologous to other DNA polymerases δ sequenced very recently (Fig. 6). They include DNA polymerase δ from *Schizosaccharomyces pombe* (Pignède et al. 1991), from *Plasmodium falciparum* (Ridley et al. 1991), calf (Zhang et al. 1991) and human (Chung et al. 1991; Yang et al. 1992). The comparison made so far includes the middle and the COOH terminal part of DNA polymerase δ which are considered to be more conserved than the NH_2-terminus (for details see Yang et al. 1992). The homology to DNA polymerase α from *Saccharomyces cerevisiae* and human cells was 2–4 times lower. In addition the homology within the DNA polymerase δ is higher than within DNA polymerase α, indicating that DNA polymerase δ might be the most conserved replicative DNA polymerase in nature. An extreme conservation of DNA polymerase δ has already been suggested based on biochemical data since DNA polymerase δ holoenzyme could be replaced by the DNA polymerase holoenzyme of the *Escherichia coli* bacteriophage T4 in the SV40 *in vitro* replication system (Tsurimoto et al. 1990).

3 Conclusion

The existence of active and isolatable DNA polymerase δ and ε holoenzyme forms from the calf thymus tissue is advantageous since this gland is a very rich and cheap source for obtaining large quantities of the factors and enzymes described. Biochemical analysis to test the model that two DNA polymerases act at the lagging strand (Nethanel and Kaufmann 1990) appear to be possible now. Functional experiments at model replication forks with DNA polymerase δ and ε holoenzyme, the DNA polymerase α/primase in the presence of RF-A and possibly alpha accessory factor (Goulian and Heard 1990) and one or more of the four cellular DNA helicases described here might tell us how three replicative DNA polymerases might share their roles at a mammalian replication fork. Finally, DNA polymerase δ appears to be the evolutionarily most conserved replicative DNA polymerase in nature.

Acknowledgements. We thank the following institutions for their support: Swiss National Science Foundations (grants 31.28592.90 and 31.30298.90), Schweizerische Krebsliga, Bonizzi-Theler Stiftung, EMDO-Stiftung Zürich, Ciba-Geigy Jubiläumsstiftung and the Kanton of Zürich.

References

Alberts BM (1987) Prokaryotic DNA replication mechanisms. Philos Trans R Soc Lond B 317:395–420

Araki H, Ropp PA, Johnson AL, Johnston LH, Morrison A & Sugino A (1992) DNA polymerase II, the probable homolog of mammalian DNA polymerase ε, replicates chromosomal DNA in the yeast *Saccharomyces cerevisiae*. EMBO J 11:733–740

Blanco L, Bernad A, Blasco MA & Salas M (1991) A general structure for DNA-dependent DNA polymerases. Gene 100:27–38

Boulet A, Simon M, Faye G, Bauer GA & Burgers PMJ (1989) Structure and function of the *Saccharomyces cerevisiae* CDC2 gene encoding the large subunit of DNA polymerase III. EMBO J 8:1849–1854

Brill SJ & Stillman B (1989) Yeast replication factor-A functions in the unwinding of the SV40 origin of DNA replication. Nature 342:92–95

Burgers PMJ, Bambara RA, Campbell JL, Chang LMS, Downey KM, Hübscher U, Lee MYWT, Linn SM, So AG & Spadari S (1990) Revised nomenclature for eukaryotic DNA polymerases. Eur J Biochem 191:617–618

Burgers PMJ (1991) *Saccharomyces cerevisiae* replication factor C. II. Formation and activity of complexes with the proliferating cell nuclear antigen and DNA polymerases δ and ε. J Biol Chem 266:22698–22706

Chung DW, Zhang J, Tan C-K, Davie EW & So AG (1991) Primary structure of the catalytic subunit of human DNA polymerase δ and chromosomal location of the gene. Proc Natl Acad Sci USA 88:11197–11201

Dornreiter I, Hoss A, Arthur AK & Fanning E (1990) SV40 T antigen binds directly to the large subunit of purified DNA polymerase α. EMBO J 9:3329–3336

Fairman MP & Stillman B (1988) Cellular factors required for multiple stages of SV40 DNA replication in vitro. EMBO J 7:1211–1218

Fanning E (1992) Simian Virus 40 Large T Antigen: the Puzzle, the Pieces, and the Emerging Picture. J Virol 66:1289–1293

Fien K & Stillman B (1992) Identification of replication factor C from *Saccharomyces cerevisiae*: a component of the leading-strand DNA replication complex. Mol Cell Biol 12:155–163

Focher F, Spadari S, Ginelli B, Hottiger M, Gassmann M & Hübscher U (1988) Calf thymus DNA polymerase δ: purification, biochemical and functional properties of the enzyme after

its separation from DNA polymerase α, a DNA-dependent ATPase and proliferating cell nuclear antigen. Nucl Acids Res 16:6279–6295
Focher F, Gassmann M, Hafkemeyer P, Ferrari E, Spadari S & Hübscher U (1989) Calf thymus DNA polymerase δ independent of proliferating cell nuclear antigen (PCNA). Nucl Acids Res 17:1805–1821
Focher F, Verri A, Maga G, Spadari S & Hübscher U (1990) Effect of divalent and monovalent cations on calf thymus PCNA-independent DNA polymerase δ and its 3'– > 5' exonuclease. FEBS Lett 259:349–352
Goulian M & Heard CJ (1990) The mechanism of action of an accessory protein for DNA polymerase α/primase J Biol Chem 265:13231–13239
Hübscher U & Stalder H-P (1985) Mammalian DNA helicase. Nucl Acids Res. 13:5471–5483
Hübscher U, Gassmann M, Spadari S, Brown NC, Ferrari E & Buhk H-J (1987) Mammalian DNA polymerase α: a replication competent holoenzyme form from calf thymus. Philos Trans R Soc Lond B 317:421–428
Hübscher U & Thömmes P (1992) DNA polymerase ε: in search of a function. Trends Biochem Sci 17:55–58
Kornberg A & Baker TA (1992) DNA replication. published by WH Freeman and Co, San Francisco, Cal
Lee, S-H, Kwong AD, Pan Z-Q & Hurwitz J (1991) Studies on the activator 1 protein complex, an accessory factor for proliferating cell nuclear antigen-dependent DNA polymerase δ. J Biol Chem 266:594–602
Lohman TM (1992) *Escherichia coli* DNA helicases: mechanisms of DNA unwinding. Molec Microbiol 6:5–14
Maki H & Kornberg A (1988) DNA polymerase III holoenzyme of *Escherichia coli*. II. A novel complex including the γ subunit essential for processive synthesis. J Biol Chem 263:6555–6560
Matson SW, Tabor S & Richardson CC (1983) The gene 4 protein of bacteriophage T7. Characterization of helicase activity. J Biol Chem 258:14017–14024
Matson SW (1991) DNA helicases of *Escherichia coli*. Prog Nucl Acid Res Mol Biol 40:289–326
Matsumoto T, Eki T & Hurwitz J (1990) Studies on the initiation and elongation reactions in the simian virus 40 DNA replication system. Proc Natl Acad Sci USA 87:9712–9716
Morrison A, Araki H, Clark AB, Hamatake RK & Sugino A (1990) A third essential DNA polymerase in *S. cerevisiae*. Cell 62:1143–1151
Murakami Y, Eki T & Hurwitz J (1992) Studies on the initiation of simian virus 40 replication in vitro: RNA primer synthesis and its elongation. Proc Natl Acad Sci USA 89:952–956
Nethanel T & Kaufmann G (1990) Two DNA polymerases may be required for synthesis of the lagging DNA strand of simian virus 40. J Virol 64:5912–5918
O'Donnell, M & Studwell PS (1990) Total reconstitution of DNA polymerase III holoenzyme reveals dual accessory protein clamps. J Biol Chem 265:1179–1187
Ottiger HP, Frei P, Hässig M & Hübscher U (1987) Mammalian DNA polymerase α: a replication competent holoenzyme form from calf thymus. Nucl Acids Res 15:4789–4807
Pignède G, Bouvier D, deRecondo A & Baldacci G (1991) Characterization of the POL3 gene product from *S. pombe* indicates inter-species conservation of the catalytic subunit of DNA polymerase δ. J Mol Biol 222:209–218
Podust V, Georgaki A, Strack B & Hübscher U (1992) Calf thymus RF-C as an essential componet for DNA polymerases δ and ε holoenzymes function: RF-C as an essential component for holoenzyme function. Nucl Acids Res 20:4159–4165
Prelich G, Kostura M, Marshak DR, Mathews MB & Stillman B (1987) The cell-cycle regulated proliferating cell nuclear antigen is required for SV40 DNA replication in vitro. Nature 326:471–475
Ridley RG, White JH, McAleese SM, Goman M, Alano P, deVries E & Kilbey BJ (1991) DNA polymerase δ: gene sequences from *Plasmodium falciparum* indicate that this enzyme is more highly conserved than DNA polymerase α. Nucl Acids Res 19:6731–6736
Senn B (1992) Funktionelle Formen von Säugetier-DNA-Polymerasen. Thesis, University of Zürich

So AG & Downey KM (1992) Eukaryotic DNA replication. Crit Rev Biochem Mol Biol 27:129–155

Stillman B (1989) Initiation of eukaryotic DNA replication in vitro. Annu Rev Cell Biol 5:197–245

Studwell PS & O'Donnell M (1990) Processive replication is contingent on the exonuclease subunit of DNA polymerase III holoenzyme. J Biol Chem 265:1171–1178

Studwell-Vaughan PS & O'Donnell M (1991) Constitution of the twin polymerase of DNA polymerase III holoenzyme. J Biol Chem 266:19833–19841

Tan C-K, Castillo C, So AG & Downey KM (1986) An auxiliary protein for DNA polymerase δ from fetal calf thymus. J Biol Chem 261:12310–12316

Thömmes P & Hübscher U (1990a) DNA helicase from calf thymus. J Biol Chem 265:14347–14354

Thömmes P & Hübscher U (1990b) Eukaryotic DNA replication. Enzymes and proteins acting at the fork. Eur J Biochem 194:699–712

Thömmes P & Hübscher U (1992) Eukaryotic DNA helicases: essential enzymes for DNA transactions. Chromosoma 101:467–473

Thömmes P, Ferrari E, Jessberger R & Hübscher U (1992) Four different DNA helicases from calf thymus. J Biol Chem 267:6063–6073

Tsurimoto T & Stillman B (1989a) Multiple replication factors augment DNA synthesis by the two eukaryotic DNA polymerases, α and δ. EMBO J 8:3883–3889

Tsurimoto T & Stillman B (1989b) Purification of a cellular replication factor, RF–C, that is required for coordinated synthesis of leading and lagging strands during simian virus 40 DNA replication in vitro. Mol Cell Biol 9:609–619

Tsurimoto T & Stillman B (1990) Functions of replication factor C and proliferating cell nuclear antigen: functional similarity of DNA polymerase accessory proteins from human cells and bacteriophage T4. Proc Natl Acad Sci USA 87:1023–1027

Tsurimoto T, Melendy T & Stillman B (1990) Sequential initiation of lagging and leading strand synthesis by two different polymerase complexes at the SV40 DNA replication origin. Nature 346:534–539

Weiser T, Gassmann M, Thömmes P, Ferrari E, Hafkemeyer P & Hübscher U (1991) Biochemical and functional comparison of DNA polymerases α, δ and ε from calf thymus. J Biol Chem. 266:10420–10428

Wessel R, Schweizer J & Stahl H (1992) Simian virus 40 T-antigen DNA helicase is a hexamer which forms a binary complex during bidirectional unwinding from the viral origin of replication. J Virol 66:804–815

Wobbe CR, Weissbach L, Borowiec JA, Dean FB, Murakami Y, Bullock P & Hurwitz J (1987) Replication of simian virus 40 origin-containing DNA in vitro with purified proteins. Proc Natl Acad Sci USA 84:1834–1838

Wold MS & Kelly TJ (1988) Purification and characterization of replication protein A, a cellular protein required for in vitro replication of simian virus 40 DNA. Proc Natl Acad Sci USA 85:2523–2527

Yang C-L, Chang L-S, Zhang P, Hao H, Zhu L, Toomey NL & Lee MYWT (1992) Molecular cloning of the catalytic subunit of human DNA polymerase δ. Nucl Acids Res 20:735–745

Yoder BL & Burgers PMJ (1991) *Saccharomyces cerevisiae* replication factor C. I.: purification and characterization of its ATPase activity. J Biol Chem 266:22689–22697

Zhang J, Chung DW, Tan C-K, Downey KM, Davie EW & So AG (1991) Primary structure of the catalytic subunit of calf thymus DNA polymerase δ: sequence similarities with other DNA polymerases. Biochemistry 30:11742–11750

Protein-DNA Interaction at Yeast Replication Origins: An ARS Consensus Binding Protein

J. F. X. Hofmann[1, 2], M. Cockell[1], and S. M. Gasser[1]

1 Introduction

Eukaryotic DNA replication initiates at many sites on each chromosome during the S phase of the cell cycle. The identification of yeast chromosomal DNA sequences that act in *cis* to allow the extrachromosomal maintenance of plasmids (ARS or autonomously replicating sequences, Stinchcomb et al. 1976) in yeast led to the suggestion that they function as origins of replication. Proof that initiation of DNA synthesis occurs at or near ARS elements comes from two-dimensional gel electrophoresis systems (Brewer and Fangman 1987; Huberman et al. 1987) that allow the separation of replication intermediates and the subsequent mapping of initiation sites for DNA replication.

The DNA sequence requirements of ARS elements have been reviewed recently (Newlon 1988; Umek et al. 1989). In general, ARS elements have a significantly higher A + T content than the bulk chromosomal DNA and they all contain one or more copies of an 11 nucleotide consensus sequence: 5'–(A/T)TTTAT(A/G)TTT (A/T)–3'. The essential regions for ARS function include the consensus sequence and a variable number of nucleotides 3' of the T-rich strand of the consensus (Celniker et al. 1984; Kearsey 1984; Palzkill and Newlon 1988). Extensive mutational analysis of the C2G1 ARS addressed the functional contribution of each nucleotide in the ARS consensus (Van Houten and Newlon 1990). Mutations that reduced or abolished ARS activity occurred at each position in the consensus sequence, demonstrating that each position of this sequence contributes to ARS function. Mutations that created alternative perfect matches to the consensus sequence had no major effect on ARS function (see summary, Fig. 2B). In addition, the region required for full ARS efficiency extends beyond the core consensus in both directions.

Recently attempts were made to revert *cis*-acting ARS mutations by in vitro mutagenesis (Kipling and Kearsey 1990). Surprisingly, most *cis*-acting mutations that restored ARS activity were found to create improved close matches to the ARS consensus (or additional matches to the ARS consensus) in bacterial vector sequences (summarized in Fig. 2B). This suggests that the ARS consensus and perhaps near matches to it are essential for ARS function and that the sequence requirements of the *S. cerevisiae* DNA replication apparatus for initiation are not very complex. Fur-

[1] Swiss Institute for Experimental Cancer Research (ISREC), Ch. des Boveresses 155, CH-1066 Epalinges/Lausanne, Switzerland.
[2] Current address: Cold Spring Harbor Laboratories, P.O. Box 100, Cold Spring Harbor, NY 11724, USA.

43. Colloquium Mosbach 1992
DNA Replication and the Cell Cycle

thermore, the specificity of effect that point mutations have on ARS function is compatible with the role of the ARS consensus as a recognition site for a protein(s) that is involved in the initiation of DNA synthesis.

In contrast to the studies of the core consensus, sequence comparisons and mutational analyses have been less revealing about the role of the 3' flanking sequence. It exhibits little primary sequence homology from one ARS to another. In one hypothesis it was suggested that the flanking sequence requires a repeated sequence of near matches to the core consensus sequence (Palzkill and Newlon 1988). These were found at the 9 ARS elements analyzed, and were generally in an inverted orientation to the perfect consensus. Only the perfect match at the core consensus sequence exhibited sensitivity to point mutations, however. Linker substitution mutations inserted along 180 bp of the yeast ARS1 confirmed that the 11/11 ARS consensus is essential for replication (Marahrens and Stillman 1992). Insertions into three other short sequences in the 3' flank caused a significant drop in the efficiency of replication. One such region is adjacent to the perfect ARS consensus, the second corresponds to a 9/11 near match to the ARS consensus in the opposite orientation, while the third is the binding site of ABF1, a transcription factor that enhances ARS efficiency. The effect of disruption of the only inverted 9/11 ARS consensus would appear to support the proposal of Palzkill and Newlon.

Another explanation of the ARS 3' flanking region has been proposed by Umek and Kowalski (1988). They suggested that DNA unwinding is the primary role of the ARS flanking sequence. Stable origin unwinding in vitro proved to be temperature dependent and certain large deletions of the ARS flanking sequence generate cold sensitive ARS mutations in vivo (Umek and Kowalski 1990). This demonstrates that thermal energy contributes to origin unwinding in vivo. In addition, insertion mutants, in which the 3' flanking region is moved significantly downstream, are replication deficient (Palzkill et al. 1986; Diffley and Stillman 1988). We will suggest below that perhaps both the unwinding propensity and near ARS matches are important features of the 3' flanking region.

Considerable effort has been put into the identification of *trans*-acting factors that bind yeast ARS elements (reviewed by Newlon 1988). An abundant nuclear protein called ABF-1 has been purified and cloned by several groups, based on its affinity for a site that is often found 3' of the ARS consensus (Diffley and Stillman 1988; Eisenberg et al. 1988; Sweder et al. 1988). Deletion of the ABF-1 binding site at ARS1 slightly reduces the efficiency of the ARS element in vivo, but the site is clearly not essential for plasmid replication. Although neither ABF-1 nor other previously identified factors bind directly to the essential ARS consensus sequence, protein-DNA interaction at the yeast ARS consensus was proposed based on in vivo footprinting data at the chromosomal copy of ARS-1 (Lohr and Torchia 1988).

In this paper we present further characterization of a yeast protein that binds in vitro to the ARS consensus and some close matches to it in the 3' flanking region (Hofmann and Gasser 1991). This ARS consensus binding protein (ACBP) is unusual in that it specifically binds the T-rich single-strand of its recognition sites.

2 Identification of a Yeast Factor that Binds to the T-rich Strand of ARS Consensus Oligonucleotides

The ARS element downstream of the copy-I Histone H4 gene has been extensively characterized in order to define the sequences required for H4 ARS function (Bouton and Smith 1986; Holmes and Smith 1989). A minimal region of at least 66 bp permits autonomous replication of non-replicating plasmids in yeast. This minimal element contains an 11/11 bp ARS consensus sequence and several 9/11 bp close matches to the ARS core in the 3' flanking domain. ARS unwinding and origin function have similar sequence requirements in the 3' flanking region of the H4 ARS consensus (Umek and Kowalski 1988).

Recently it was shown that DNA binding activities from yeast extracts interact with single stranded sequences that contain the ARS consensus (Hofmann and Gasser 1991; Kuno et al. 1990; Schmidt et al. 1991). One of these binding activities has been purified by DNA affinity chromatography using single stranded oligonucleotides that contain the T-rich strand of the histone H4 ARS consensus (Hofmann and Gasser 1991). The ARS consensus binding factor (called ACBP) purifies as a 67 kD protein.

It is interesting to note that more than one factor in yeast preferentially interacts with T-rich single stranded DNA. When nuclear or whole cell extracts from yeast are used in gel retardation assays, a second activity (distinct from ACBP) can be cross-linked to the T-rich strand of the H4 ARS by exposure to UV light, even in the presence of an excess of single stranded *E. coli* DNA (Fig. 1, labelled complex II). The presence of two T-strand binding factors has also been noted by Kuno et al. 1991. The complexes can be differentiated by several criteria: the more rapidly migrating complex (I) shows a more pronounced sequence specificity, and copurifies with a nuclear scaffold fraction (Fig. 1). The T-rich strand of the ARS consensus competes well for complex I formation, while the complementary A-rich strand and the oligonucleotide dA_{14} do not. Both dT_{14} and an oligo containing a point mutation in the C2G1 ARS consensus, converting the T at position 10 to a G, appear to have lower affinities than the intact ARS consensus (Fig. 1). The upper complex (II) is less sensitive to point mutations in the consensus and is absent from nuclear scaffold extracts. UV crosslinking studies show that the protein forming complex II has an apparent molecular weight around 40 kDa, and is distinct from ACBP by both size and immunological criteria (data not shown).

In order to determine the sequence specificity of purified ACBP, we quantified the ability of the wild type ARS consensus sequence to compete for the binding of purified ACBP to mutant forms of the ARS consensus. The semisynthetic C2G1 ARS was chosen for this study, since an exhaustive analysis of the effect of point mutations in the C2G1 ARS consensus on ARS activity has been published (Van Houten and Newlon 1990; summarized in Fig. 2B). Three of the four point mutations tested significantly reduce the affinity of the ARS binding factor for the ARS consensus (Fig. 2A). Two of these alter thymidine residues (positions 3 and 10) to C and G residues, respectively; both of these modifications also eliminate ARS activity at the C2G1 ARS (Fig. 2B). In one oligonucleotide the guanidine at position 7 is changed to a thymidine, thus enriching the consensus for T residues. This, nonetheless, reduced the ability of this sequence to compete for binding to the ARS factor.

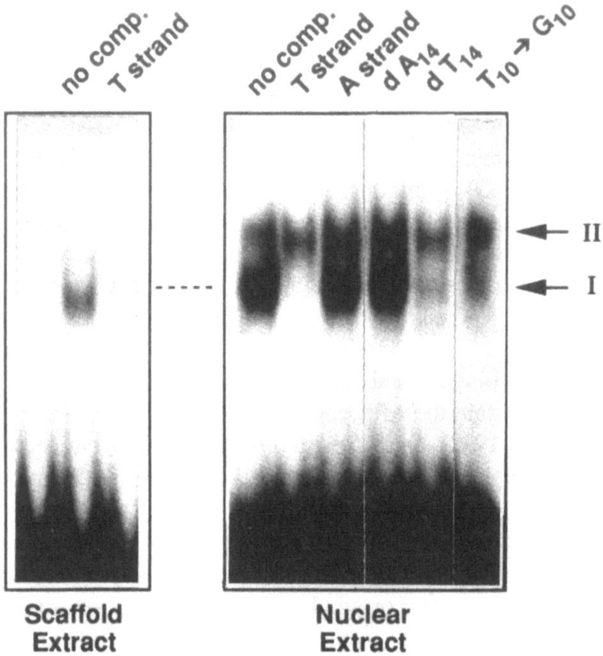

Fig. 1. ACBP is a scaffold factor that binds the T-rich strand of the H4 ARS consensus. Band-shift studies were done with a 26nt oligonucleotide containing the H4 ARS consensus in the presence of 1 μg single-stranded *E. coli* DNA and 1 μg polyd(IC) (see Hofmann and Gasser 1991). Either a soluble extract from yeast nuclear scaffolds or a nuclear transcription extract (Verdier et al. 1991) was used as a source of yeast nuclear proteins. Two complexes are formed with the nuclear fraction (*I* and *II*) while only complex *I* is formed with scaffold proteins. To examine specificity the following competitors were added in a 100-fold molar excess: *T-strand* T-rich H4 ARS oligo, same as labelled probe; *A-strand* complementary strand of the H4 ARS oligo; dA_{14} oligo dA of 14 nucleotides in length; dT_{14} same but oligo dT; $T_{10}G_{10}$ indicates the T-rich strand of the C2G1 ARS oligonucleotide (see Fig. 2A) but with a G substitution at position 10 of the consensus

One of the point mutations tested that inactivates the C2G1 ARS (A to T in position 5) appears to enhance the affinity of ACBP for the consensus in vitro (Fig. 2A). This may indicate that increased affinity to the single-strand consensus is also detrimental to ARS function. The significant variation of binding affinities due to single nucleotide changes nonetheless demonstrates that the ACBP-DNA interaction is sequence specific. In three of four cases the drop in affinity correlates with loss of ARS function (Fig. 2B).

3 ACBP Recognizes the ARS Consensus at HMR-E and at ARS1

An important consideration at this point is whether ACBP binds other endogenous ARS elements and, if so, at which sites. To investigate this, ACBP binding sites were

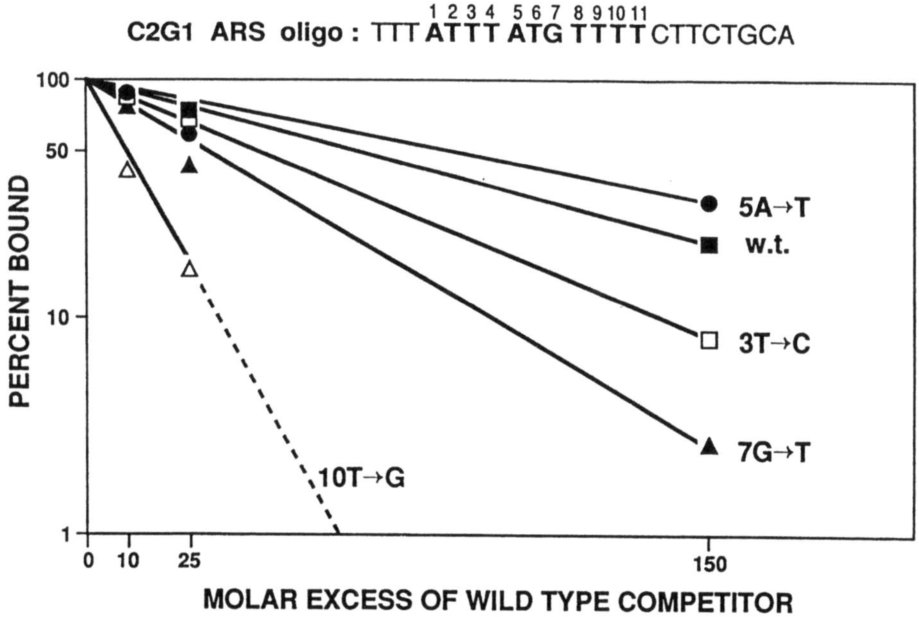

Fig. 2. **A** Single point mutations in the C2G1 ARS consensus affect ACBP binding. Gel retardation assays were done with a 22nt oligonucleotide that contains the ARS consensus of the C2G1 ARS and authentic flanking sequences of the semisynthetic ARS element that was used for functional analysis of point mutants therein (Van Houten and Newlon 1990). The ARS consensus is shown in *bold letters* and the individual positions of the consensus are indicated. The single-stranded oligonucleotides of the wild-type sequence and four sequences with single nucleotide changes at the indicated positions were end-labelled and were incubated in 20 μl bandshift buffer for 20 min at room temperature in the presence of 2 μg single-stranded competitor DNA and 100 ng purified ACBP. The complexes were subsequently loaded on a 5% native polyacrylamide gel. Binding reactions for the C2G1 ARS and the indicated point mutants were also done in the presence of increasing amounts of the wild-type semisynthetic C2G1 ARS oligonucleotide. Competition for complex formation was evaluated by densitometric scanning. The complex formation using the 10T to G mutation in the presence of 150-fold excess of wild-type ARS consensus was below detectable levels, as indicated by the *dotted line*. This figure is reproduced with permission from Hofmann and Gasser (1991). **B** A summary of point mutations in the ARS consensus affecting ARS function and ACBP binding. A summary of important bases in the yeast autonomously replicating sequence (ARS) consensus sequence. Mutagenesis studies (Van Houten and Newlon 1990), reversion studies (Kipling and Kearsey 1990), and ARS consensus-binding protein (ACBP) studies (Hofmann and Gasser 1991) have contributed to the evaluation of the relative importance of the different residues in the yeast ARS consensus. The wild-type consensus sequence is indicated and the position of each residue is *numbered*. For the C2G1 ARS, mutations that created an inactive ARS consensus are indicated by a *black box*, those that weaken but do not eliminate ARS function are *hatched*, and those that are still functional are written in *white boxes* (Van Houten and Newlon 1991). Reversion studies in the M13 vector show that, for functional ARS consensus, 6 out of 11 bases are invariably wild type (Kipling and Kearsey 1991). These are given here in *bold type*. Interaction of ACBP with the wild-type consensus is greatly reduced by point mutations (*bold type*) at positions *3*, *7* and *10*. There is one permitted mutation (*encircled*) at position 5 where thymidine replaces the consensus adenosine

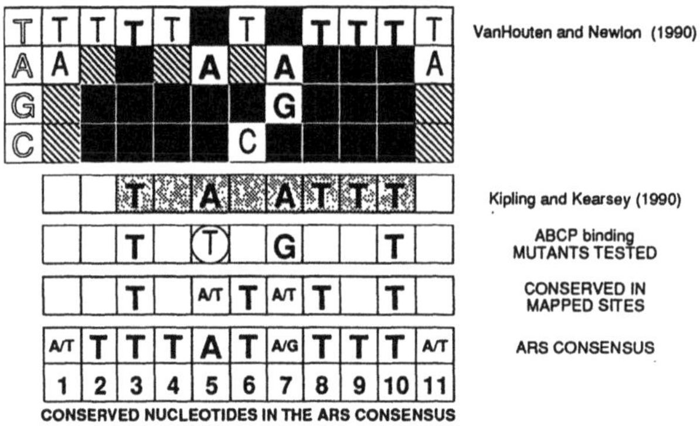

Fig. 2. B

mapped on ARS1, and on the silencer sequence HMR-E which has been shown to contain ARS function (Brand et al. 1987). Purified ACBP binds to the T-rich strand of the HMR-E ARS consensus in the presence of *E. coli* single-stranded competitor, and this binding can be competed by oligonucleotides containing the H4 ARS consensus (Hofmann and Gasser 1991). A 7 bp deletion in the ARS core that abolishes ARS function of the 138 bp AhaIII-AluI silencer fragment (Brand et al. 1987), but recreates a 9/11 bp close match (5'TTTTAGGTATT3') to the consensus, reduces the binding efficiency of purified ARS binding factor by approximately 65%.

Several binding sites for this same factor could be mapped on the two strands of ARS1. A single site is present on the top strand of the sequence as depicted in Fig. 3A. This binding site is found between the PstI and the HindIII sites of ARS1, a fragment that contains the T-rich strand of the only 9/11 bp ARS consensus found inverted with respect to the perfect consensus. Gel retardation experiments using the HindIII-RsaI fragment of the bottom strand show two shifted complexes with different mobility (Fig. 3A). We interpret this to be due to the presence of two potential binding sites for ACBP: probably the T-rich strand of the ARS consensus (between the BglII and the RsaI site) and the T-rich strand of a near match situated between the HinfI and the PstI site. Cleavage of the HindIII-RsaI fragment at the BglII site eliminated one complex, and cleavage at HinfI removed the second (Fig. 3A).

The two binding sites in ARS1 are found in domains that play dominant roles in ARS function, i.e. domain A, that contains the essential ARS consensus; and the proximal portion of domain B, 3' to the ARS consensus, which significantly affects the efficiency of plasmid replication (Diffley and Stillman 1988). As mentioned above, linker mutagenesis into the inverted 9/11 consensus caused at least a four-fold drop in mitotic stability of a centromere-containing plasmid (Marahrens and Stillman 1992).

4 ACBP Binds the H4 ARS Consensus and a 9/11 Near Match to it in the 3' Flanking Domain

Since the 3' domain of the Histone H4 ARS is known to be important for ARS function, it was determined whether there are binding sites for ACBP in this region. Several close matches to the ARS consensus are present in the 3' domain of the ARS core (Holmes and Smith 1989). Binding sites were mapped by gel retardation assays using single-stranded restriction fragments of the H4 ARS. Purified ACBP was bound to these restriction fragments in the presence of single-stranded competitor DNA. Binding was competed with an oligonucleotide containing the H4 ARS consensus (Hofmann and Gasser 1991). A summary of the mapping data is shown in Fig. 3B. In addition to the binding site at the ARS core consensus, a single binding site of ACBP is present at a T-rich region on the opposite strand in the 3' domain of the ARS core consensus, in which there are two overlapping 9/11 bp matches to the consensus.

The borders of the two protein binding sites at the H4 ARS precisely correspond to the borders of the minimal fragment that confers autonomous replication on non-replicating plasmids (Holmes and Smith 1989). Linker insertion, but not point mutations, in the 9/11 bp ARS consensus match to which ACBP binds, increases the loss rate of the mutant H4 ARS on plasmids, whereas deletion of the near match element in the minimal ARS 3' flanking domain results in loss of high frequency of transformation (Holmes and Smith 1989). It is not known if this is the result of deleting an ACBP binding site or another, undefined DNA structure. Two other 9/11 matches to the ARS consensus in the 3' flank of the H4 ARS are not recognized by ACBP in vitro (indicated in Fig. 3B).

5 ACBP Copurifies with Leading Strand Polymerases

In view of the single-stranded binding activity of ACBP it was tested whether or not the purified protein was related to the large subunit of Replication Factor A (RF-A), and whether ACBP has helicase or unwinding activity in the presence of double stranded DNA. Using an anti-peptide antibody to the C-terminus of the 70 kD single strand binding protein of yeast RF-A (kindly supplied by T. Kelly), we found that ACBP is not recognized. In addition, the 70 kD protein recognized by the RF-A antibody in yeast nuclear extracts, migrates more slowly than the protein recognized by anti-ACBP (F. Palladino, data not shown). Both an unwinding assay using double-stranded ARS1 fragments, and a helicase assay using an M13 template with annealled primers were negative for the purified ARS binding protein (Hofmann 1991).

The fact that ACBP can bind specifically to replication origins in vitro suggested that it might nonetheless interact with characterized components of the DNA replication complex. In addition, we observed that wash fractions from the DNA affinity chromatography step of the purification were able to stimulate the DNA binding activity of purified ACBP. Thinking that these stimulatory factors might be DNA polymerases, we tested purified fractions of both the leading strand DNA polymerases II* (ε) and III (δ), and the lagging strand polymerase (α) for the ability to stimulate

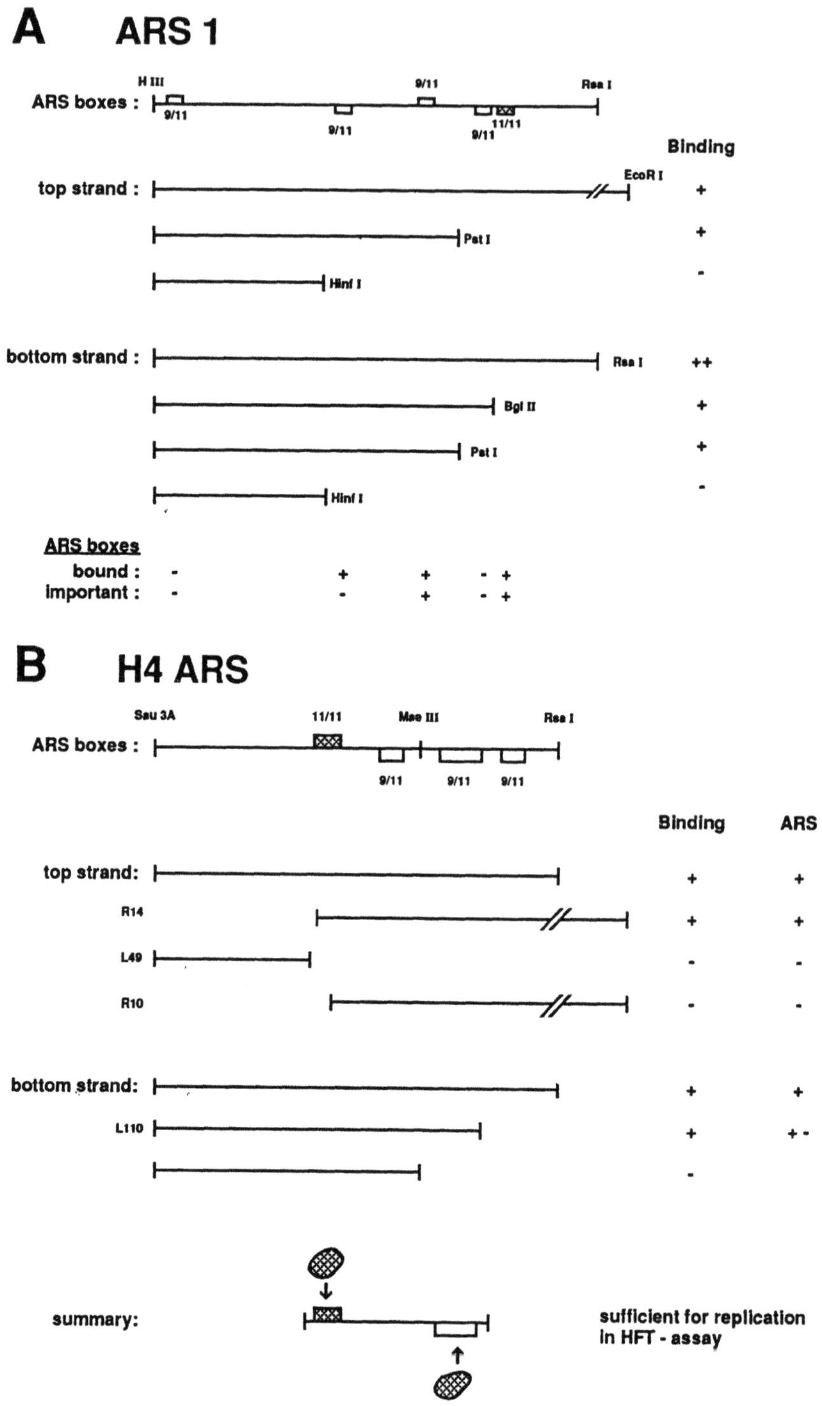
A ARS 1
H III
9/11
Rsa I
ARS boxes :
9/11
9/11
11/11
9/11
Binding
top strand :
EcoR I
+
Pst I
+
Hinf I
-
bottom strand :
Rsa I
++
Bgl II
+
Pst I
+
Hinf I
-
ARS boxes
bound : - + + - +
important : - - + - +
B H4 ARS
Sau 3A
11/11
Mae III
Rsa I
ARS boxes :
9/11
9/11
9/11
Binding
ARS
top strand:
+ +
R14
+ +
L49
- -
R10
- -
bottom strand:
+ +
L110
+ +-
-
summary:
sufficient for replication
in HFT - assay

ACBP binding, and for the presence of ACBP in the purified fractions. We found that the two putative leading strand polymerase (δ and ε) fractions contain a 67 kD protein which binds specifically to the H4 ARS oligo, and which is probably identical to ACBP (see Fig. 4). Moreover, addition of purified yeast DNA polymerases δ and ε (generous gifts of Akio Sugino) stimulates ACBP binding efficiency three-fold (Fig. 4, lanes δ and ε, +). Contrary to this, yeast topoisomerase II and DNA polymerase α fractions contain no endogenous ACBP activity and do not stimulate binding of ACBP.

6 Highly Conserved Epitopes in the ARS Consensus Binding Protein

A polyclonal serum was raised against the purified yeast ACBP. The serum reacts in Western blots with a single polypeptide of 67 kD in whole cell extracts and in purified nuclear fractions from *S. cerevisiae* (Fig. 5). The same serum reacts with proteins of similar sizes in whole cell extracts of *Schizosaccharomyces pombe*, Hela cells, mouse keratinocytes and *E. coli*. Smaller cross-reactive bands at roughly 30 kD in mammalian extracts (open arrow) are recognized by the preimmune serum, while the reactions with proteins between 60 and 65 kD are stimulated in the immune serum (F. Palladino, pers. communication). This argues for conservation of at least one epitope of ACBP throughout evolution. The association of ACBP with DNA polymerases during purification also appears to be a conserved feature of the protein: purified fractions of DNA polymerases α, δ and ε from calf thymus (gifts of U. Hubscher, Zürich), all contain a 65 kD protein that strongly cross-reacts with the anti-ACBP

Fig. 3. The binding of ARS factor to the histone H4 ARS and ARS1 correlates with ARS activity. **A** Binding sites of the ARS consensus binding protein on the two opposite single strands of the Histone H4 ARS and ARS1 were mapped by gel retardation (Hofmann and Gasser 1991). For ARS1, restriction sites indicated at the *bottom* are the genomic sites of the restriction fragments used. In this case the bottom strand contains the 11/11 perfect match of the ARS consensus, and the 3' flank is *to the left of the hatched box. ARS boxes*: boxes above or below the straight line indicate the presence of the T-rich strand of an ARS sequence in the top or bottom strand, respectively. The *stippled box* represents the 11/11 bp ARS consensus sequence. The *empty boxes* represent 9/11 matches to the ARS consensus. *Top strand* and *bottom strand* indicate individual single strands. Complex formation in the gel retardation are indicated by + in the panel "binding". ++ stands for two shifted bands of different mobility in the gel retardation assay, probably representing two ACBP binding sites on the fragment. At the *bottom*, binding sites (Hofmann and Gasser 1991) are compared with data on replicative activity of linker insertion mutations at the indicated ARS consenses (data from Marahrens and Stillman 1992). **B** For *H4 ARS*, the *top* and *bottom strand* refer to the strands of the published sequence (Smith and Andrésson 1983; Bouton and Smith 1986). The *top strand* contains the T-rich strand of the ARS core consensus. *Symbols* are as described above, with the *wider empty box* representing two overlapping 9/11 matches in the 3' flanking region (Holmes and Smith 1989). *Top strand* and *bottom strand* indicate individual single strands. Complex formation in the gel retardation are indicated by + in the panel "binding". If the corresponding double-stranded fragment confers high frequency of transformation on a non-replicating plasmid (Bouton and Smith 1986), this is indicated by + in the panel "ARS". (A and B after Hofmann and Gasser 1991)

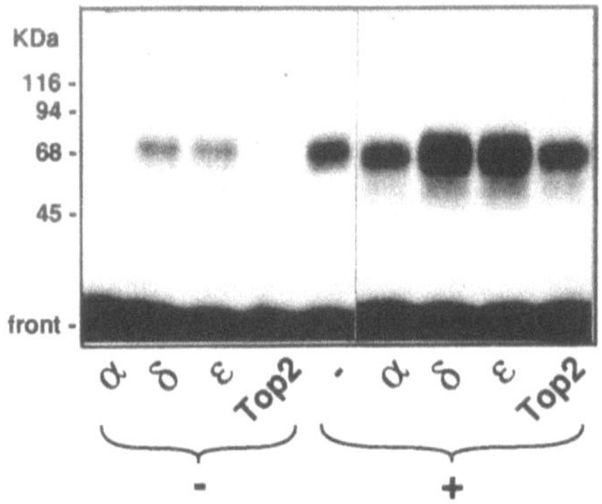

Fig. 4. Detection of endogenous ARS binding activity in yeast DNA polymerase fractions and stimulation of ARS binding activity by polymerases. 0.5 μl (1 μg) aliquots of the following protein samples were tested in the UV crosslinking assay with the T-rich strand of the H4 ARS oligonucleotide, either in the absence (–) or presence (+) of a 0.1 μl aliquot of affinity purified ACBP: Yeast DNA polymerase α (α); yeast DNA polymerase δ (δ), yeast DNA polymerase ε (ε), yeast topoisomerase II (Top2). The fifth lane contained only purified ACBP. In all assays an excess of *E. coli* single-stranded DNA and polyd(IC) were present as described in Hofmann and Gasser (1991). The crosslinked reactions were run on an SDS gel and autoradiographed. Quantification of crosslinked material was done by liquid scintillation of the excised bands

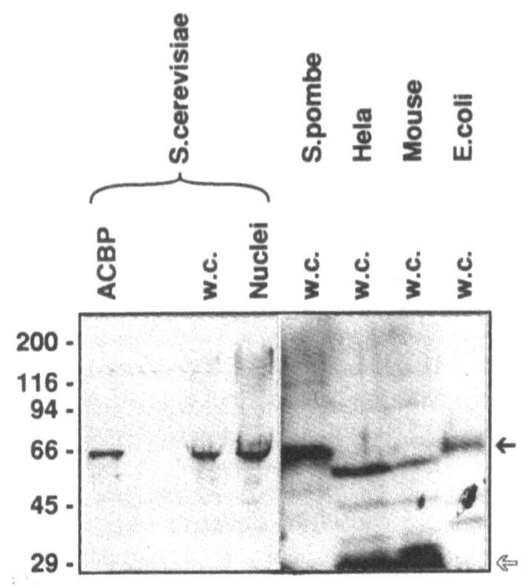

Fig. 5. Characterization of the α-ACBP antiserum. The antiserum was used in Western blots at a dilution of 1:1000 to react with the following extracts: extracts from *Saccharomyces cerevisiae*: a whole cell extract (*w.c.*), a nuclear extract (*Nuclei*), and the purified ARS-binding protein (*ACBP*); a whole cell extract from *Schizosaccharomyces pombe*; a whole cell extract from HeLa cells; a whole cell extract from mouse keratinocytes; and a cell lysate from *E. coli*. The *filled arrow* indicates reactivity stimulated in the immune serum, while reaction with the preserum identifies a protein of roughly 30 kDa in mammalian extracts (data not shown)

serum (data not shown). The association with DNA polymerases argues for a conserved function as well as a conserved epitope in the ACBP protein.

7 Discussion

The data mapping ACBP binding sites are in good agreement with proposed requirements for ARS activity (Palzkill and Newlon 1988). Based on extensive deletion analysis at the yeast genomic ARS C2G1, the authors proposed a requirement for one or several close matches to the ARS core consensus present 3' of the T-rich strand of the 11/11 bp ARS consensus and preferentially in an inverted orientation. Such an organization corresponds to the ACBP binding sites mapped in this paper at the H4 ARS and ARS1.

What might be the role of the ARS binding factor in initiation of DNA replication? The preference of ACBP for one single strand of the ARS consensus provides a potential mechanism for specifying the template for leading strand synthesis and for targetting the appropriate polymerase to this site. The apparent interaction of ACBP with DNA polymerases would be consistent with this proposal. A simple model would predict that bidirectional DNA replication requires the presence of two single-strand binding sites for the ARS binding factor on opposite strands with a 3' orientation to each other, in order to initiate two diverging replication forks. Due to the 5' to 3' directionality of the replicating enzymes, the positioning of two binding sites on opposite strands 5' to each other would result in two converging replication forks which might finally collide and result in a non-functional origin.

Experiments by Holmes and Smith (1989) are consistent with our proposed requirement for an ACBP-binding site on each strand. Inversion of the perfect ARS consensus at the H4 ARS, which places the T-rich sequence on the opposite strand, results in an inactive ARS element. A similar requirement for the correct orientation of the ARS consensus to its 3' flanking domain has been found at ARS1 (Diffley and Stillman 1988). These results may either reflect the need for binding sites on opposite strands, or the disruption of the ACBP binding site by the inversion. By using these constructs and others with mutant 3' domains for bidirectional DNA replication on plasmids and for ACBP binding, this model for ACBP function can be readily tested.

It is clear that origins of replication must contain readily unwound sequences in order to form single stranded templates for primase or helicase insertion. It is not clear why the unwinding region is located 3' of the core consensus. Assuming that the 3' flank provides a second ARS consensus as a protein binding site, as proposed in our model, it is still unclear why this second consensus should be less highly conserved than the core consensus. Although ACBP binds preferentially to the ARS consensus, its requirements for binding in vitro are less stringent than the sequence requirements observed at the core ARS consensus (see Fig. 2A). Thus the requirement for one perfect consensus at each origin may reflect constraints placed on the sequence by another factor, one which perhaps recognizes the DNA sequence in double-stranded form. Further evidence implicating ACBP directly in DNA replication awaits cloning and disruption of the gene.

References

Abraham J, Nasmyth K, Strathern KA, Klar AJS & Hicks JB (1984) Regulation of mating type information in yeast. J Mol Biol 176:307–331

Bouton AH & Smith M (1986) Fine structure analysis of the DNA sequence requirements for autonomous replication of *Saccharomyces cerevisiae* plasmids. Mol Cell Biol 6:2354–2363

Brand AH, Micklem G & Nasmyth K (1987) A yeast silencer contains sequences that can promote autonomous plasmid replication and transcriptional activation. Cell 51:709–719

Brewer BJ & Fangman WL (1987) The localization of replication origins on ARS plasmids in *S. cerevisiae*. Cell 51:463–471

Bouton AH & Smith M (1986) Fine structure analysis of the DNA sequence requirements for autonomous replication of *S. cerevisiae* plasmids. Mol Cell Biol 6:2354–2363

Celniker SE, Sweder K, Srienc F, Bailey JE & Campbell JL (1984) Deletion mutations affecting autonomously replicating sequence ARS1 of *Saccharomyces cerevisiae*. Mol Cell Biol 4:2455–2466

Diffley JFX & Stillman B (1988) Purification of a yeast protein that binds to origins of DNA replication and a transcriptional silencer. Proc Natl Acad Sci USA 85:2120–2124

Eisenberg S, Civalier D & Tye B-K (1988) Specific interactions between a *S. cerevisiae* protein and a DNA element associated with certain autonomously replicating sequences. Proc Natl Acad Sci USA 85:743–746

Hofmann JFX & Gasser SM (1991) Identification and purification of a protein that binds the yeast ARS consensus sequence. Cell 64:951–960

Hofmann JFX (1991) Components of the yeat nuclear scaffold: RAP-1 and an ARS consensus binding protein. PhD. Thesis, University of Lausanne, Lausanne, Switzerland.

Holmes SG & Smith M (1989) Interaction of the H4 ARS consensus sequence and its 3' flanking domain. Mol Cell Biol 9:5464–5472

Huberman JA, Spotila LD, Nawotka KA, El-Assouli SM & Davis LR (1987) The in vivo replication origin of the yeast 2 um plasmid. Cell 51:473–481

Kearsey S (1984) Structure requirements for the function of a yeast chromosomal replicator. Cell 37:299–307

Kipling D & Kearsey SE (1990) Reversion of autonomously replicating sequence mutations in *Saccharomyces cerevisiae*: creation of a eucaryotic replication origin with procaryotic vector DNA. Mol Cell Biol 10:265–272

Kuno K, Murakami S & Kuno S (1990) Single-strand-binding factors which interact with ASR1 of *Saccharomyces cerevisiae*. Gene 95:73–77

Kuno K, Kuno S, Matsushima K & Murakami S (1991) Evidence for the binding of at least two factors, including T-rich strand binding factors, to the single-stranded ARS1 sequence in *Saccharomyces cerevisiae*. Mol Gen Genet 230:45–48

Lohr D & Torchia T (1988) Structure of the chromosomal copy of ARS1. Biochemistry 27:3961–3965

Marahrens Y & Stillman B (1992) A yeast chromosomal origin of DNA replication defined by multiple functional elements. Science 255:817–823

Newlon CS (1988) Yeast chromosome replication and segregation. Microbiol Rev 52:568–601

Palzkill TG & Newlon CS (1988) A yeast replicating origin consists of multiple copies of a small conserved sequence. Cell 53:441–450

Palzkill TG, Oliver SG & Newlon CS (1986) DNA sequence analysis of ARS elements from chromosome III of *Saccharomyces cerevisiae*: Identification of a new conserved sequence. Nucleic Acid Res 14:6247–6263

Schmidt AMA, Heterich SU & Krauss G (1991) A single-stranded DNA binding protein from *Saccharomyces cerevisiae* specifically recognizes the T-rich strand of the core consensus of ARS elements and discriminates mutant sequences. EMBO J 10:981–985

Smith MM & Andrésson OS (1983) DNA-sequences of yeast H3 and H4 histone genes from two non allelic sets encode identical H3 and H4 proteins. J Mol Biol 169:641–661

Stinchcomb DT, Struhl K & Davis RW (1976) Isolation and characterization of a chromosomal replicator. Nature 282:39–43
Sweder K, Rhode PR & Campbell JL (1988) Purification and characterization of proteins that bind to yeast ARSs. J Biol Chem 263:17270–17277
Umek RM & Kowalski D (1988) The ease of unwinding as a determinant of initiation at yeast replication origins. Cell 52:559–567
Umek RM & Kowalski D (1990) Thermal energy suppresses mutational defects in DNA unwinding at a yeast replication origin. Proc Natl Acad Sci USA 87:2486–2490
Umek RM, Linskens MHK, Kowalski D & Huberman JA (1989) New beginnings of eucaryotic DNA replication origins. Biochim et Biophys Acta 1007:1–14
Van Houten JV & Newlon CS (1990) Mutational analysis of the consensus sequence of a replication origin from yeast chromosome III. Mol Cell Biol 10:3917–3925

Eukaryotic Origins of DNA Replication

M. L. DePamphilis[1], W. C. Burhans[1], L. T. Vassilev[1] and Z.-S. Guo[1]

1 Introduction

Initiation of DNA replication acts as a failsafe point in the cell proliferation cycle, because it is a commitment to cell division and, therefore, central to control of the eukaryotic cell cycle (Laskey et al. 1989). In addition, initiation of DNA replication is coupled to programmed changes in gene expression that occur during the life of a single cell as well as during the development of multicellular organisms (Villarreal 1991). Therefore, understanding, at the molecular level, the processes by which DNA replication is initiated is crucial to understanding regulation of cell proliferation, its linkage to gene expression, and its aberrations that lead to diseases such as cancer.

Regulation of DNA replication can be viewed as the interaction of four primary components (Fig. 1): (1) *cis*-acting DNA sequences that functions as the "origin of DNA replication" (reviewed in Burhans et al. 1992), (2) proteins that recognize and bind to the origin (e.g. T-antigen, reviewed in Burhans et al. 1992), (3) proteins that replicate DNA (e.g. DNA primase-DNA polymerase-α, reviewed in Stillman 1989, Challberg and Kelly 1989; Hurwitz et al. 1990; Thömmes and Hübscher 1990), and (4) proteins that regulate the activity of these two groups of proteins (e.g. phosphorylation state of T-antigen and RF-A, reviewed in Prives 1990; Fanning 1992; Din et al. 1990). Thus, one question that is central to understanding how DNA replication is

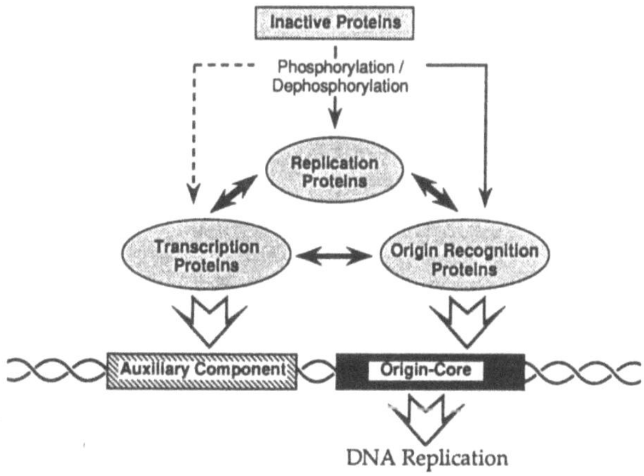

Fig. 1. Basic components involved in regulation of initiation of DNA replication in eukaryotic chromosomes

[1] Roche Institute of Molecular Biology, Roche Research Center, Nutley, NJ 07110, USA.

43. Colloquium Mosbach 1992
DNA Replication and the Cell Cycle

regulated throughout the biological world is: "What is an origin of replication and how does it work?"

2 Coincidence Between Genetic and Functional Origins of Replication

Origins of replication are recognized either genetically as unique, *cis*-acting DNA sequences required for initiation of DNA replication (*ori*), or functionally as those sites where replication begins (reviewed in Burhans et al. 1992). Methods for identification and characterization of eucaryotic origins have been summarized by Vassilev and DePamphilis (1992). The most precise method for mapping functional origins that exist within a chromosome (as opposed to initiation at the ends of linear chromosomes) is to identify the transition from discontinuous to continuous DNA synthesis on each template strand by locating the positions of RNA-primed nascent DNA chains. This defines the *O*rigin of *B*idirectional *R*eplication (OBR). The OBR has been mapped at the resolution of single nucleotides in SV40, PyV, mitochondrial DNA (mtDNA), *E. coli* and bacteriophage λ. In each case, the OBR lies adjacent to or within *ori*. The locations of DNA primase recognition sites at the bidirectional origin in bacteriophage T7, and protein priming sites at the unidirectional origins in the adenovirus (Ad) and bacteriophage ϕ29 families (reviewed in Salas 1991) also reveal coincidence between genetic and functional origins. Mapping replication bubbles using the 2-D gel electrophoresis method on bovine papillomavirus (BPV), Epstein-Barr virus (EBV), and yeast plasmid DNA has placed the OBR at or near *ori*. Thus, in those examples of origins that are most well characterized, a strong coincidence exists between the cis-acting sequences that determine origin activity and the actual site where replication begins.

3 Organization and Function of Origin Components

Based on studies of animal virus and mitochondrial genomes, most, perhaps all, eukaryotic *ori* sequences consist of at least two principal components organized in a highly polarized manner (reviewed in DePamphilis 1988; Burhans et al. 1992):

1. The origin core (*ori*-core) component contains all the *cis*-acting sequence information required to initiate DNA replication. *Ori*-core is the minimal *cis*-acting sequence required to initiate DNA replication under all conditions; it is analogous to a transcription promoter. *Ori*-core provides the binding site for "origin recognition proteins". Origin recognition proteins can participate in DNA replication directly by initiating DNA unwinding at *ori* (e.g. SV40 and PyV T-antigen) or indirectly by associating with replication proteins and thus guiding them to the site where DNA synthesis is to be initiated. For example, SV40 T-antigen binds specifically to SV40 *ori*-core as well as to the catalytic subunit of DNA polymerase-α, the enzyme required to initiate DNA synthesis (Dornreiter et al. 1990; Gannon and Lane 1990). Ad pre-terminal protein associates with Ad DNA polymerase and the resulting complex binds specifi-

cally to Ad *ori*-core (Challberg and Kelly 1989; Salas 1991). Thus, *ori*-core determines where replication begins in the chromosome and in which animal species replication occurs.

2. Origin auxiliary (*ori*-auxiliary) components facilitate the activity of *ori*-core. *Ori*-auxiliary components are dispensible under some conditions without altering the mechanism of replication, and with little if any effect on the efficiency of origin activity (e.g. Ad, BPV, SV40 and PyV). Therefore, they serve an auxiliary function, analogous to transcription enhancers, rather than a required function. *Ori*-auxiliary components consist of transcription factor binding sites (e.g. Ad, BPV, SV40, PyV, EBV, HSV, yeast ARS, and mtDNA) or easily unwound DNA sequences (yeast ARS). The purpose of *ori*-auxiliary components may be to regulate DNA replication by determining under what conditions replication can be initiated. For example, the PyV (PyV) *ori* functions only in those mouse cell types that can activate its enhancer. Transcription factors may also regulate timing of replication during a single S-phase. In mammalian chromosomes, active genes are replicated early during S-phase while silent genes are replicated late.

3. Origins of replication are highly polarized. Sequence motifs and protein binding sites that comprise *ori*-core and *ori*-auxiliary components are frequently organized with strict spatial and orientation rules. Moreover, the site where replication begins is frequently located asymmetrically with respect to the various sequence motifs that comprise *ori*, and RNA-primed and protein-primed DNA synthesis initiation events are found only on one strand of *ori*-core. Therefore, only one of the two template strands in *ori* is actually used to initiate DNA synthesis. Examples are SV40, PyV, Ad, mtDNA (Figs. 2, 3; Guo et al. 1992; Challberg and Kelly 1989; Clayton 1991). The highly polarized nature of replication origins makes it unlikely that two replication forks are initiated simultaneously from the center of the origin. More likely is the "replication zone" model (DePamphilis et al. 1988) in which one fork progresses beyond the origin before initiation begins in the opposite direction.

4 Role of Origin Auxiliary Components

Transcriptional elements were first recognized as *ori*-auxiliary components in the PyV and SV40 origins of DNA replication, and subsequently, a variety of transcriptional elements were identified as components of most, perhaps all, origins of eucaryotic DNA replication. The purpose of *ori*-auxiliary components may be to regulate DNA replication by determining under what conditions replication can be initiated, and to link DNA replication to gene expression.

4.1 Ori-Auxiliary Components Excercise Their Activity Through Transcription Factors

The evidence is now compelling that transcriptional elements that function as *ori*-auxiliary components must bind specific transcription factors in order to facilitate

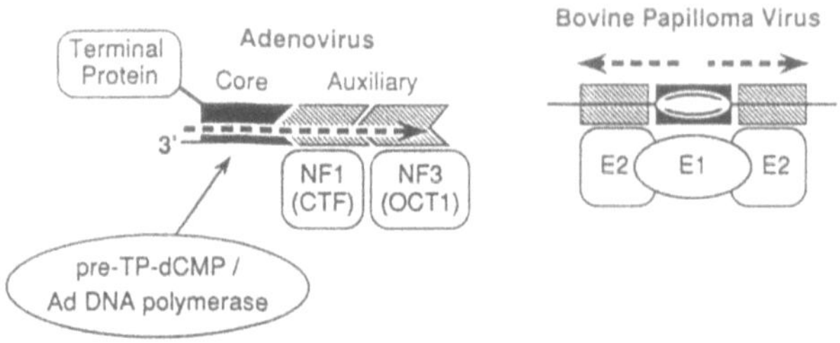

Fig. 2. Specific transcription factors are required to faciliate the simian virus 40 (SV40) and polyomavirus (PyV) origins of DNA replication. Polarity and spacial requirements for origin components are indicated

origin activity. EBV and BPV encode proteins that function both as enhancer activation proteins and origin activation proteins. Tandem arrays of either the EBV EBNA-1 or BPV E2 protein binding sites can stimulate transcription from a variety of promoters, including ones not present in their cognate genomes. However, EBV *ori*-core is fully active only when associated with a tandem array of EBV EBNA-1 protein binding sites and provided with EBNA-1 protein (Wysokenski and Yates 1989; Ambinder et al. 1990) and BPV *ori*-core is fully active only in the presence of BPV E2 protein which binds to sites flanking the BPV *ori*-core (Ustav et al. 1991; Yang et al. 1991).

Cellular transcription factors can also stimulate viral origins. The Ad2 *ori* (Fig. 2) consists of a core component flanked at one end by binding sites for cellular transcription factors NF1(CTF) and NF3(OCT1). Both of these transcription factors have been shown to stimulate Ad2 DNA replication, although only their DNA binding domains are required (Bosher et al. 1990; Verrijzer et al. 1990). SV40 and PyV origins are stimulated by several cellular transcription factors as well as the viral-encoded transcription factor, T-ag, but in contrast to Ad2, stimulation of initiation requires a specific transcription activation domain as well as a DNA binding domain (Fig. 3; re-

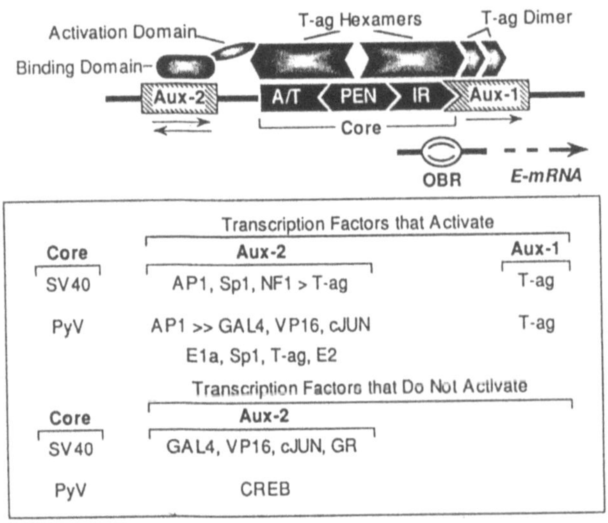

Fig. 3. Transcription factors can facilitate binding of origin recognition proteins. Examples are adenovirus type 2 (Ad2) and bovine papilloma virus (BPV)

viewed in Guo et al. 1992). Given the wide variety of transcription factors that bind to SV40 and PyV enhancers, it is likely that this list will increase.

Two important conclusions emerge from these studies: First, transcription factors that activate one origin do not necessarily activate another, and second, the ability of a transcription factor to stimulate transcription at promoters does not necessarily reflect its ability to stimulate replication activity at origins. Presumably, this reflects the need for specific protein-protein interactions between transcription factors and origin recognition proteins. In the case of Ad2, the fact that NF1 binds to Ad DNA polymerase (Bosher et al. 1990; Chen et al. 1990) suggests that the "activation domain" used to stimulate Ad *ori* activity simply differs from the one used to stimulate RNA polymerase II transcription.

4.2 Specific Transcription Factors Stimulate SV40 and PyV Origins of DNA Replication

SV40 and PyV *aux*-1 consist of a binding site for the viral encoded protein T-ag, which functions as a transcription factor as well as an origin recognition protein. This binding site must reside in a precise orientation and distance with respect to one end of *ori*-core, and cannot be replaced by other sequences, including ones that promote DNA unwinding (Guo et al. 1991). This would occur if the T-ag hexamer responsible for initiating DNA unwinding was facilitated only by association with the T-ag dimer that binds to *aux*-1 (Deb and Deb 1989). Unfortunately, T-ag is also a viral DNA replication protein, making it difficult to alter T-ag protein in order to demonstrate this interaction directly. However, in the case of *aux*-2, it has been shown that specific activation domains are required for transcription factors to stimulate SV40 and PyV *ori*-core, suggesting that only transcription factors that can interact with the T-ag associated with *ori*-core can facilitate *ori*-core activity (summarized in Fig. 3). Association between proteins bound to *aux*-2 and the A/T-rich domain would account for the DNA binding site observed by Yamaguchi and DePamphilis (1986).

Aux-2 can be completely replaced by synthetic oligonucleotides that bind specific transcription factors. The natural SV40 *aux*-2 contains three Sp1 and three T-ag binding sites. A synthetic oligonucleotide that bound Sp1 but not T-ag provided ~75% of *aux*-2 activity, while a sequence that bound T-ag but not Sp1 provided ~20% of *aux*-2 activity (Guo et al. 1992). Therefore, the combined effects of these two proteins could account for SV40 *aux*-2 activity in vivo. The stimulatory activity of SV40 and PyV *aux*-2 requires binding of transcription factors with a specific activation domain. A DNA sequence containing AP1 binding sites completely substitutes for either SV40 or PyV *aux*-2 (Guo et al. 1992), and NF1 binding sites can substitute for SV40 *aux*-2 (Cheng and Kelly 1989). Mutations in the AP1 or NF1 binding sites that eliminate binding of the appropriate transcription factor as well as its ability to stimulate transcription also eliminate the ability of these sequences to stimulate DNA replication. Moreover, synthetic AP1 binding sites can completely replace PyV *aux*-2 in mouse embryonic F9 cells, but only when these cells are supplemented with cFOS and cJUN (Murakami et al. 1991). F9 cells are naturally deficient in cFOS and cJUN.

In other cells, however, the JUN activation domain was less effective. cJUN or vJUN activation domains fused to the *E. coli* LexA DNA binding domain stimulated PyV *ori*-core about 10-fold in mouse fibroblasts when PyV *aux*-2 was replaced with LexA binding sites (Wasylyk et al. 1990). Similar experiments using yeast Gal4 DNA binding sites revealed that the GAL4 DNA binding domain fused to various portions of cJUN could stimulate PyV *ori*-core about 20-fold in mouse fibroblasts, but could not stimulate SV40 *ori*-core in monkey fibroblasts (Guo et al. 1992). These fusion proteins readily stimulated transcription from an appropriate reporter plasmid in the same cells used to test viral DNA replication. Therefore, different members of the AP-1 family must be involved in activating these two origins. Furthermore, the fact that cJUN is at least ten-fold better at stimulating PyV *ori*-core in undifferentiated F9 cells than in differentiated fibroblasts suggests the need for cell-specific "coactivator" proteins to mediate the interaction between transcription factors and replication proteins as well as between transcription factors and an RNA polymerase initiation complex. Part of the difficulty in reproducing the effects of *ori*-auxiliary sequences in vitro may reflect the fact that some transcription factors and their coactivator proteins are either low in abundance or unstable in cell extracts (Guo et al. 1991, 1992).

The need for specific activation domains in proteins binding to *ori*-auxiliary sequences was confirmed by the fact that GAL4 binding sites provided PyV *aux*-2 activity only in the presence of proteins containing both a DNA binding domain as well as an appropriate activation domain. The GAL4 DNA binding domain itself had little, if any, activity in stimulating DNA replication or transcription. However, GAL4 and its fusion proteins strongly stimulated transcription in each of the cell lines tested for SV40 or PyV DNA replication. Wild-type GAL4 protein stimulated PyV *ori*-core 20- to 30-fold, and this stimulation required both the DNA binding domain as well as an activation domain such as that of GAL4, HSV VP16, Ad E1a, pseudorabies virus IE protein, BPV E2 or cJUN (Baru et al. 1991; Bennett-Cook and Hassell 1991; Nilsson et al. 1991; Guo et al. 1992). The effects of AP1 and GAL4 proteins required that the binding sites be placed in close proximity to the A/T-rich element of *ori*-core, and that PyV T-ag be present. However, the activity observed with GAL4 and LEXA fusion proteins was only 10% to 30% as much as that observed with the natural PyV *aux*-2 region ($\alpha + \beta$ enhancers). Furthermore, the effects of GAL4 and its fusion proteins on the SV40 *ori*-core were undetectable, suggesting that SV40 and PyV T-ag molecules interact with different transcription factor activation domains. This would explain why the natural SV40 *aux*-2 region stimulated PyV *ori*-core less than 10% as well as the natural PyV *aux*-2 (Bennett et al. 1989). Some natural transcription factors also failed to stimulate replication. The glucocorticoid receptor (GR) does not replace SV40 *aux*-2 (O'Connor and Subramani 1988) and the cAMP response element-binding protein (CREB) does not replace PyV *aux*-2 (Murakami et al. 1991).

5 Mechanisms by Which Transcription Factors Facilitate the Origin Core Component

1. Transcription factors can promote transcription through *ori*. The only example of this mechanism in eukaryotes is mitochondrial DNA (mtDNA) *oriH* where an upstream promoter directs transcription through three highly conserved sequence blocks where some of these transcripts are cleaved by a site-specific endoribonuclease and used to prime DNA synthesis (Clayton 1991). This mechanism appears unique to mtDNA *oriH*, because SV40 (Li and Kelly 1984; Decker et al. 1986), PyV (C. Prives, pers. comm.) and BPV (Yang et al. 1991) *ori* activities in cell lysates are not affected by α-amanitin, a specific inhibitor of RNA polymerases II and III. Moreover, promoters in Ad and PyV are at least 200 bp away from *ori*-core and direct transcription away from *ori*.

2. Transcription factors can facilitate the assembly of a replication initiation complex. Two examples of this mechanism are the adenovirus (Ad) and bovine papilloma virus (BPV) origins of replication (Fig. 2). NF1 interacts with Ad DNA polymerase to facilitate binding of subsaturating concentrations of Ad2 preterminal protein/DNA polymerase complex (pTP-pol) to Ad *ori*-core; stimulation is 25-fold greater at low pTP-pol concentrations than at high concentrations (Bosher et al. 1990; Chen et al. 1990). Facilitation by NF1 does not involve chromatin structure since stimulation by NF1 is as great with purified proteins and DNA as it is in vivo. Nevertheless, both NF1 and NF3 binding sites are dispensible when replication proteins encoded by Ad4 are substituted for those of Ad2; in Ad4 *ori*, the NF1 site is absent and the NF3 site has no effect (Temperley et al. 1991), demonstrating that Ad4 pTP-pol interacts with *ori*-core more effectively than Ad2 pTP-pol. These results strongly support a role for the NF1 protein in facilitating assembly of a replication initiation complex. The same role seems to apply to BPV E2 protein. Initiation of BPV DNA replication requires two viral encoded proteins, E1 and E2. BPV replication requires both BPV E1 and E2 proteins. E1 is similar to SV40 and PyV T-ag, and E1 alone can bind to *ori*. However, the affinity of E1 for *ori* is increased ~ten-fold in the presence of E2, and both E2 binding sites are protected from nuclease digestion (Ustav et al. 1991; Yang et al. 1991).

3. Transcription factors can facilitate the activity of a replication initiation complex. Two examples are yeast and SV40. The yeast ARS core component is an 11 bp consensus sequence that replicates poorly in the absence of its flanking sequences. All ARS elements are strongly stimulated by "domain B" which must lie proximal to the 3'-end of the T-rich strand of *ori*-core (Holmes and Smith 1989). However, domain B does not appear to require a specific sequence, simply one that unwinds easily, and, therefore, functions as a "DNA unwinding element" (Umek and Kowalski 1990). Whether or not the activity of this "DNA unwinding element" is facilitated by binding specific transcription factors (Marahrens and Stillman 1992) remains to be demonstrated.

An analogous role for *ori*-auxiliary sequences has been proposed for SV40 (Gutierrez et al. 1990), except that *ori*-auxiliary function clearly requires binding of specific transcription factors (Fig. 4). Using extracts of SV40-infected CV-1 cells under conditions that were optimized for SV40 *ori*-auxiliary activity (Guo et al. 1989),

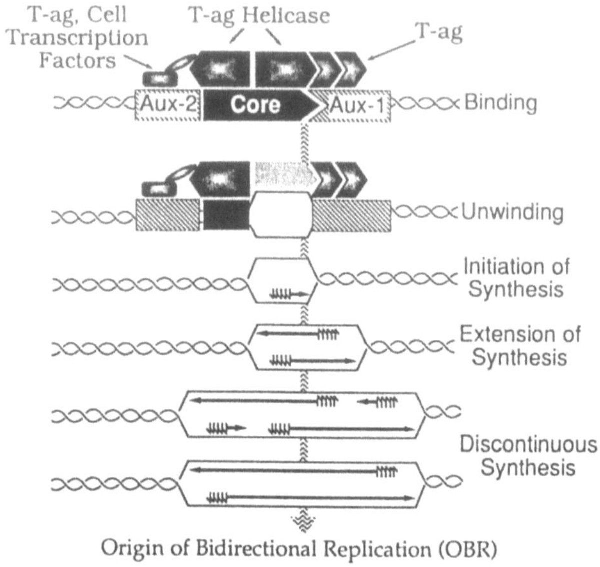

Fig. 4. Transcription factors can facilitate unwinding of DNA at origins. SV40 T-ag helicase binds strongly and specifically to *ori*-core when represented as dsDNA, but weakly and non-specifically to *ori*-core when represented as ssDNA. This change is indicated by change in *shading* of one of the two T-ag hexamers. Therefore, association with transcription factors that bind to flanking sequences (*aux*-1 and *aux*-2) may stabilize the unwound intermediate until DNA synthesis is initiated. Easily unwound DNA sequences ("DUE") appear to provide the same function in yeast ARS elements (Umek and Kowalski 1990)

it was observed that the replicationally active form of T-ag in these extracts has a strong affinity for *ori*-core, but a weak, although specific, affinity for *ori*-auxiliary sequences (Gutierrez et al. 1990). Nevertheless, titration experiments revealed that the activity of *ori*-auxiliary sequences required interaction with T-ag (Guo et al. 1989). In fact, T-ag binding sites alone function as effectively in vitro as they do in vivo when used as *aux*-1 (Guo et al. 1991) or *aux*-2 (Guo et al. 1992). Since saturating amounts of T-ag did not reduce the stimulatory role of *ori*-auxiliary sequences, they must promote initiation of replication at some step following T-ag binding to *ori*. This step appears to be the unwinding of *ori*-core DNA, because deletion of *aux*-1 and *aux*-2 reduced the efficiency of *ori*-specific, DNA unwinding by either T-ag in cell extracts or purified T-ag to the same extent it reduced DNA replication (Gutierrez et al. 1990). Therefore, *aux*-1 and *aux*-2 facilitate T-ag dependent DNA unwinding after the T-ag preinitiation complex has bound to *ori*-core. As T-ag proceeds to unwind *ori*-core, its strong sequence-specific affinity for duplex DNA is lost. During this transition from DNA unwinding by T-ag to initiation of DNA synthesis by DNA primase-DNA polymerase-α, T-antigen's moderate affinity for *aux*-1 and *aux*-2 may promote DNA unwinding by stabilizing the initial unwound intermediate.

Transcription factors can prevent repression from chromatin structure. Ori-auxiliary sequences can also facilitate formation of initiation complexes indirectly by preventing chromatin structure from interfering with binding of replication proteins (Fig. 5). Preassembly of nucleosomes onto a DNA substrate can interfere with binding of

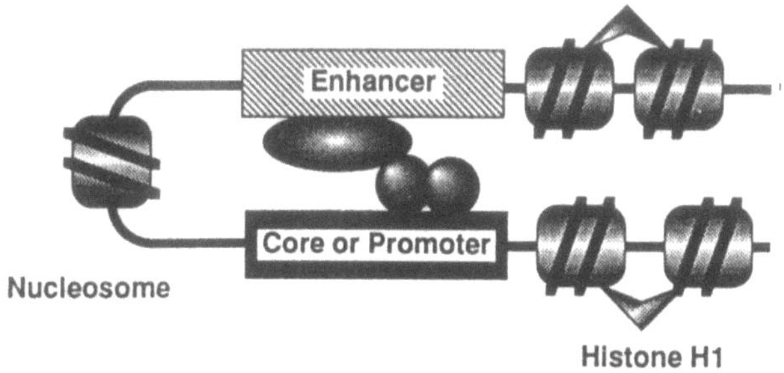

Fig. 5. Transcription factors can prevent repression of replication origins and transcription promoters by chromatin structure. This mechanism appears to apply to polyomavirus (PyV) and Epstein-Barr virus (EBV)

transcription factors to promoters or T-ag to *ori*-core (Cheng and Kelly 1989). In the case of SV40, this interference can be prevented by prebinding of NF1 to a synthetic *aux*-2 sequence prior to assembly of the DNA into nucleosomes in *Xenopus* extracts (Cheng and Kelly 1989). Since NF1 binding sites could replace *aux*-2 in monkey cells, but failed to stimulate replication in HeLa cell extracts, it was suggested that the role of *aux*-2 is to prevent repression of *ori*-core by chromatin structure. However, subsequent studies using Gal4 binding sites to replace *aux*-2 demonstrated that this in vitro chromatin reconstitution system does not accurately model *aux*-2 activity in vivo. Prebinding of GAL4:VP16 to DNA in vitro prevented repression of SV40 DNA replication in vitro (Cheng et al. 1992), but Gal4 binding sites in the presence of GAL4:VP16 did not stimulate SV40 DNA replication when transfected into COS-1 cells, although they did stimulate a TATA box promoter at least 300-fold in COS-1 cells (Guo et al. 1992). Furthermore, the same result can also be achieved by prebinding of T-ag to *ori* (J. Sogo, pers. comm.), suggesting that T-ag alone can prevent interference from chromatin structure. The question remains as to whether or not proteins bound to *aux*-2 facilitate T-ag in preventing repression during nucleosome assembly. NF1 may have failed to stimulate SV40 *ori*-core in HeLa extracts because appropriate coactivator proteins were not present in sufficient amount. Moreover, repression in chromatin with normal nucleosome spacing is mediated primarily by histone H1 (Laybourn and Kadonaga 1991), and histone H1 is absent from chromatin assembled in *Xenopus* oocytes (Wolffe 1989). Therefore, preassembly of chromatin in extracts of *Xenopus* oocytes may not provide an appropriate model for repression by chromatin in mammalian cells.

Nevertheless, the ability of enhancers to prevent repression by chromatin structure appears to be a valid model for *aux*-2 activity in PyV and EBV, as well as the effect of enhancers on promoters. In the case of PyV, *ori*-core no longer requires an enhancer when chromatin structure is either absent or altered. The PyV enhancer (*aux*-2) is not required to activate PyV *ori*-core in cytoplasmic extracts (Prives et al. 1987), conditions under which chromatin assembly does not occur (Gruss et al. 1990). Enhancers also are not required to activate either an *ori*-core or a promoter encoded by plasmids injected into the male pronucleus of fertilized mouse eggs (Martínez-Salas et al. 1988, 1989; Wiekowski et al. 1991) where repression by chromatin structure is absent (M. Wiekowski and M. DePamphilis, submitted). Enhancers appear to prevent chromatin formation over an origin or promoter by linking the enhancer DNA region with either the origin or promoter DNA region through association of proteins bound specifically to the enhancer with proteins bound specifically to either the *ori*-core (DePamphilis et al. 1988) or promoter (Felsenfeld 1992). This close association of two regions of duplex DNA will prevent nucleosome assembly over the origin or promoter region.

In summary, *ori*-auxiliary components display multiple functions, and the extent to which they are needed may simply reflect the relative abilities of proteins that initiate replication such as T-ag and pTP-pol to bind *ori*-core and initiate DNA unwinding or DNA synthesis. NF1 and E2 proteins may facilitate binding of origin recognition proteins to *ori*-core. However, since Py T-ag binding to Py *ori*-core is much weaker than T-ag binding to SV40 *ori*-core, chromatin structure may interfere more strongly with Py *ori*-core activity than with SV40 *ori*-core activity. For example, transcription factors can differ significantly in their ability to bind DNA organized into nucleosomes. Therefore, enhancers may provide a specific mechanism for preventing repression by chromatin structure. This would account for the ability to dispense with the need for enhancers to activate origins or promoters under certain conditions. SV40 and Py *aux*-1 as well as yeast ARS domain B appear to promote DNA unwinding, although perhaps by different mechanisms. This would account for the consistently strong effect of SV40 *aux*-1 both in vivo and in vitro.

6 DNA Replication in Mammalian Chromosomes Begins at Specific Sites

Given the great wealth of information about origins of DNA replication in procaryotic cells, their bacteriophage and plasmids, animal viruses and mitochondrial DNA, and even the simple eukaryote, yeast, it is surprising that one of the major questions remaining in this field is whether or not members of the animal kingdom utilize similar sequences to initiate DNA replication in their chromosomes. Attempts to identify *A*utonomously *R*eplicating *S*equence (ARS) elements in mammalian or insect cells has generated contradictory as well as controversial results (Linskens and Huberman 1990; Burhans et al. 1992). It is fair to state that at this time, there is no compelling evidence that metazoan chromosomes utilize ARS elements analogous to those found in yeast. In fact, experiments with eggs and egg extracts from frogs, sea urchins or fish reveal that these cells can initiate semiconservative replication in virtually any DNA molecule, suggesting that initiation of replication does not require unique *cis*-acting sequences. Moreover, the number of initiation sites for DNA replication in preblastula embryos of amphibians and insects is significantly greater than in differentiated cells of the same animal (McKnight and Miller 1977), suggesting that many DNA sequences are capable of acting as initiation sites. On the other hand, injection of DNA into preblastula embryos of mammals revealed that only those DNA molecules replicated that contained a known origin of replication such as those from polyomavirus or SV40, and that the requirements for initiation of viral replication were those expected from experiments in mammalian differentiated cells

(DePamphilis et al. 1988). The only exception was replication or expression of genes within the paternal pronucleus of one-cell mouse embryos. Under these conditions, enhancers were no longer required to activate either the PyV origin core component or a transcription promoter (Wiekowski et al. 1991). These results suggest that all mammalian cells require unique *cis*-acting sequences to initiate either replication or transcription, although no nonviral DNA sequence has yet been demonstrated to function as an origin when injected into mouse embryos. Therefore, we set about to identify a true origin of replication that functions in mammalian chromosomes.

Heintz and Hamlin (1982) and later Burhans et al. (1986a, b) identified a DNA fragment downstream of the dihydrofolate reductase (DHFR) gene in Chinese hamster ovary (CHO) cells that was rapidly labeled with DNA precursors (Fig. 6), suggesting the presence of a replication origin in this region. However, these experiments suffered from the fact that they were carried out on the amplified DHFR locus in CHO C400 cells, a cell line containing 500 tandem copies of an approximately 200 kilobases (kb) long genomic region. Furthermore, cells were synchronized at their G1/S boundary by amino acid starvation and treatment with aphidicolin, a drug that inhibits specifically DNA polymerases α and δ. Thus, this putative replication origin may have been specific for amplified DNA, or the synchronization procedure may have contributed to preferential labeling of specific regions of the genome. Therefore, Vassilev et al. (1990) searched for an origin using a method that is sensitive enough to detect single copy sequences, and that requires neither synchronization of cells nor metabolic inhibitors.

6.1 Nascent Strand Method for Mapping Origins

In an asynchronous population of proliferating mammalian cells, the size of replication bubbles generated by each origin of replication will vary from small bubbles that had just been initiated to large bubbles that are about to complete replication. If one can identify the shortest nascent DNA strands that contain a unique DNA sequence, then one can map the origin of replication for those nascent strands. If replication is bidirectional, the origin of replication is at the center of the strands. If replication is unidirectional, then the origin is at one end. To map the DHFR origin, three unique DNA segments were selected that were distributed across the putative origin region. Nascent DNA chains were briefly radiolabeled in vivo with ^{3}H-dC and BrdU. Nascent DNA was then fractionated according to length under denaturing conditions, and newly replicated ^{3}H-BrdU-DNA in each fraction was purified from contaminating unreplicated DNA by immunoprecipitation with anti-BrdU antibodies. PCR amplification was used to increase the concentration of each DNA segment to levels easily detected by blotting-hybridization using sequence-specific ^{32}P-probes.

Depending on their length at the time when ^{3}H-dC and BrdU were added, nascent DNA strands will contain one or all of the unique sequences selected. If the origin lies between two segments, then the shortest nascent chains will contain only that segment closest to the origin, while the longer nascent chains will contain the segment closest to the origin plus one or more of the other segments selected. If replication is bidirectional, then nascent DNA chains will contain DNA segments from both sides of the origin with the ratio of origin proximal segment to origin distal segments decreasing as shorter chains grow into longer chains. The size of the shortest DNA fraction that contains a given segment is determined, and from this distance, the location of the origin is calculated. Results from this method demonstrated that replication originated bidirectionally about 17 kb downstream from the 3'-end of the DHFR gene within a 2.5 kb locus that is coincident with the initiation region identified by early labeled DNA fragments (Fig. 6). Therefore, this region must contain a true mammalian origin that functions in single-copy sequences during normal cell proliferation.

6.2 Okazaki Fragment Distribution Method for Mapping Origins

Two questions remained. First, did replication from this DHFR origin utilize the replication fork mechanism, and second did replication begin within a small region of DNA comparable in size to origins found in the genomes of other organisms? Replication can occur in three ways: (1) displacement of one template strand with concomitant DNA synthesis on the other template (ϕ29 phage, Ad, and mtDNA), (2) displacement of both template strands followed by random initiation of synthesis on both templates (suggested for preblastula *Xenopus* embryos), and (3) the replication fork mechanism (chromosomes of bacteria, bacteriophage T4, T7 and λ, animal viruses SV40 and PyV, *Drosophila* and plasmids in bacteria and yeast). In the replication fork mechanism, DNA synthesis occurs concomitantly on both DNA template strands as rapidly as they are unwound. Since the two template strands are antiparallel, and all DNA polymerases synthesize DNA only in the 5' to 3' direction, the direction of synthesis on one template must be opposite that of fork movement (retrograde). This is accomplished by the repeated initiation of short nascent DNA chains ("Okazaki fragments") that are eventually joined to the 5'-ends of long nascent DNA chains (discontinuous DNA synthesis). When DNA replication proceeds bidirectionally from specific sites, the resulting transition from discontinuous to continuous DNA synthesis that occurs on each strand of DNA defines an *O*rigin of *B*idirectional *R*eplication (OBR, Fig. 4). Thus, the replication fork model predicts that Okazaki fragments will anneal predominantly, if not exclusively, to the retrograde template under all conditions of labeling. This permits identification of an OBR based on template specificity of Okazaki fragment synthesis. Such a strategy was applied to the DHFR initiation region (Burhans et al. 1990).

Cells were synchronized at their G1/S border in order to accumulate a large number of replication forks within the initiation zone. In an asynchronous population of mammalian cells, the quantity of Okazaki fragments that could be isolated from a reasonable number of cells is insufficient to detect a hybridization signal. Okazaki fragments were labeled for 1.5 min by treating the cells with NP-40 to make them permeable to dNTPs, and then releasing them into S-phase in the presence of [α-^{32}P]dATP and BrdUTP under conditions that permit faithful continuation of DNA replication events initiated in vivo downstream of the DHFR gene (Burhans et al. 1986a; Heintz and Stillman 1988). To obtain highly purified populations of Okazaki fragments, DNA was fractionated according to its size by gel electrophoresis, recovered by electroelution, and ^{32}P-DNA separated from unlabeled parental DNA by immunoprecipitating single-stranded Br-DNA with anti-BrdU antibodies. Nascent DNA was then hybridized to individual complementary template strands that had been cloned into single-stranded bacteriophage M13 DNA and immobilized on membranes.

Okazaki fragments can be distinguished from DNA repair products in four ways (DePamphilis and Wassarman 1980; 1982; DePamphilis 1987): (1) Okazaki fragments are short nascent DNA fragments with a mean length of 100 to 150 nucleotides; (2) Okazaki fragments are rapidly joined to long growing nascent DNA strands; (3) Okazaki fragments contain a short oligoribonucleotide of 8 to 12 nucleotides covalently attached to their 5'-ends (the "RNA primer"); (4) Okazaki fragments anneal predominantly, if not exclusively, to DNA templates representing the retrograde arms of replication forks. All four criteria have been demonstrated for the ^{32}P-DNA fragments used to locate the OBR in viral (Hay and DePamphilis 1982; Hendrickson et al. 1987) and cellular (Burhans et al. 1990, 1991) genomes.

Results from this approach have identified an OBR within a 0.45 kb region 17 kb downstream from the DHFR gene in both single copy CHO cells and CHO C400 cells containing amplified DHFR genes (Fig. 6; Burhans et al. 1990). At least 80% of the replication forks within the 27-kb region examined emanated from this OBR. Since this OBR lies within the initiation locus first identified by Heintz and Hamlin (1982) and referred in later papers by Hamlin and coworkers as ori-β, we refer to this OBR as OBR-β. The same strategy for locating an OBR, combined with cutting the Okazaki fragment ^{32}P-DNA:template DNA hybrids at a unique restriction site, has been used to map the OBR with single nucleotide resolution in SV40, PyV, *E. coli* and bacteriophage λ. Thus, initiation of DNA replication in mammalian chromosomes appears to use the same replication fork mechanism previously described in a variety of prokaryotic and eukaryotic genomes, suggesting that mammalian chromosomes also utilize specific *cis*-acting sequences as origins of replication.

6.3 Imbalanced DNA Synthesis Method for Mapping Origins

Mapping an OBR by measuring the distribution of Okazaki fragments between the two arms of replication forks takes advantage of a naturally occurring asymmetry resulting from discontinuous synthesis on retrograde arms. In an effort to confirm the results of Burhans et al. (1990), we set out to determine whether or not the same populations of replication forks could be recognized by mapping the direction of continuous DNA synthesis on forward arms of forks. Such a method had been reported by Handeli et al. (1989). They analyzed DNA synthesis on forward arms of replication forks after treating exponentially proliferating cells with emetine, a general inhibitor of protein synthesis. Earlier studies had suggested that, in the absence of histone synthesis, histone octamers in front of replication forks segregate exclusively to forward arms of replication forks, leaving nascent DNA on retrograde arms unprotected by histones and therefore sensitive to nonspecific endonucleases such as micrococcal nuclease. If this were true, then an OBR could be recognized by the transition from nuclease protected to nuclease sensitive nascent DNA on each template strand, analogous to measuring the transition from continuous to discontinuous DNA synthesis. When Handeli et al. (1989) applied this approach to the DHFR locus in CHO cells, they concluded that the bulk of replication forks downstream of the DHFR gene emanated from an OBR somewhere within a 14-kb region that included OBR-β (Burhans et al. 1990) and the initiation zone identified by Vassilev et al. (1990). We have confirmed that this strategy does map the OBR. However, we have also found that the primary reason that this method works results from preferential inhibition of Okazaki fragment synthesis, and not from conservative nucleosome segregation. When protein synthesis is inhibited in vivo, discontinuous synthesis is rapidly inhibited while continuous synthesis is slowly inhibited. Mechanistically, this suggests that some component of DNA primase-DNA polymerase-α complex that is required for initiation of Okazaki fragments undergoes rapid turnover in vivo, while DNA polymerase-δ complex that is used to extend DNA chains is stable (see Fig. 4). Thus, emetine induces "imbalanced DNA synthesis" leading to DNA synthesis on forward arms in the absence of synthesis on retrograde arms.

The "imbalanced DNA synthesis" method begins by incubating exponentially proliferating mammalian cells with emetine to inhibit protein synthesis and BrdU to label DNA synthesized in the absence of protein synthesis. Cellular DNA is then isolated and sonicated in order to separate labeled (heavy) from unlabeled (light) regions on the same DNA stand. Heavy DNA is separated from light DNA by equilibrium centrifugation in alkaline Cs_2SO_4 gradients. Heavy DNA and light DNA are then immobilized on a membrane and hybridized with strand-specific ^{32}P-labeled RNA probes representing specific segments of DNA throughout the genomic region of interest. The OBR is recognized by the transition from forward to retrograde template on each DNA strand: On one side of the OBR, heavy DNA representing continuous DNA synthesis on forward arm templates will anneal to the "Watson" strand while on the other side of the OBR, it will anneal to the "Crick" strand.

Burhans et al. (1991) observed that, in the presence of emetine, nascent DNA on forward arms of replication forks in hamster cell lines containing either single or amplified copies of the DHFR gene region was enriched five- to seven-fold over nascent DNA on retrograde arms. This forward arm bias was observed on both sides of OBR-β, consistent with at least 85% of replication forks within this region emanating from OBR-β (Fig. 6). These results were consistent with those of Handeli et al. (1989). However, the replication fork asymmetry induced by emetine does not result from conservative nucleosome segregation, as previously believed (Handeli et al. 1989), but from preferentially inhibiting Okazaki fragment synthesis on retrograde arms of forks to produce "imbalanced DNA synthesis". Three lines of evidence support this conclusion. First, the bias existed in long nascent DNA strands prior to nuclease digestion of non-nucleosomal DNA. Second, the fraction of RNA-primed Okazaki fragments was rapidly diminished. Third, electron microscopic analysis of SV40 DNA replicating in the presence of emetine revealed forks with single-stranded DNA on one arm, and nucleosomes randomly distributed to both arms. Thus, in the presence of either cycloheximide (Cusick et al. 1984; Sogo et al. 1986) or emetine, nucleosome segregation is distributive. Cycloheximide also preferentially inhibits Okazaki fragment synthesis (J. Sogo, unpubl. data).

7 Metazoan Origins of Replication Appear to Function Only in the Context of a "Real" Chromosome

Based on the studies cited above and summarized in Fig. 6, we conclude that DNA replication in mammalian chromosomes begins at specific DNA sites and proceeds bidirectionally using the replication fork mechanism. However, this region of hamster DNA does not function as an ARS element in plasmids (5 kb) or cosmids (32 kb) when transfected or electroporated into several different cell lines; the level of replication observed in DNA molecules containing the CHO DHFR OBR-β region was indistinguishable from DNA molecules of similar size and shape that did not contain this sequence. The same was true when plasmids were incubated with cell extracts capable of replicating SV40 DNA, or were injected into the nuclei of mouse one-cell and two-cell embryos. However, Handeli et al. (1989) reported that a 14-kb sequence containing OBR-1 functioned as an origin of replication when randomly integrated into CHO cell chromosomes, suggesting that cellular origins, in contrast to viral origins, may function only in the context of a natural chromosome.

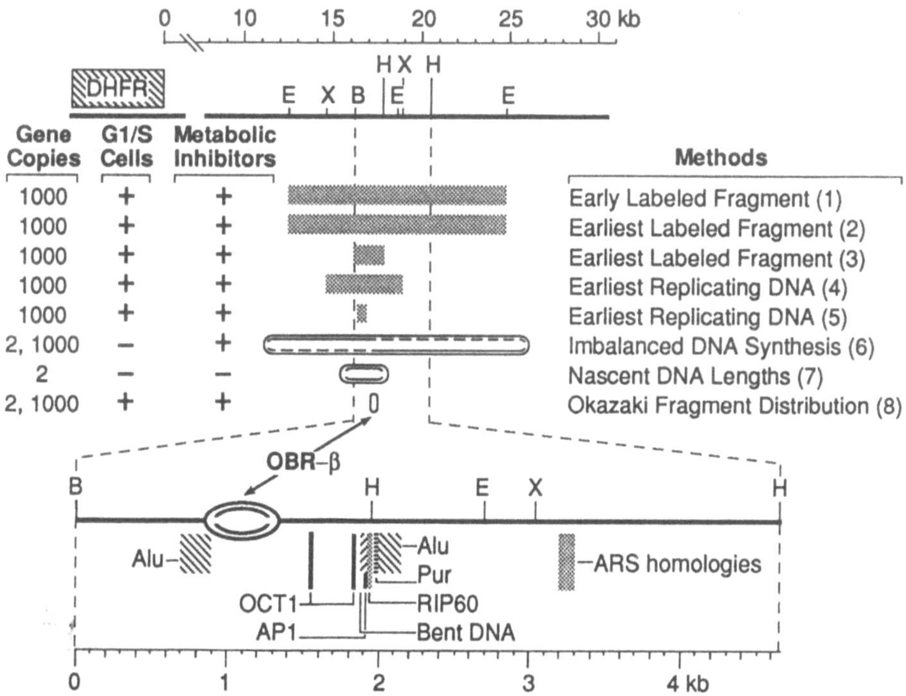

Fig. 6. Replication in mammalian chromosomes occurs bidirectionally from specific DNA sites. The clearest example is found downstream of the hamster DHFR gene [*O*rigin of *B*idirectional *R*eplication (OBR-1)]. "Methods" are described in (*1*) Heintz and Hamlin (1982), (*2, 3*) Burhans et al. (1986a, b), (*4, 5*) Leu and Hamlin (1989), Anachkova and Hamlin (1989), (*6*) Handeli et al. (1989), Burhans et al. (1991), (*7*) Vassilev et al. (1990), and (*8*) Burhans et al. (1990). Restriction enzyme sites are Eco RI (*E*), Xba I (*X*), Bam H1 (*B*), and HindIII (*H*). Details of the OBR region are described by N. Heintz (this Vol.)

All of the analyses of newly synthesized DNA chains that were identified by incorporation of density-labeled and radio-labeled DNA precursors led to the conclusion that DNA replication in the DHFR region begin at a site-specific OBR. Surprisingly, analysis of replicating DNA structures by 2-D gel elctrophoresis led to the opposite conclusion. In this method, the total population of cellular DNA is cut into unique restriction fragments that are then fractionated by two-dimensional gel electrophoresis on the basis of their size and shape. Unique DNA sequences are identified by hybridization with specific ^{32}P-DNA probes. Replication intermediates are identified on the basis of theoretical considerations together with analysis of simple model systems such as replicating plasmid DNA. When this method was applied to the DHFR locus in CHO C400 and CHO cells, DNA replication appeared to be initiated randomly throughout the 55-kb region downstream of the DHFR gene with forks traveling in both directions at equal frequency (Vaughn et al. 1990; Dijkwel et al. 1992; P. Dijkwel, pers. comm.). However, if this interpretation of the 2-D gel data is correct, then it should not be possible to observe asymmetries due to replication forks: Okazaki fragments as well as long nascent DNA chains would anneal equally well to both templates throughout this genomic region!

There are three ways that one can reconcile the 2-D gel results with those based on analysis of nascent DNA strands. First, it may be difficult to quantify the number of initiation events in any given DNA region of the gel, so that while some initiation events do occur throughout a large DNA region, most initiation events occur at a specific site (i.e. OBR-β). The second possibility is that artifacts might arise in the 2-D gel method when applied to complex eukaryotic genomes such as those found in mammalian cells (discussed in Vassilev and DePamphilis 1992). The 2-D gel method searches for structures that migrate as replication intermediates regardless of whether or not they contain nascent DNA. It is possible to imagine artifacts in which unusual DNA structures could arise from the increased sequence complexity of mammalian genomes or from the procedures used to enrich for replication bubbles may be misidentified as replication intermediates. Finally, two theoretical models have been suggested for chromosome replication in cells from multicellular organisms with complex differentiated organization (Metazoa).

Linskens and Huberman (1990) have suggested that replication begins at many sites throughout a large DNA region, but that replication events initiated outside the OBR identified by analysis of nascent DNA are restricted to unidirectional replication away from the OBR. In this way, all of the active replication forks on one side of the OBR would progress in the opposite direction of the active replication forks on the opposite side of the OBR. We suggest an alternative model in which many DNA sequences can potentially serve as replication origins, but only a small subset of those sequences act as replication origins in the context of a natural chromosome (Fig. 7). Thus, when cells are presented with bare DNA molecules introduced by transfection, injection, or electroporation, or when DNA is added to cell extracts, each DNA molecule can initiate at one of many possible sites. This creates the perception that initiation of replication is independent of sequence specificity. However, when DNA is introduced into cytoplasm of *Xenopus* eggs, it will not replicate until it has first been organized into chromatin and then the chromatin organized into "pseudo-nuclei" (Laskey et al. 1989). If nuclei assembly is blocked, DNA replication does not occur.

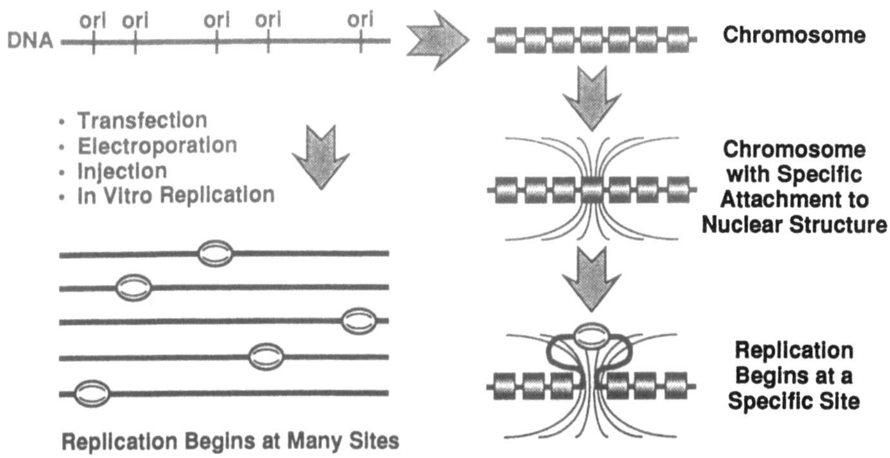

Fig. 7. The 'Jesuit Model' for initiation of DNA replication in animal chromosomes. Cells from multicellular organisms with complex differentiated organization (Metazoa) can initiate DNA replication at many sequences when presented with relatively small DNA molecules by transfection, injection, or electroporation, or when DNA is added to cell extracts. However, when the same DNA is first organized into nuclei, some aspect of nuclear organization limits replication in these "real chromosomes" to a small subset of potential origin sequences. Thus, selection of initiation sites in cellular chromosomes is reminiscent of the Jesuit's dictum that "many are called but few are chosen"

Therefore, some aspect of nuclear organization may limit replication in chromosomes to a small subset of potential origin sequences. In effect, nuclear structure designates one of many potential origins as the primary origin of replication (i.e. the OBR). The appearance of additional replication bubbles at other sites along the DNA may represent either low frequency initiation events or aborted initiation events at the many alternative sites. Since each "pseudo-nucleus" formed in frog egg extract appears to select a different initiation site (Hyrien and Méchali 1992; Mahbubani et al. 1992), we suggest that all nuclei derived from the same progenitor nucleus must use the same origin, and that this is the origin (OBR) that is mapped in cultured cell lines. Thus, selection of initiation sites in cellular chromosomes is reminiscent of the Jesuit's dictum that "many are called but few are chosen". If, in fact, a mammalian origin of replication is not simply a unique DNA sequence, but a sequence associated with a particular nuclear structure, then initiation of replication will automatically disassemble this structure and thus erase the origin until it is once again reassembled during G2 or M phases of the cell cycle. In that way, a mammalian cell could accomplish one of its major objectives: to replicate all its DNA by initiating replication at thousands of different origins without ever replicating the same region twice in a single S-phase.

References

Anachkova B & Hamlin JL (1989) Replication in the amplified dihydrofolate reductase domain in CHO cells may initiate at two distinct sites, one of which is a repetitive sequence element. Mol Cell Biol 9:532–540
Baru M, Shlissel M & Manor H (1991) The yeast GAL4 protein transactivates the polyomavirus origin of DNA replication in mouse cells. J Virol 65:3496–3503
Bennett-Cook ER & Hassell JA (1991) Activation of polyomavirus DNA replication by yeast GAL4 is dependent on its transcriptional activation domains. EMBO J 10:959–969
Bosher J, Robinson EC & Hay RT (1990) Interactions between the adenovirus type 2 DNA polymerase and the DNA binding domain of nuclear factor I. New Biol 2:1083–1090
Burhans WC, Selegue JE & Heintz NH (1986b) Replication intermediates formed during initiation of DNA synthesis in methotrexate-resistant CHOC 400 cells are enriched for sequence derived from a specific, amplified restriction fragment. Biochemistry 25:441–449
Burhans WC, Selegue JE & Heintz NH (1986a) Isolation of the origin of replication associated with the amplified Chinese hamster dihydrofolate reductase domain. Proc Natl Acad Sci USA 83:7790–7794
Burhans WC, Ortega JM & DePamphilis ML (1992) Origins of DNA replication in eukaryotic cells: What are they and how do they work? Crit Rev Biochem Mol Biol (in press)
Burhans WC, Vassilev LT, Wu J, Sogo JM, Nallaseth F & DePamphilis ML (1991) Emetine allows identification of origins of mammalian DNA replication by imbalanced DNA synthesis, not through conservative nucleosome segregation. EMBO J 10:4351–4360
Burhans WC, Vassilev LT, Caddle MS, Heintz NH & DePamphilis ML (1990) Identification of an origin of bidirectional DNA replication in mammalian chromosomes. Cell 62:955–965
Challberg MD & Kelly TJ (1989) Animal virus DNA replication. Ann Rev Biochem 58:671–717
Chen M, Mermod N & Horwitz M (1990) Protein-protein interactions between adenovirus DNA polymerase and nuclear factor I mediate formation of the DNA replication preinitiation complex. J Biol Chem 265:18634–18642
Cheng L & Kelly TJ (1989) Transcriptional activator nuclear factor I stimulates the replication of SV40 minichromosomes in vivo and in vitro. Cell 59:541–551
Cheng L, Workman JL, Kingston RE & Kelly TJ (1992) Regulation of DNA replication in vitro by the transcriptional activation domain of GAL4-VP16. Proc Natl Acad Sci USA 89:589–593
Clayton DA (1991) Replication and transcription of vertebrate mitochondrial DNA. Annu Rev Cell Biol 7:453–478
Deb SP & Deb S (1989) Preferential binding of SV40 T-antigen dimers to origin region I. J Virol 63:2901–2907
Decker RS, Yamaguchi M, Possenti R & DePamphilis ML (1986) Initiation of SV40 DNA replication in vitro: aphidicolin causes accumulation of early replicating intermediates and allows determination of the initial direction of DNA synthesis. Mol Cell Biol 6:3815–3825
DePamphilis ML (1988) Transcriptional elements as components of origins of eukaryotic DNA replication. Cell 52:635–638
DePamphilis ML, Martínez-Salas E, Cupo DY, Hendrickson EA, Fritze CE, Folk WR & Heine U (1988) Initiation of polyomavirus and SV40 DNA replication, and the requirements for DNA replication during mammalian development, pp 165–175. In: Stillman B, Kelly T (eds) Eukaryotic DNA replication. Cancer cells, vol 6. Cold Spring Harbor Laboratory, Cold Spring Harbor
Dijkwel PA, Vaughn JP & Hamlin JL (1990) Mapping of replication initiation sites in mammalian genomes by two-dimensional gel analysis: stabilization and enrichment of replication intermediates by isolation on the nuclear matrix. Mol Cell Biol 11:3850–3859
Din S, Brill SJ, Fairman MP & Stillman B (1990) Cell cycle regulated phosphorylation of DNA replication factor A from human and yeast cells. Genes Dev 4:968–977

Dornreiter I, Höss A, Arthur AK & Fanning E (1990) SV40 T-antigen binds directly to the large subunit of purified DNA polymerase-a. EMBO J 9:3329–3336

Fanning E (1992) SV40 large T-antigen: the puzzle, the pieces, and the emerging picture. J Virol 66:1289–1293

Gannon JV & Lane DP (1990) The New Biologist 2:84–92

Gruss C, Gutierrez C, Burhans WC, DePamphilis ML, Koller Th & Sogo JM (1990) Nucleosome assembly in mammalian cell extracts before and after DNA replication. EMBO J 9:2911–2922

Guo Z-S, Gutierrez C, Heine U, Sogo JM & DePamphilis ML (1989) Origin auxiliary sequences can facilitate initiation of simian virus 40 DNA replication in vitro as they do in vivo. Mol Cell Biol 9:3593–3602

Guo Z-S & DePamphilis ML (1992) Specific transcription factors stimulate both simian virus 40 and polyomavirus origins of DNA replication. Mol Cell Biol (in press)

Guo Z-S, Heine U & DePamphilis ML (1991) T-antigen binding to site I facilitates initiation of SV40 DNA replication but does not affect bidirectionality. Nucl Acids Res 19:7081–7088

Gutierrez C, Guo Z-S, Roberts J & DePamphilis ML (1990) Simian virus origin auxiliary sequences weakly facilitate binding of T-antigen, but strongly facilitate initiation of DNA unwinding. Mol Cell Biol 10:1719–1728

Handeli S, Klar A, Meuth M & Cedar H (1989) Mapping replication units in animal cells. Cell 57:909–920

Hay RT & DePamphilis ML (1982) Initiation of simian virus 40 DNA replication in vivo: location and structure of 5'-ends of DNA synthesized in the *ori* region. Cell 28:767–779

Heintz NH & Hamlin JL (1982) An amplified chromosomal sequence that includes the gene for dihydrofolate reductase initiates replication within specific restriction fragments. Proc Natl Acad Sci USA 79:4083–4087

Heintz NH & Stillman BW (1988) Nuclear DNA synthesis in vitro is mediated via stable replication forks assembled in a temporally specific fashion in vivo. Mol Cell Biol 8:1923–1931

Hendrickson EA, Fritze CE, Folk WR & DePamphilis ML (1987) The origin of bidirectional DNA replication in polyoma virus. EMBO J 6:2011–2018

Hurwitz JF, Dean B, Kwong AD & Lee S-H (1990) In vitro replication of DNA containing the SV40 origin. J Biol Chem 265:18043–18046

Hyrien O & Méchali M (1992) Plasmid replication in *Xenopus* eggs and egg extracts: a 2D gel electrophoretic analysis. Nucl Acids Res 20:1463–1469

Laskey RA, Fairman MP & Blow JJ (1989) S phase of the cell cycle. Science 246:609–614

Laybourn PJ & Kadonaga JT (1991) Role of nucleosomal cores and histone H1 in regulation of transcription by RNA polymerase II. Science 254:238–245

Leu T-H & Hamlin JL (1989) High-resolution mapping of replication fork movement through the amplified dihydrofolate reductase domain in CHO cells by in-gel renaturation analysis. Mol Cell Biol 9:523–531

Li JJ & Kelly TJ (1984) Simian virus 40 DNA replication in vitro. Proc Natl Acad Sci USA 81:6973–6977

Linskens MHK & Huberman JA (1990) The two faces of higher eukaryotic DNA replication origins. Cell 62:845–847

Mahbubani HM, Paull T, Elder JK & Blow JJ (1992) DNA replication initiates at multiple sites on plasmid DNA in *Xenopus* egg extracts. Nucl Acids Res 20:1457–1462

Marahrens Y & Stillman B (1992) A yeast chromosomal origin of DNA replication defined by multiple functional elements. Science 255:817–823

Martínez-Salas E, Cupo DY & DePamphilis ML (1988) The need for enhancers is acquired upon formation of a diploid nucleus during early mouse development. Genes Dev 2:1115–1126

Martínez-Salas E, Linney E, Hassell J & DePamphilis ML (1989) The need for enhancers in gene expression first appears during mouse development with formation of a zygotic nucleus. Genes Dev 3:1493–1506

McKnight SL & Miller OL (1977) Electron Microscope Analysis of Chromatin Replication in the Cellular Blastoderm *Drosophila melanogaster* Embryo. Cell 12:795–804

Murakami Y, Satake M, Yamaguchi-Iwai Y, Sakai M, Muramatsu M & Ito Y (1991) The nuclear proto-oncogenes c-*fos* and c-*jun* as regulators of DNA replication. Proc Natl Acad Sci USA 88:3947–3951

Nilsson M, Forsberg M, You Z, Westin G & Magnusson G (1991) Enhancer effect of bovine papillomavirus E2 protein in replication of polyomavirus DNA. Nucl Acids Res 19:7061–7065

O'Connor DT & Subramani S (1988) Do transcriptional enhancers also augment DNA replication? Nucl Acids Res 16:11207–11222

Prives C (1990) The replication functions of SV40 T-antigen are regulated by phosphorylation. Cell 61:735–738

Prives C, Murakami Y, Kern FJ, Folk W, Basilico C & Hurwitz J (1987) DNA sequence requirements for replication of polyomavirus DNA in vivo and in vitro. Mol Cell Biol 7:3694–3704

Salas M (1991) Protein-priming of DNA replication. Annu Rev Biochem 60:39–71

Stillman B (1989) Initiation of eukaryotic DNA replication in vitro. Annu Rev Cell Biol 5:197–245

Temperley SM, Burrow CR, Kelly TJ & Hay RT (1991) Identification of two distinct regions within the adenovirus minimal origin of replication that are required for adenovirus type 4 DNA replication in vitro. J Virol 65:5037–5044

Thömmes P & Hübscher U (1990) Eukaryotic DNA replication. Enzymes and proteins acting at the fork. Eur J Biochem 194:699–712

Umek RM & Kowalski D (1990) Thermal energy suppresses mutational defects in DNA unwinding at a yeast replication origin. Proc Natl Acad Sci USA 87:2486–2490

Ustav M, Ustav E, Szymanski P & Stenlund A (1991) Identification of the origin of replication of bovine papillomavirus and characterization of the viral origin recognition factor E1. EMBO J 10:4321–4329

Vassilev LT & DePamphilis ML (1992) Guide to identification of origins of DNA replication in eukaryotic cell chromosomes. Crit Rev Biochem Mol Biol (in press)

Vassilev LT, Burhans WC & DePamphilis ML (1990) Mapping an origin of DNA replication at a single copy locus in exponentially proliferating mammalian cells. Mol Cell Biol 10:4685–4689

Vaughn JP, Dijkwel PA & Hamlin JL (1990) Replication initiates in a broad zone in the amplified CHO dihydrofolate reductase domain. Cell 61:1075–1087

Verrijzer CP, Kal AJ & der Vliet PCV (1990) The DNA binding domain (POU domain) of transcription factor oct-1 suffices for stimulation of DNA replication. EMBO J 9:1883–1889

Villarreal LP (1991) Relationship of eukaryotic DNA replication to committed gene expression: general theory for gene control. Microbiol Rev 55:512–542

Wasylyk C, Schneikert J & Wasyly B (1990) Oncogene v-jun modulates DNA replication. Oncogene 5:1055–1058

Wiekowski M, Miranda M & DePamphilis ML (1991) Regulation of gene expression in preimplantation mouse embryos: effects of the zygotic clock and the first mitosis on promoter and enhancer activities. Dev Biol 147:403–414

Wirak DO, Chalifour LE, Wassarman PM, Muller WJ, Hassell JA & DePamphilis ML (1985) Sequence-dependent DNA replication in preimplantation mouse embryos. Mol Cell Biol 5:2924–2935

Wolffe AP (1989) Dominant and specific repression of *Xenopus* oocyte 5S RNA genes and satellite I DNA by histone H1. EMBO J 8:527–537

Wysokenski DA & Yates JL (1989) Multiple EBNA1 binding sites are required to form an EBNA1 dependent enhancer and to activate a minimal replicative origin within *oriP* of Epstein-Barr virus. J Virology 63:2657–2666

Yamaguchi M & DePamphilis ML (1986) DNA binding site for a factor(s) required to initiate SV40 DNA replication. Proc Natl Acad Sci USA 83:1646–1650

Yang L, Li R, Mohr IJ, Clark R & Botchan MR (1991) Activation of BPV-1 replication in vitro by the transcription factor E2. Nature 353:628–632

Protein-Induced Alterations in DNA Structure at the *dhfr* Origin of Replication

P. Held[1], E. Soultanakis[1], L. Dailey[2], T. Kouzarides[3], N. Heintz[2], and N. H. Heintz[1]

1 Introduction

To identify the *trans*-acting factors that participate in initiation of DNA synthesis in mammalian cells, we have studied the interaction of nuclear proteins with the DNA sequences that encompass the origin of replication associated with the Chinese hamster dihydrofolate reductase *(dhfr)* gene. A variety of experimental approaches suggests that replication of amplified *dhfr* genes in the methotrexate-resistant variant of CHO cells, CHOC 400, initiates within a 4.3 kb Xba I fragment located 14 kb 3' to the *dhfr* gene (Heintz and Hamlin 1982; Burhans et al. 1986a, b; Anachakova and Hamlin 1988; Leu and Hamlin 1989). By mapping the strand specificity of Okazaki fragment synthesis throughout the *dhfr* origin region, an origin of bidirectional DNA synthesis (OBR) has been located to a 450 bp segment of the 4.3 kb Xba I fragment (Burhans et al. 1990). The DNA sequences that encompass the *dhfr* OBR include a region of stably bent DNA composed of 4 tracts of $A_{(3-4)}$ residues spaced precisely 10 bp apart (Fig. 1; Caddle et al. 1990a, b). The 3' end of the bent DNA region includes a ATT-rich motif that binds a 60 kDa nuclear protein named RIP60 (Dailey et al. 1990). RIP60 enhances DNA bending upon binding to ATT-rich repeats (Caddle et al. 1990b). Located 14 bp upstream of the RIP60 binding site between bend elements B3 and B4 is a perfect match to the AP-1 consensus response element (TGACTCA) that binds c-fos/c-jun heterodimers, one of the transcription factors that mediate changes in gene expression by the tumor promoter TPA (Angel and Karin 1991). Since transcription factors may participate in the regulation of initiation of DNA synthesis in eukaryotes (Heintz 1992), we have used DNase I footprinting to examine the effect of c-fos/c-jun heterodimers and c-jun homodimers on the structure of the bent DNA region in the presence and absence of RIP60. Here, we show that both c-fos/c-jun and c-jun/c-jun complexes bind to the AP-1 site within the bent DNA region, and provide preliminary evidence that suggests c-fos/c-jun heterodimers inhibit the binding of RIP60 to the ATT-rich repeats that abut the bent DNA motif. Because c-fos/c-jun heterodimers bend DNA in a direction opposite to that of c-jun/c-jun homodimers (Kerppola and Curran 1991), these proteins may act with RIP60 to dictate alternate DNA structures at the *dhfr* OBR during the onset of the S phase. We also report that the DNA helicase activity that cofractionates with RIP60 (Dailey et al. 1990)

[1] Department of Pathology, University of Vermont, College of Medicine Soule Medical Alumni Building, Burlington, VT 05405, USA.
[2] Laboratory of Molecular Biology, Rockefeller University, New York, NY 10021 USA.
[3] Wellcome/CRC Institute, Cambridge, UK.

43. Colloquium Mosbach 1992
DNA Replication and the Cell Cycle

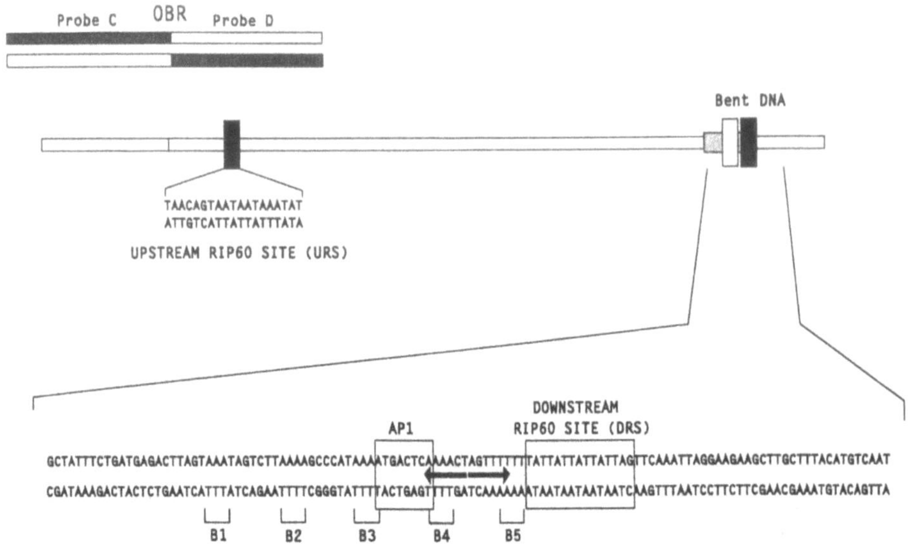

Fig. 1. Schematic representation of the bent DNA motif associated with the *dhfr* origin of replication. The bent DNA motif defined by bend elements B1–B5 *(brackets)* is shown relative to the origin of bidirectional DNA replication (OBR) defined by the strand specificity of Okazaki fragment synthesis (see Burhans et al. 1990). The bent DNA sequences encompass a consensus AP-1 site and a binding site for the 60 kDa origin-binding protein, RIP60. The *bold arrows* indicate the perfect inverted repeat that separates the AP-1 and RIP60 sites in the bent DNA region. A second RIP60 site is located upstream of the bent DNA region within the sequences defined by probe D

travels 3' to 5' on single-stranded (ss) DNA, and is stimulated by the ssDNA binding proteins RF-A and *E. coli* SSB.

2 Materials and Methods

2.1 Proteins and DNase I Footprinting

RIP60 was purified approximately 8800-fold from HeLa cell nuclear extract as described previously (Dailey et al. 1990). Peptides encoding the DNA binding and dimerization domains of c-fos (fos C7) and c-jun (junP) were expressed in bacteria and purified as described (Bannister et al. 1991). DNase I footprinting and gel shift probes were prepared by PCR amplification of the bent DNA region in the presence of an ^{32}P-end labeled oligonucleotide primer for either the top strand (p24) or the bottom strand (p23). Gel mobility shift experiments and DNase I footprinting were performed as described by Dailey et al. (1990). For footprinting of gel shift complexes, labeled probes that spanned the entire bent DNA region were incubated for 20 min with various combinations of test proteins under conditions described by Dailey et al. (1990). The samples were then treated with DNase I for 1 min, after which the reaction was terminated by the addition of EDTA to 10 mM and loaded immediately

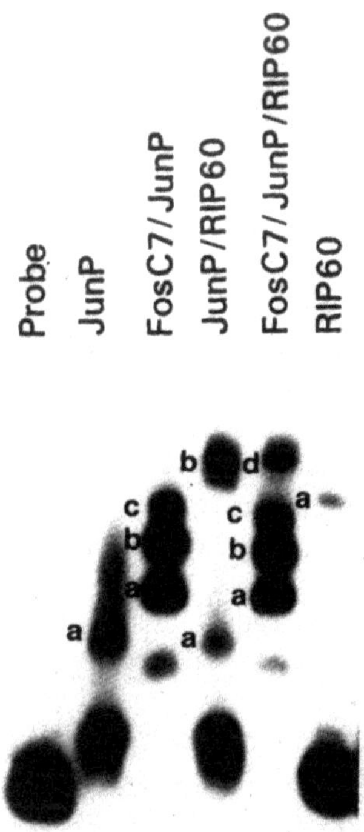

Fig. 2. Interaction of c-fos/c-jun heterodimers, c-jun/c-jun homodimers, and RIP60 with the bent DNA region. A 130 bp end labeled probe spanning the bent DNA region was incubated with the indicated combinations of fosC7, junP, and RIP60, treated briefly with DNase I, and the resulting protein/DNA complexes were resolved on a 4% polyacrylamide gel. After autoradiography, the DNA in the indicated complexes was recovered, concentrated, and resolved on denaturating sequencing gels to evaluate the position of the protein/DNA interactions (see Fig. 3). Shown is the autoradiography of the gel for the probe labeled on the top strand; an identical pattern was obtained for the probe labeled on the bottom strand (not shown)

on a neutral 4% polyacrylamide gel. After electrophoresis, the wet gel was exposed to Kodak X-Omat film and the positions of the gel shift complexes were determined by autoradiography. The DNA from each complex (as indicated in Fig. 2) and the free probe were eluted from the gel, concentrated by precipitation with ethanol, and resolved on denaturing 10% polyacrylamide sequencing gels. The position of specific protein footprints was determined by comparing the DNase I cleavage patterns with one another and with chemical sequencing reactions of the probe.

2.2 DNA Helicase Assays

DNA helicase activity was assessed by a standard oligonucleotide displacement assay as described previously (Lahue and Matson 1988; Dailey et al. 1990). A ^{32}P-end labeled 21 base oligonucleotide complementary to M13 DNA was annealed to single-stranded (ss) M13 DNA in 1 mM EDTA, 10 mM Tris-HCl (pH 7.9), 50 mM NaCl, and the oligo/M13 hybrid was purified by CL4B chromatography as described (Dailey et al. 1990). Reaction products were separated on a 10% neutral polyacrylamide gel, the gel was dried, and thc relative amount of displaced oligonucleotide was determined by 2D scintillation counting on the dried gel in a Betascope Analyzer (Betagen). *E. coli* SSB was obtained from USBiochemicals (Cleveland); RF-A was a gift from Bruce Stillman (Cold Spring Harbor Laboratory, NY).

3 Results and Discussion

3.1 The AP-1 Site Within the Bent DNA Motif from the dhfr Origin Region Binds fos/jun Heterodimers and jun/jun Homodimers

Heterodimers of c-fos and c-jun bind the AP1 consensus sequence (TGACTCA) with at least 30-fold higher affinity than do homodimers of c-jun (reviewed by Angel and Karin 1991). C-fos is not able to form homodimers efficiently, and has no demonstrable DNA binding activity. To determine if c-fos and c-jun are able to recognize the consensus AP-1 binding site located within the *dhfr* bent DNA motif, a 65 bp oligonucleotide spanning the AP-1 and RIP60 sites (ORI) was used as a probe in gel shift assays with various combinations of purified polypeptides encoding the dimerization and DNA binding domains of c-fos (fosC7) and c-jun (junP). These experiments showed that while fosC7 alone failed to bind the ORI probe, both fosC7/junP heterodimers and junP/junP homodimers formed specific complexes with this region of the bent DNA sequences (data not shown). In these experiments, while a single junP/junP protein/ORI DNA complex was observed, three distinct protein-DNA complexes were observed with the fosC7/junP protein preparation. To ascertain the composition of these complexes, various combinations of proteins were mixed with an end labeled probe, treated with DNase I, and the reactions were resolved on 4% neutral polyacrylamide gels as described in Section 2.1. The DNA from each complex indicated in Fig. 2 was isolated from the mobility shift gel, and resolved on denaturing sequencing gels.

DNase I footprinting of the single junP complex, junP (a in Fig. 2, lane 2), shows that the consensus AP-1 site within the bent DNA motif is protected by junP/junP homodimers (lane 3, panels A and B, Fig. 3). Several DNase I cleavage products that map in the bottom strand of the AP-1 are suppressed in the presence of junP (Fig. 3A), as is the single cleavage product that maps in the top strand of the probe (Fig. 3B). No other alterations in the DNase I cleavage patterns are observed, suggesting that the interaction of junP with the DNA is limited to the consensus AP-1 sequence.

With fosC7/junP, several DNA/protein complexes were observed (labeled a–c in Fig. 2, lane 3). Analysis of the fosC7/junP (a) and (b) gel shift complexes show that both of these complexes display an identical DNase I footprint (cf. lanes 4 and 5, Fig. 3A, B). Like junP, both the (a) and (b) fosC7/junP complexes reduce the signal from the single band that arises from cleavage of the AP-1 sequence in the top strand (Fig. 5B) as well as the several cleavage products that arise from the bottom strand of the AP-1 site (Fig. 3A). The footprints from the fosC7/junP (a) and (b) gel shift complexes are identical, suggesting that the c-fos/c-jun interactions with DNA in these two complexes is similar, and that the slower mobility of complex (b) is due to the binding of additional molecules of fosC7 and junP by protein-protein interactions. In other experiments (data not shown), the fosC7/junP (c) complex was shown to be identical with those of the fosC7/junP (a) and (b) complexes. These experiments show that both junP/junP homodimers and fosC7/junP heterodimers bind to a single site within the bent DNA region, and that this site corresponds to the consensus AP-1 site. Because five-fold more junP/junP than fosC7/junP protein was used in these gel shift

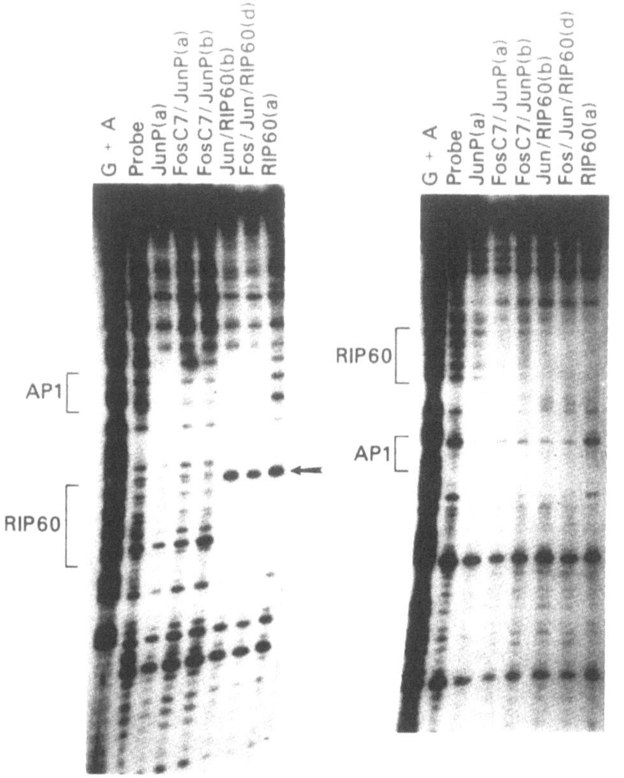

Fig. 3. DNase I footprint analysis of c-fos/c-jun and c-jun/c-jun binding to the bent DNA region in the presence and absence of RIP60. The DNA from the protein/DNA complexes in Fig. 2 were purified and analyzed on 10% denaturing polyacrylamide sequencing gels. *G* and *G+A* denote chemical sequencing reactions of the probe for the bottom strand (*left hand panel*) and the top strand (*right hand panel*). *Brackets* indicate the position of the AP-1 and RIP60 binding sites; the *arrow* indicates the characteristic DNase I hypersensitive site that is induced in the bottom strand of the bent DNA region by RIP60 binding to the ATT-rich repeats

experiments, it is also evident that fosC7/junP heterodimers bind the origin AP-1 site far more efficiently than do junP/junP homodimers (cf. lanes 2 and 3, Fig. 2).

While circular permutation assays show that both fos/jun heterodimers and jun/jun homodimers bend DNA, phase analysis shows that these protein complexes bend DNA in opposite directions (Kerrpola and Curran 1991). C-jun/c-jun homodimers bend DNA toward the minor groove, and therefore when the AP-1 site placed in phase with poly dA tracts spaced 10 bp apart, c-jun enhances intrinsic DNA bending (Kerrpola and Curran 1991). In contrast, c-fos/c-jun heterodimers bend DNA toward the major groove, and therefore tend to counteract intrinsic DNA bending. How these proteins affect DNA bending when the AP-1 site is located within a bent DNA molecule, as is the case for the *dhfr* origin region, is unknown. Using high resolution DNase I and chemical footprinting at a wide range of protein concentrations, we have observed several distinct differences in the c-fos/c-jun and c-jun/c-jun footprints on the bent DNA motif that indicate these two complexes affect the structure of the bent DNA region in different ways (E. Soultanakis, L. Dailey, N. Heintz, and N. H. Heintz, in prep.)

3.2 FosC7/junP Heterodimers and junP/junP Homodimers Have Different Effects on the Binding of RIP60

Because the c-fos/c-jun and c-jun/c-jun complexes bend DNA differently, we then asked how these complexes affect the binding of RIP60 to the nearby ATT-rich repeats located 14 bp downstream from the AP-1 site. Although circular permutation assays have shown that RIP60 also bends DNA (Caddle et al. 1990b), the direction of DNA bending by RIP60 has not been determined. Note that twelve of the 14 bp that comprise the spacer region between these two sites form a perfect inverted repeat (Fig. 1, bold arrows).

When RIP60 is added to gel shift reactions containing junP, nearly all of the junP/junP (a) complex disappears and a new supershifted complex is observed (complex b, lane 4, Fig. 2). Footprinting of this complex shows that it contains both junP/junP homodimers bound to the AP-1 site and RIP60 bound to the ATT-rich repeats (lane 6, Fig. 3A, B). Occupancy of the RIP60 site is evident from protection of DNase I cleavage within the ATT-repeats as well as from the induction of DNase I hypersensitivity at the junction between he poly-dA/dT tract and the ATT-rich repeats (cf. lanes 6 and 8, Fig. 3A). At equivalent concentrations of RIP60, the gel shift signal from the junP/RIP60/DNA complex exceeds that obtained with RIP60/DNA complex alone, suggesting that junP/junP does not interfere with, and may even stabilize, the binding of RIP60 to the nearby ATT repeats. In contrast, addition of RIP60 to the fosC7/junP reactions results in a very small portion of supershifted complex (complex d, lane 5, Fig. 2). Footprinting shows that, as in the junP/RIP60/DNA complex, both the AP-1 and the RIP60 site are occupied by protein in the supershifted fosC7/junP/RIP60 complex (cf lanes 6 and 7, Fig. 3A, B). The results of these low resolution experiments suggest that the binding of junP/junP homodimers and fosC7/junP heterodimers to the AP-1 site embedded within the origin bent DNA motif may influence the binding of RIP60 at the nearby ATT repeats differently.

Using both truncated and full length forms of fos and jun (a gift from T. Kerrpola and T. Curran, Roche Institute), we have titrated these factors against fixed concentrations of RIP60 in both gel shift and high resolution footprinting experiments. The results of a representative footprinting experiment with truncated c-junP and RIP60 is shown in Fig. 4. In the presence of increasing concentrations of RIP60 and junP, both the AP-1 site and the RIP60 site become occupied (lanes 7–8, 10–11, Fig. 4A, B). In the presence of both factors prominent DNase I hypersensitivity in the bottom strand is observed in the palindromic region at the junction between the poly-dA/dT tract and the RIP60 site (Fig. 4A, lanes 7–8, 10–11). Note that in this experiment binding of RIP60 increases the DNase I sensitivity of the bottom strand of the AP-1 site. In other experiments, we have shown that saturating concentrations of heterodimers composed of full length c-fos and c-jun quantitatively eliminate occupancy of the RIP60 site (E. Soultanakis, L. Dailey, N. Heintz, and N. H. Heintz, in prep.). Using

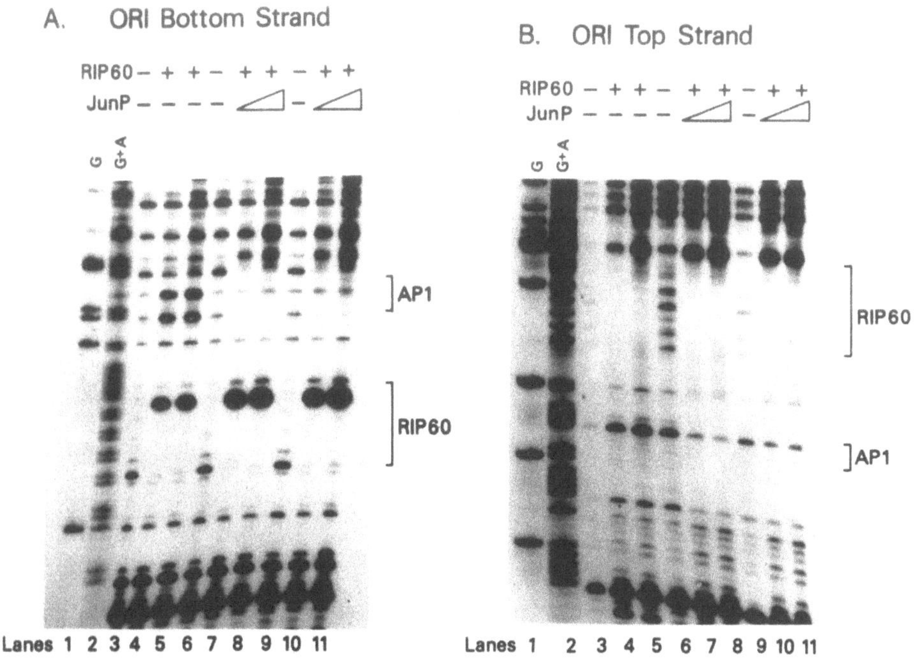

Fig. 4. DNase I footprint analysis of c-jun binding in the presence and absence of RIP60. Increasing concentrations of two preparations of junP (lanes 7–8, 9–10) were incubated with end labeled probes for the bottom strand (*A*) or top strand (*B*) of the bent DNA region in the presence or absence of RIP60. *G* and *G+A* indicate chemical sequencing reactions of the probes; the AP-1 and RIP60 sites are indicated by *brackets*. Note that RIP60 alters the DNase I sensitivity of the bottom strand of the AP-1 site

phase analysis vectors, we are presently determining if differences in DNA bending accounts for the differences in the DNA binding activity of RIP60 that are observed with c-fos/c-jun heterodimers and c-jun/c-jun homodimers.

In summary, the bent DNA motif located in the *dhfr* origin region binds at least three protein factors: RIP60, c-fos/c-jun heterodimers, and c-jun/c-jun homodimers. Different combinations of these factors, all of which bend DNA, may dictate alternative structures in the bent DNA region and thereby affect the structure of the palindromic sequence that separates the AP-1 and RIP60 binding sites. Electron microscopic examination of RIP60 protein/DNA complexes show that in vitro RIP60 mediates DNA looping between the bent DNA region and an upstream RIP60 binding site (P. Held, L. Dailey, I. Mastrangelo, and P. Hough, personal communication). Because of the proximity of the AP-1 site to the RIP60 site in the bent DNA region, we are particularly interested in determining if c-fos and c-jun affect the ability of RIP60 to mediate DNA looping between the upstream and downstream sites (see Fig. 1). As is found for other origins of replication in viruses and in yeast (reviewed in Heintz 1992), transcription factors such as AP-1 may act in concert with other origin recognition proteins to foster the assembly of a specialized nucleoprotein complex at the *dhfr* origin of replication during the onset of DNA synthesis.

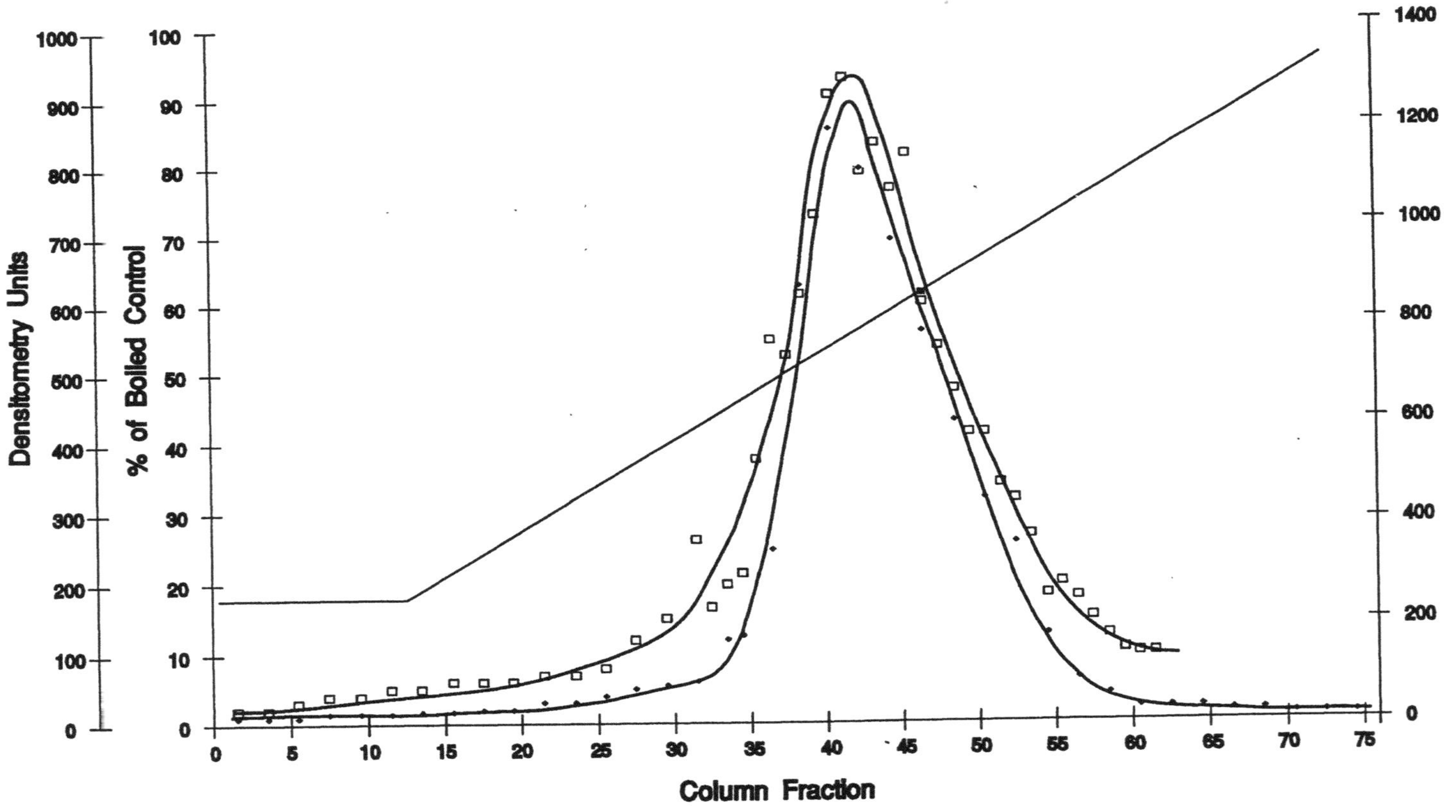

Fig. 5. Co-elution of RIP60 and a DNA helicase activity from the ATT-oligonucleotide affinity column. The final step in the purification of RIP60 is affinity chromatography on an ATT-oligonucleotide column (see Dailey et al. 1990). In this experiment, a linear KC1 gradient was applied to the column and each fraction was assayed for RIP60 DNA binding activity by gel mobility shift analysis (*filled diamonds*) and DNA helicase activity by an oligonucleotide displacement assay (*open boxes*). The helicase activity is expressed as percent of the radioactivity displaced relative to boiled substrate in the absence of protein. The relative amount of DNA binding activity was assessed by densitometric scanning of autoradiographs from the gel mobility shift assays

3.3 *RIP60 Cofractionates with a DNA Helicase Activity*

RIP60 has been purified from HeLa cell nuclear extract by a combination of conventional ion exchange and DNA affinity chromatography (Dailey et al. 1990). Using a standard oligonucleotide displacement assay, fractions from the oligonucleotide affinity column that were rich in RIP60 DNA binding activity were also shown to contain an ATP(or dATP)-dependent DNA helicase activity; the DNA helicase activity appeared to cofractionate with a 100 kDa polypeptide (RIP100) that could be labeled in vitro with ATP (Dailey et al. 1990). Quantitation of the DNA helicase activity and RIP60 DNA binding activity shows that both coelute from the ATT-oligo affinity column (Fig. 5). Further purification of the helicase activity contained in the peak fractions from the ATT-oligonucleotide column has shown that the helicase activity can be separated from RIP100 (L. Dailey, unpubl. observ.). Thus, our original assignment of the helicase activity to the 100 kDa protein we call RIP100 was erroneous. Further investigation of the properties of the helicase indicates that it is retained on the ATT-oligonucleotide column due to specific interactions with the ATT-rich sequence or RIP60 protein, or both (L. Dailey, unpubl. observ.). Assignment of

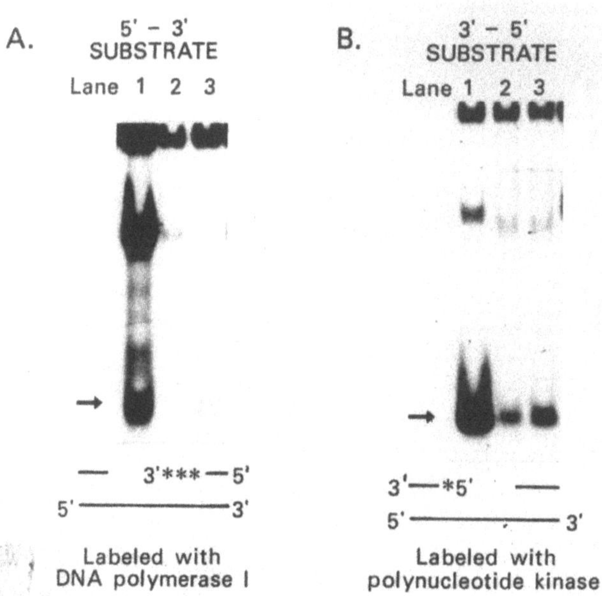

Fig. 6. The DNA helicase associated with RIP60 translocate 3' to 5' on ssDNA. Partial duplexes containing oligonucleotides radiolabeled at either the 3' end with DNA polymerase I (*A*) or the 5' end with polynucleotide kinase (*B*) were incubated under standard conditions with 1 (*lane 2*) or 2 (*lanes 3*) µl of the most active fraction from the ATT-oligonucleotide column, and the reaction products were resolved by electrophoresis on a neutral polyacrylamide gel. As a control, each substrate was boiled for 10 min prior to electrophoresis (*lane 1*). *Arrows* indicate the labeled products expected from a DNA helicase that translocates on ssDNA in the 5' to 3' (*A*) or 3' to 5' (*B*) direction

the helicase activity to a specific polypeptide (or polypeptides) awaits purification of this enzyme to homogeneity.

Using the peak fractions of helicase activity recovered from the ATT-oligonucleotide column, we have begun investigating the properties of this enzyme (or enzymes). To determine the direction that the enzyme travels on ssDNA, substrates containing partial duplexes with uniquely labeled ends were prepared (see Fig. 6). For the 3' labeled substrate, an oligonucleotide that spans the polylinker region of M13 was annealed to ssM13 DNA and extended briefly with *E. coli* DNA polymerase I in the presence of ^{32}P-dNTPs (Fig. 6A). For the 5' labeled substrate, the oligonucleotide

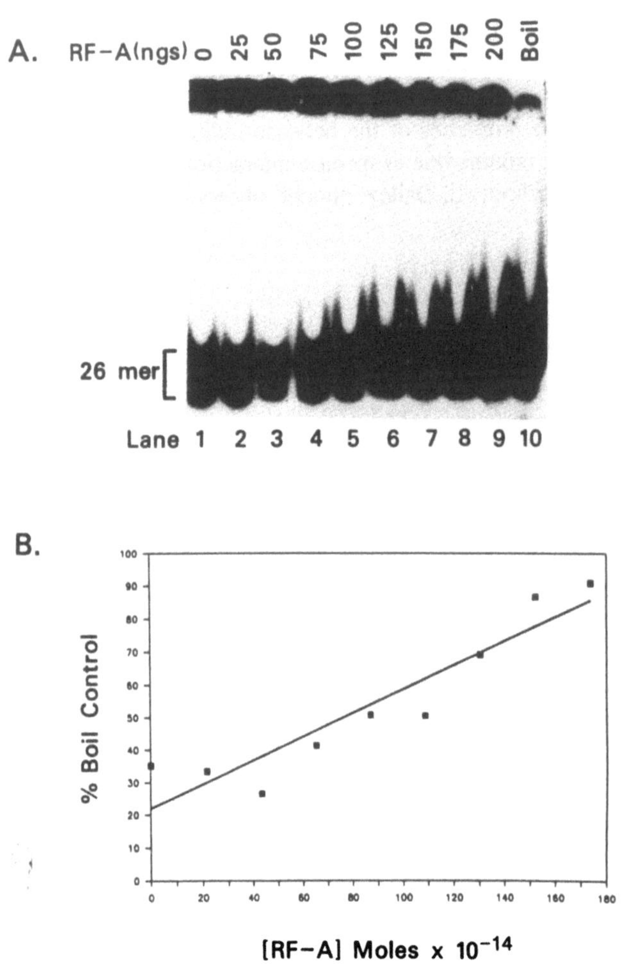

Fig. 7. The DNA helicase is stimulated by human SSB, RF-A. A partial duplex composed of a 5' end labeled oligonucleotide and ssM13 DNA was incubated with a fixed amount of DNA helicase activity and increasing concentrations of RF-A under standard reaction conditions. Reaction products were resolved by electrophoresis *(A)* and the amount of displaced oligonucleotide relative to boiled substrate *(B)* was determined by 2D scintillation counting in a Betascope Analyzer (Betagen)

was end labeled with ^{32}P-ATP and polynucleotide kinase prior to annealing to M13 ssDNA (Fig. 6B). Both substrates were purified by exclusion chromatography and digested with a restriction enzyme that cuts M13 DNA within the polylinker sequences. In the presence of ATP, only the fragment labeled at the 5' end with polynucleotide kinase is displaced from the partial duplex (Fig. 6B, lanes 2 and 3). These data indicate that the predominant helicase activity that cofractionates with RIP60 DNA binding on the ATT-oligo affinity column translocates in a 3' to 5' direction on ssDNA.

The strand displacement activity of several DNA helicases has been shown to be stimulated by ssDNA binding proteins (SSBs). Stimulation may arise from specific protein-protein interactions between the SSB and the helicase, or by coating of the ssDNA regions of the substrate such that the effective concentration of the primer/template junction is increased. To test the effects of SSBs on the helicase ac-

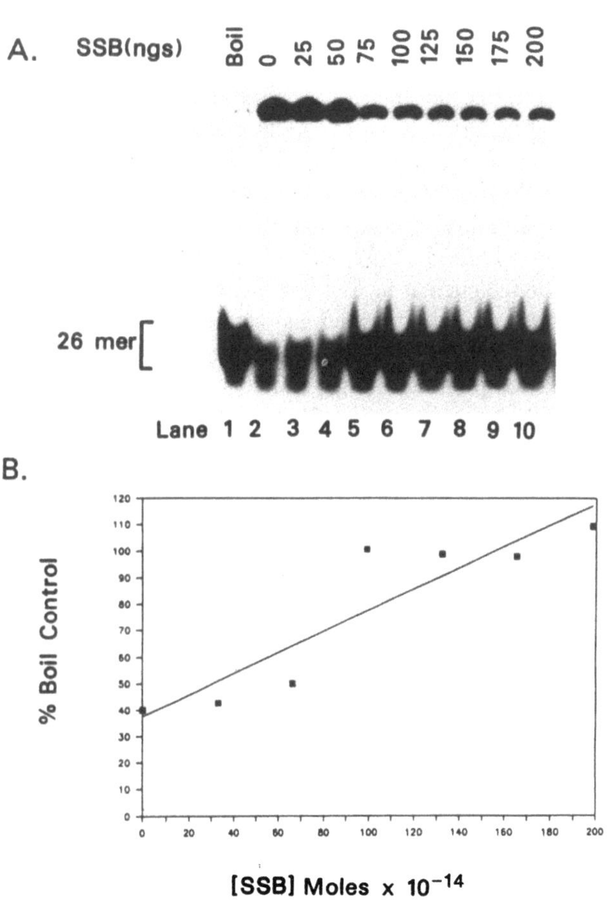

Fig. 8. The DNA helicase is stimulated by the *E. coli* SSB. As in Fig. 7, a partial duplex composed of a 5' end labeled oligonucleotide and ssM13 DNA was incubated with a fixed amount of DNA helicase activity and increasing concentrations of *E. coli* SSB under standard reactions conditions. Reactions products were resolved by electrophoresis (*A*) and the amount of displaced oligonucleotide relative to boiled substrate (*B*) was determined by 2D scintillation counting in a Betascope Analyzer (Betagen)

tivity, we incubated a fixed amount of the peak fractions containing the helicase with increasing concentrations of the human SSB, RF-A (a gift from Bruce Stillman) or *E. coli* SSB in the standard oligonucleotide displacement assay. Both RF-A (Fig. 7) and *E. coli* SSB (Fig. 8) stimulated the helicase activity in a dose-dependent fashion, with a maximal 10-fold increase in activity at near saturating amounts of SSB. Because stimulation was observed with both RF-A and *E. coli* SSB, these results suggest that stimulation of the helicase activity was due to increasing the effective concentration of the primer/template junction, and not due to specific protein-protein interactions.

4 Summary and Future Directions

Based on studies in bacteria, viruses, and yeast, we expect that initiation of DNA replication in mammalian cells is mediated by the assembly of a specific nucleoprotein complex at origins of replication. Because there is at present no reliable assay for the DNA sequences or *trans*-acting factors that participate in initiation of DNA synthesis in multicellular organisms, our approach has been to search for origin-binding proteins that display the general properties of other well characterized initiation factors. Because DNA bending has been implicated in a number of metabolic processes involving protein-DNA interactions (Trifonov 1992, and references therein), to date we have concentrated our efforts on the nuclear factors that interact with the *dhfr* origin bent DNA motif. In this respect, we have purified and studied the properties of RIP60, a 60 kDa protein that binds to an ATT-rich repeat that demarcates the 3' end of the bent DNA region. DNase I footprinting and gel mobility shift analysis show that RIP60 binds to two sites within the origin region (P. Held, L. Dailey, and N. H. Heintz, unpubl. data); the upstream site is located about 720 bp 5' to the site that abuts the bent DNA region (see Fig. 1). While examination of RIP60/DNA complexes by electron microscopy shows that in vitro RIP60 mediates looping between the two binding sites, this observation does not directly implicate DNA looping in initiation at the *dhfr* origin. Rather, the DNA looping studies demonstrate that RIP60, like many other DNA binding proteins, has distinct protein domains that are able to participate in protein multimerization and DNA binding.

While several viral initiation factors contain both origin DNA binding and DNA helicase activity, in mammalian cells these initiation functions may reside in distinct polypetides. The chromatographic behavior of the DNA helicase activity suggests that this factor cofractionates with RIP60 on the ATT-oligonucleotide column because of specific interactions with RIP60 and/or the ATT-rich sequences (L. Dailey, unpublished data). Hence, the DNA helicase may represent one of several activities that are recruited to origin sequences by specific protein-protein interactions. If either RIP60 or the DNA helicase can be shown to play a role in initiation of DNA synthesis, it will be interesting to determine how these and other factors assemble on the *dhfr* origin sequences.

In yeast, an understanding of origin structure and function is emerging. Yeast origins of replication are multipartite regulatory elements (Marahrens and Stillman 1992) that include binding sites for transcription factors. The participation of transcription factors in initiation of DNA synthesis suggests a model in which initiation is regulated by the concerted action of core replication proteins that function at all (or many) origins and accessory *trans*-acting factors that display origin-specific effects (Heintz 1992); this model provides a ready mechanism for coordinating these transcription and replication processes during the cell cycle. Of the transcription factors that have been shown to interact with the *dhfr* origin region in vitro, AP-1 binding proteins and oct1 may prove particularly appropriate for coordinating replication and transcription during the cell cycle. Oct1 is multiply phosphorylated in a cell cycle-specific fashion (Roberts et al. 1991), and one function of this protein is the regulation of histone H2B transcription during the S phase (N. Heintz 1991). AP-1 DNA binding activity is required for initiation of DNA replication in mammalian cells (Riabowol et al. 1992), and c-jun in particular has been shown to be required for entry into the S phase (Kovary and Bravo 1991). Interestingly, c-jun alone is able to enhance polyoma virus DNA replication through the agency of an origin AP-1 enhancer sequence (Murakami et al. 1991).

Here we have provided preliminary evidence that in vitro c-jun homodimers and c-fos/c-jun heterodimers may differentially affect the interaction of RIP60 with the bent DNA motif from the *dhfr* origin region. To ascertain if these interactions are physiologically relevant, we have begun examining the organization of the bent DNA region in synchronized cells by genomic footprinting techniques with the long range goal of reconstructing the in vivo footprint with purified factors in vitro. Once an inventory of the proteins that interact with the *dhfr* origin in whole cells has been established, studies to define the biochemical roles of these proteins in initiation of DNA synthesis can begin in earnest.

Acknowledgements. We appreciate the assistance of Judy Kessler in preparing the figures, and Rhoda Rowell in preparing the text. This work was supported by grants from the NIH to N. H. Heintz, N. Heintz, and L. Dailey.

References

Anachakova B & Hamlin JL (1989) Replication in the amplified dihydrofolate reductase domain in CHO cells may initiate at two distinct sites, one of which is a repetitive sequence element. Mol Cell Biol 9:532–540

Angel P & Karin M (1991) The role of jun, fos, and the AP-1 complex in cell-proliferation and transformation. Biochim Biophys Acta 1072:129–157

Bannister AJ, Cook A & Kouzarides T (1991) In vitro DNA binding activity of fos/jun and B2LF1 but not C/EBP is affected by redox changes. Oncogene 6:1243–1250

Burhans WC, Selegue JE & Heintz NH (1986a) Isolation of the origin of replication associated with the amplified Chinese hamster dihydrofolate reductase domain. Proc Natl Acad Sci USA 83:7790–7794

Burhans WC, Selegue JE & Heintz NH (1986b) Replication intermediates formed during initiation of DNA synthesis in methotrexate-resistant CHCO 400 cells are enriched for sequences derived from a specific, amplified restriction fragment. Biochemistry 25:441–449

Burhans WC, Vassilev LT, Caddle MS, Heintz NH & DePamphilis ML (1990) Identification of an origin of bidirectional DNA replication in mammalian chromosomes. Cell 62:955–965

Caddle MS, Lussier RH & Heintz NH (1990a) Intramolecular DNA triplexes, bent DNA and DNA unwinding elements in the initiation region of an amplified dihydrofolate reductase replicon. J Mol Biol 211:19–33

Caddle MS, Dailey LD & Heintz NH (1990b) RIP60, a mammalian origin binding protein, enhances DNA bending near the dihydrofolate reductase origin for replication. Mol Cell Biol 10:6236–6243

Dailey L, Caddle MS, Heintz N & Heintz NH (1990) Purification of RIP60 and RIP100, mammalian proteins with origin-specific DNA-binding and ATP-dependent DNA helicase activities. Mol Cell Biol 10:6225–6235

Heintz N (1991) The regulation of histone gene expression during the cell cycle. Biochim Biophys Acta 1088:327–339

Heintz NH (1992) Transcription factors and the control of DNA replication. Current Opinion Cell Biol (in press)

Heintz NH & Hamlin JL (1982) An amplified chromosomal sequence that includes the gene for dihydrofolate reductase initiates replication within specific restriction fragments. Proc Natl Acad Sci USA 79:4083–4087

Kerrpola TK & Curran T (1991) Fos-jun heterodimers and jun homodimers bend DNA in opposite orientations: implications for transcription factor cooperativity. Cell 66:317–326

Kovary K & Bravo R (1991) The jun and fos protein families are both required for cell cycle progression in fibroblasts. Mol Cell Biol 11:4466–4472

Lahue EE & Matson SW (1988) *Escherichia coli* DNA helicase I catalyzes a unidirectional and highly processive unwinding reaction. J Biol Chem 263:3208–3215

Leu T-H & Hamlin JL (1989) High resolution mapping of replication fork movement through the amplified dihydrofolate reductase domain in CHO cells by in-gel renaturation analysis. Mol Cell Biol 9:1026–1033

Marahrens Y & Stillman B (1992) A yeast chromosomal origin of DNA replication defined by multiple functional elements. Science 255:817–823

Murakami Y, Satake M, Yamaguchi-Iwai, Sakai M, Muramatsu M & Ito Y (1991) The nuclear protooncogenes c-jun and c-fos as regulators of DNA replication. Proc Natl Acad Sci USA 88:3947–3951

Riabowol K, Schiff J & Gilman MZ (1992) Transcription factor AP-1 activity is required for initiation of DNA synthesis and is lost during cellular aging. Proc Natl Acad Sci USA 89:157–161

Roberts SB, Segil N & Heintz N (1991) Oct-1 is multiply phosphorylated during the cell cycle. Science 253:1022–1024

Trifonov EN (1991) DNA in profile. TIBS 16:467–470

Mechanism and Control of Cellular DNA Replication

B. Stillman[1]

1 Introduction

Our understanding of DNA replication in bacteria has relied heavily upon studies on the replication of bacteriophage and plasmid chromosomes and the ability to compare and contrast the mechanisms of DNA synthesis and the functions of the replicative enzymes. In much the same way, studies of the mammalian DNA viruses have contributed to an understanding of the enzymology of cellular DNA replication. The choice model system to date for understanding the enzymology of eukaryotic DNA replication has been the study of Simian Virus 40 (SV40) DNA replication. This is due to the fact that SV40 DNA exists in the cell as a small, circular chromosome that can be thought of as the equivalent of a single replicon within the larger cellular chromosomes. Furthermore, SV40 DNA exists in a chromatin structure that resembles the structure of cellular chromatin and therefore could be a useful model for chromosome, as well as DNA replication (Stillman 1986; DePamphilis and Bradley 1986; Cheng and Kelly 1989; Smith and Stillman 1989).

The overall mechanism of SV40 DNA replication has been elucidated from studies on the replication of the DNA in infected cells, mostly by pulse labeling experiments followed by gel electrophoresis, as well as by electron microscopy (reviewed by DePamphilis and Bradley 1986; Challberg and Kelly 1989). The virus genome is a circular, double-stranded genome of 5243 base pairs that contains a single origin of DNA replication located in the noncoding region that also constitutes the bidirectional promoter for both early and late gene transcription. Initiation of DNA replication occurs at the origin and thereafter, proceeds in both directions from the origin. The two divergent replication forks meet when they replicate approximately half of the genome and termination involves resolution of the two daughter molecules into two completely replicated chromosomes. This latter step involves topoisomerases I and II (Sundin and Varshavsky 1980; 1981; Weaver et al. 1985; Yang et al. 1987; Snapka et al. 1988).

SV40 encodes six proteins that are required for the various stages of viral infection, but only one of these, the SV40 large tumor (T) antigen, is required for the replication of SV40 DNA. Consequently, the bulk of the proteins that replicate SV40 DNA in the infected cell are cellular proteins and it is believed that these proteins also function to replicate cellular DNA during the S phase of the cell cycle in uninfected cells. The development of extracts from primate, and particularly human cells, that can support the replication of SV40 DNA when purified SV40 T antigen is added was

[1] Cold Spring Harbor Laboratory, Cold Spring Harbor, New York 11724, USA.

43. Colloquium Mosbach 1992
DNA Replication and the Cell Cycle

a major breakthrough that has enabled detailed studies on the mechanism of DNA replication to proceed (Li and Kelly 1984; 1985; Stillman and Gluzman 1985; Wobbe et al. 1985). It was anticipated that identification and characterization of the cellular proteins present in these extracts would be an appropriate starting point for studying cellular DNA replication in eukaryotic cells. Below, I have briefly reviewed progress over the last six years of this endeavor and summarize the functions of the cellular replication proteins, particularly those that interact with the cellular DNA polymerases.

Ultimately, the goal of these studies is to understand how cell chromosomes replicate in the S phase of the cell cycle and how this complex process is regulated. Over the past three years, my laboratory has also begun to address this question directly by studying the replication of chromosomal DNA in the yeast *S. cerevisiae*. This genetically manipulable organism is eminently suited because it is the only eukaryote that has well defined DNA sequences that function as origins of DNA replication. By combining biochemical and genetic approaches, we have begun to study the nature of the *cis*- and *trans*-acting elements that combine to initiate the replication of cellular chromosomes. These studies, performed in parallel with the biochemical studies on SV40 DNA replication, should lead to considerable insight into a complicated cellular process.

The progress cited herein focuses primarily on developments in my own laboratory and is not meant to be an extensive review of the literature. Such reviews can be found in Challberg and Kelly (1989); Stillman (1989); Wang et al. 1989; Wang (1991); Campbell and Newlon (1991) and Fangman and Brewer (1991).

2 Initiation of SV40 DNA Replication

The SV40 origin of DNA replication contains three essential sequence elements that need to be correctly spaced. The central element corresponds to the binding site for the SV40 T antigen (site II) and is a 27 base pair palindrome containing four copies of the T antigen recognition penta-nucleotide sequence 5'-GAGGC-3' (see Deb et al. 1986). On one side of site II is a partially palindromic sequence called the early palindrome and on the other side is an A·T rich sequence that is naturally bent. When T antigen binds to site II DNA in the presence of ATP, it forms a large oligomer that most likely consists of two hexamers of T antigen (reviewed by Borowiec et al. 1990). The formation of this complex, called the T complex (Tsurimoto et al. 1989), induces the local unwinding of part of the early palindrome sequence and a structural distortion of the A·T rich DNA (Borowiec and Hurwitz 1988; see Borowiec et al. 1990 for review). The T complex is relatively stable, but addition of a single strand DNA binding protein of either prokaryotic or eukaryotic origin, causes the DNA on either side of the origin to unwind extensively (Borowiec et al. 1990). The locally unwound replication intermediate has been called the 'unwound complex' (Tsurimoto et al. 1989; Borowiec et al. 1990). Interestingly, only the single strand DNA binding protein, RPA (formerly called RF-A), that has been purified from mammalian cells will function during DNA replication in vitro. Even the structurally related RPA from

S. cerevisiae will not substitute for the mammalian protein during DNA replication, suggesting that specific protein-protein interactions must occur. These multiple stages of origin unwinding are analogous to the early stages of initiation of DNA replication at the *E. Coli ori*C and phage λ replication origins (reviewed by Bramhill and Kornberg 1988).

The SV40 T antigen, in addition to functioning as a site specific DNA binding protein, is also a DNA helicase that can unwind duplex DNA into its single strand components (Stahl et al. 1986). The helicase moves in the 3' to 5' direction on the strand to which it is bound (Goetz et al. 1988; Wiekowski et al. 1988) and would therefore translocate along the strand that is template for the continuously synthesized, leading strand at a replication fork. There is evidence that T antigen functions during the elongation of DNA replication as a DNA helicase at the replication fork (Dodson et al. 1987; Wiekowski et al. 1987; Wessel et al. 1992). The functions of these cellular proteins and how they interact with T antigen will be outlined below.

3 Cellular DNA Replication Factors

To identify cellular factors required for the complete replication of plasmid DNAs containing the SV40 origin, we have fractionated cellular extracts into multiple essential components and then purified the individual components to homogeneity. A summary of the components we have identified to date is shown in Fig. 1. T antigen was purified by an immunoaffinity procedure (Simanis and Lane 1985) from SF9 insect cells infected with a recombinant baculovirus vector (Lanford 1988). T antigen is a phospho-protein of 708 amino acids and migrates in an SDS-polyacrylamide gel with an approximate molecular weight of 97 000 (97K) (Fig. 2). The functions of T antigen in DNA replication have recently been reviewed by Fanning (1992) and briefly, have been mentioned above. All of the cellular proteins were initially purified from human 293 cells or calf thymus extracts. The topoisomerases I and II have long been known to be involved in eukaryotic DNA replication, functioning during elongation and for segregation of the newly replicated daughter chromosomes (DiNardo et al. 1984; Thrash et al. 1984; Uemura and Yanagida 1984; Holm et al. 1985) and sim-

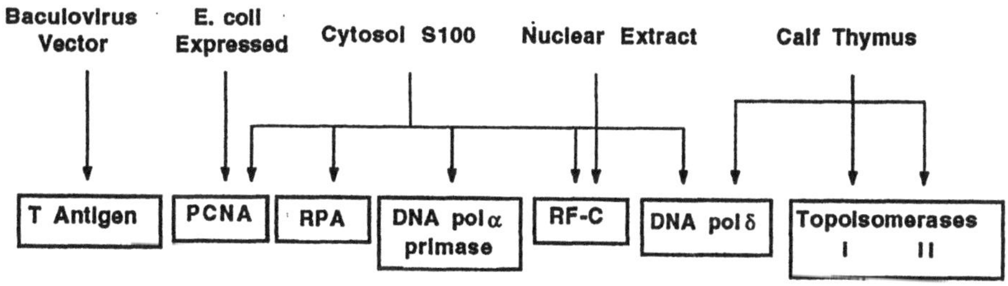

Fig. 1. A summary of the fractionation scheme for the identification of cellular proteins required for SV40 DNA replication in vitro. SV40 T antigen is purified from recombinant baculovirus infected SF9 cells. PCNA is either purified from human 293 cells or from *E. coli* containing a recombinant vector plasmid expressing the human protein. The proteins are described in the text

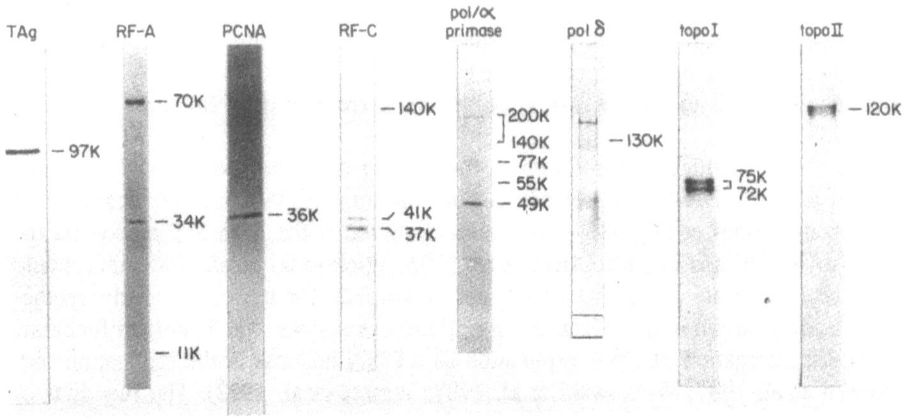

Fig. 2. Gel electrophoresis of the purified DNA replication factors. The indicated purified proteins were subjected to electrophoresis in SDS-polyaryladamide gels, which were then stained with either Coomassie Brilliant Blue or with silver. The molecular weights of the protein subunits for each replication factor are indicated. The topoisomerases I and II are both active proteolytic fragments of the calf thymus topoisomerase I and II

ilar functions have been inferred from in vitro studies with the SV40 system (Yang et al. 1987). In studies described below, a mixture of purified topoisomerases I and II (Fig. 2) have been used (Tsurimoto et al. 1989). The functions of the other cellular proteins in the SV40 cell-free system are now described.

3.1 Replication Protein A (RPA)

This protein was first purified as a factor that was essential for the initiation of DNA replication (Wobbe et al. 1987; Fairman and Stillman 1988; Wold and Kelly 1988). The replication factor has been variably called RF-A, RP-A or eukaryotic SSB (for single strand DNA binding protein), but RPA will be used henceforth. It contains protein subunits of 70K, 34K and 11K that are tightly bound to each other. The 70K subunit of RPA is a single stranded DNA binding protein that is capable of binding preferentially to single, versus double stranded DNA (Wold et al. 1989; Brill and Stillman 1989; Erdile et al. 1990). The functions of the other two subunits are not known, however antibodies directed against the 34K subunit inhibit SV40 DNA replication in vitro (Kenny et al. 1990; Erdile et al. 1990; S. Din and B. Stillman, in preparation). Interestingly, the 34K subunit is phosphorylated in a cell cycle regulated fashion (Din et al. 1990). The protein is not phosphorylated in the G_1 phase but becomes phosphorylated at the start of the S phase and is then dephosphorylated during mitosis. The cdc2 cell-cycle regulated protein kinase, with an associated cyclin B or cyclin A subunit, phosphorylates RPA from human cells on serine residues that are essential for RPA phosphorylation in vivo (Dutta and Stillman 1992). The cdc2 protein kinase and the related cdk2 kinase, play a major role in regulation of entry into the S phase of the cell cycle (reviewed by Pines and Hunter 1990). Furthermore, the cdc2-cyclin kinase complex has been shown to stimulate SV40 DNA replication and unwinding of the SV40 origin of DNA replication in the presence of RPA and cell extracts prepared from human cells that were in the G1 phase of the cell cycle (Roberts and D'Urso 1988; D'Urso et al. 1990; Dutta and Stillman 1992). This suggests that one of the activating events that promotes initiation of SV40 DNA replication is phosphorylation of the RPA 34K subunit.

RPA also has an essential role in the elongation of DNA replication. RPA stimulates the activity of DNA polymerase α (pol α) (Tsurimoto and Stillman 1989a; Kenny et al. 1989) and cooperates with two other DNA replication factors, PCNA and RF-C (see below) to stimulate the activity of DNA polymerase δ (pol δ) (Kenny et al. 1989; Tsurimoto and Stillman 1989a; Tsurimoto and Stillman 1991b; Lee et al. 1991a). Furthermore, RPA is also required to prevent abnormal replication of the leading strand template DNA by pol α (Tsurimoto et al. 1990; Tsurimoto and Stillman 1991a, b). Therefore, this multi-subunit protein plays a critical role in all stages of DNA replication.

To determine if RPA functions in the replication of cell chromosomal DNA, we have turned to the genetically manipulable organism *S. cerevisiae* to search for functional homologues of the human DNA replication factors. A protein was detected in *S. cerevisiae* as a homologue of RPA (Brill and Stillman 1989). This protein was identified and purified using an origin dependent unwinding assay in which the yeast protein cooperated with SV40 T antigen to extensively unwind a plasmid containing the SV40 replication origin. Like the mammalian RPA, the purified protein, called scRPA, had protein subunits of 69K, 36K and 13K, with the 69K subunit functioning as the single strand DNA binding protein and the 36K subunit being phosphorylated in a cell cycle dependent manner (Brill and Stillman 1989; Din et al. 1990). Therefore, in many respects, the yeast protein resembles the human protein, both structurally and functionally. The scRPA protein would not, however, substitute for the human protein in the complete DNA replication reaction because it is not capable of stimulating human pol α. All three subunits of scRPA are essential for the growth of cells (Heyer et al. 1990; Brill and Stillman 1991). Moreover, in addition to a role in DNA replication, RPA has been shown to be required for the repair of U.V. induced lesions in DNA (Coverley et al. 1991) and has been implicated in genetic recombination (Heyer et al. 1990; Moore et al. 1991).

3.2 DNA Polymerase α/Primase Complex

The pol α/primase complex was long known to be involved in replicative DNA synthesis in eukaryotic cells and temperature sensitive mutations in the gene encoding *S. cerevisiae* pol α (Pol I) arrest cells at the non-permissive temperature in S phase (Budd and Campbell 1987; Pizzagalli et al. 1988). Using an immunoaffinity purification procedure (Murakami et al. 1986), we have purified pol α/primase complex from human cells and demonstrated that it is essential for initiation of DNA replication in vitro (Tsurimoto et al. 1990). Similar results have been reported by other laboratories (Murakami et al. 1986; Ishimi et al. 1988; Wold et al. 1989; Weinberg et al. 1990). The purified protein fraction contains the pol α/primase subunits of molecular weight

180K, 77K, 55K and 49K, (Fig. 2) but this fraction also contains a few other proteins of unknown function or significance.

Recent studies have indicated that pol α/primase complex can initiate DNA replication in the presence of T antigen, RPA and the topoisomerases. Omission of pol α/primase complex from the reconstituted replication reactions resulted in the complete loss of DNA synthesis, suggesting that the first priming of DNA synthesis at the replication origin is by primase, followed by synthesis of the first nascent strands by pol α. The first strand synthesized at the origin most likely corresponds to the first Okazaki fragments for lagging strand DNA replication (Tsurimoto et al. 1990). Pol α then moves away from the replication origin to become the lagging strand DNA polymerase. At the same time, and under optimal conditions in vitro, pol α is incapable of continuing as the leading strand DNA polymerase. This implies that a second DNA polymerase complex is formed at the origin for initiation of leading strand DNA replication. This complex involves pol δ, replication factor C and PCNA.

Protein-protein interactions play a critical role in the initiation of SV40 DNA replication. T antigen is known to interact specifically with the 180K catalytic subunit of human DNA polymerase α (Smale and Tjian 1986; Gannon and Lane 1987, 1990; Dornreiter et al. 1990) and with the RPA protein complex (Dornreiter et al. 1992). Moreover, RPA also binds specifically to the primase component of the DNA polymerase α-primase complex and can, under certain conditions, stimulate its activity (Matsumoto et al. 1990; Collins and Kelly 1991; Dornreiter et al. 1992). The emerging view from these studies is that specific protein-protein interactions play a critical role in the initiation of SV40 DNA replication. The assembly of a multiprotein complex of T antigen at the origin of DNA replication provides a landing-pad for cellular DNA replication proteins to assemble into an active replication complex. Since T antigen has both origin recognition and DNA helicase functions in the assembly of the active replication complex, it is anticipated that similar origin recognition and DNA helicase activities will function during initiation of DNA replication at cellular origin sequences.

3.3 Replication Factor C (RF-C) and the Proliferating Cell Nuclear Antigen (PCNA)

The human cell DNA replication factors described above are all essential for initiation of DNA synthesis at the replication origin. Three other factors described below are essential for complete DNA replication, but unlike RPA and pol α/primase complex, they are required at a later stage. The first of these is RFC, which consists of three protein subunits of 140K, 41K and 37K molecular weight (Fig. 2; Tsurimoto and Stillman 1989b, 1990, 1991b; Lee et al. 1991a). The RFC complex is a DNA binding protein that binds specifically to single strand DNAs containing an annealed primer DNA (Tsurimoto and Stillman 1990, 1991a). RFC has also been purified from the yeast, *S. cerevisiae*, and has biochemical properties similar to the human protein (Burgers 1991; Yoder and Burgers 1991; Fien and Stillman 1992). Nuclease protection and gel-shift experiments demonstrate that RFC binds to the 3' end of the primer DNA, covering part of the duplex DNA and part of the single strand DNA (Tsurimoto and Stillman 1991a). The binding is greatly stabilized in the presence of ATP, particularly a non-hydrolyzable analog of ATP, ATPγS. Protein-DNA crosslinking experiments demonstrated that the 140K subunit of RFC is the subunit that primarily binds to DNA.

Another function of RFC that is required for its function in DNA replication is an ATPase activity (Tsurimoto and Stillman 1990, 1991a; Lee et al. 1991a; Chen et al. 1992). The 41K subunit of RFC binds to ATP in the presence of the primer DNA and is probably part of the ATPase. The ATPase activity is stimulated by the presence of DNA in the reaction, with single strand DNA containing an annealed primer being the best co-factor. The ATPase activity of RFC is stimulated by the proliferating cell nuclear antigen (PCNA), another essential replication factor (Prelich et al. 1987b). The addition of PCNA to the gel-shift experiments with RFC causes a decrease in complex formation in the absence of ATP, but a marked increase in a slower migrating complex in the presence of ATP (Tsurimoto and Stillman 1991b). Nuclease protection experiments also show that the binding of RFC to the 3' end of the primer on a primer-template DNA is stimulated in the presence of PCNA and that the protected region is extended on the duplex DNA. Therefore, RFC and PCNA form a complex on the 3' end of the primer when it is annealed to a template DNA and this complex is recognized by DNA polymerase δ to initiate replication of the leading strand template (see below).

The functions of these DNA replication factors are analogous to the functions of the DNA replication factors that function to replicate bacteriophage T4 and *E. coli* genomes (Table 1). RFC and PCNA can be equated with the phage T4 genes 44/62 and gene 45 encoded proteins, respectively, or with the *E. coli* pol III accessory proteins γ, δ, δ', τ and β, respectively. Interestingly, the phage T4 DNA polymerase has homology to the pol α like DNA polymerases (human pol α and pol δ, yeast pol α and pol δ; reviewed by Wang et al. 1989; Wang 1991).

Table 1. Conserved functions of DNA replication proteins

Human	Phage T4	*E. coli*	Functions
RPA	Gene 32	SSB	Single-stranded DNA binding Stimulates DNA polymerase Promotes origin unwinding
RFC	Gene 44/62	γδ.δ'τ	DNA-dependent ATPase Primer-template binding Stimulates DNA polymerase
PCNA	Gene 45	β (*dna* N)	Stimulates DNA polymerase Stimulates DNA-dependent ATPase
DNA pol α[a]	Gene 43	DNA pol III	DNA polymerase
DNA pol δ			
DNA primase[a]	Gene 61	*dna* G	Primase

[a] The human DNA pol α and primase activities function as a multiprotein complex.

3.4 DNA Polymerase δ

Fractionation of the cytosol S100 extract (Fig. 1) into multiple components also yielded a factor that co-purified with DNA polymerase δ, (pol δ) (Lee et al. 1989; Weinberg and Kelly 1989; Melendy and Stillman 1991). The most highly purified fraction from this purification contains a few protein bands, but included in these is a protein of 130K molecular weight that precisely co-fractionates with the activity. Previous fractionation of pol δ from various laboratories demonstrated that pol δ from either human cells, calf thymus or the yeast *S. cerevisiae*, contains protein subunits of 125K and 50K molecular weight (reviewed by Wang 1991). The calf thymus pol δ, purified by a published procedure (Lee et al. 1984) can completely substitute for the human pol δ during SV40 DNA replication (Tsurimoto et al. 1990).

Genetic and biochemical experiments in the yeast *S. cerevisiae* have clearly demonstrated a role for pol δ in replicative DNA synthesis (Bauer et al. 1988; Boulet et al. 1989; Sitney et al. 1989). The yeast *CDC2* gene encodes the DNA polymerase δ that is stimulated by yeast PCNA (Bauer and Burgers 1988) and temperature sensitive mutants in the gene affect DNA synthesis in the S phase of the cell cycle. Therefore, a role for two DNA polymerases in replicative DNA synthesis seems to be a general property of eukaryotic DNA replication. A third DNA polymerase, DNA polymerase ε, has been shown to be essential for cell cycle progression through S phase in *S. cerevisiae*, (Morrison et al. 1990; Araki et al. 1992). DNA pol ε from human cells, however, does not function in the replication of SV40 DNA in vitro (Lee et al. 1991b), so the precise role of this polymerase remains to be determined.

3.5 DNA Polymerase Accessory Factors

The two DNA polymerases, α and δ are stimulated by the replication factors in fundamentally different ways. RPA and RFC both stimulate the activity of pol α about two-fold on a primed single-stranded template DNA and increase the processivity of the polymerase so that products of 300–400 nucleotides are made (Tsurimoto and Stillman 1989b; Kenny et al. 1989). In contrast, the three replication factors, RPA, RFC and PCNA cooperate to stimulate the rate of initiation and processivity of pol δ on primed single strand DNA under DNA replication conditions (pH 7.5) and in the presence of ATP (Tsurimoto and Stillman 1989; 1990; Lee et al. 1991a). PCNA, however, will stimulate the processivity of pol δ alone on poly dA/oligo dT template/primer DNAs at pH 6.9, (Tan et al. 1986; Prelich et al. 1987b; unpubl. observ.), indicating that it can independently affect the function of both pol δ and RFC. These results suggest that pol δ functions with RFC, PCNA and RPA to synthesize long, continuously synthesized strands, whereas pol α, with RPA and RFC, synthesize relatively short DNA strands. Similar results have been reported with the *S. cerevisiae* homologues of these proteins (Burgers 1991; Fien and Stillman 1992).

3.6 Two DNA Polymerases Are Required for SV40 DNA Replication

The reconstruction of SV40 DNA replication with the proteins described above directly demonstrates that two DNA polymerases are required for the complete DNA replication (Tsurimoto et al. 1990; Weinberg et al. 1990; Lee et al. 1991a). To ascertain the role that these polymerases and their accessory factors play during DNA replication, we have examined the 'in vitro phenotype' of replication reactions when each of the factors have been omitted from the replication reactions (Tsurimoto et al. 1990). As stated above, RPA and pol α/primase complex are required for the synthesis of the first nascent DNA strands at the replication origin and are therefore essential. In contrast, omission of either RFC, PCNA or pol δ reduced DNA replication by approximately three-fold. Examination of the products of such reactions indicates that the leading strand template is not replicated, but the lagging strand template is replicated (Prelich and Stillman 1988; Tsurimoto and Stillman 1989b; Tsurimoto et al.

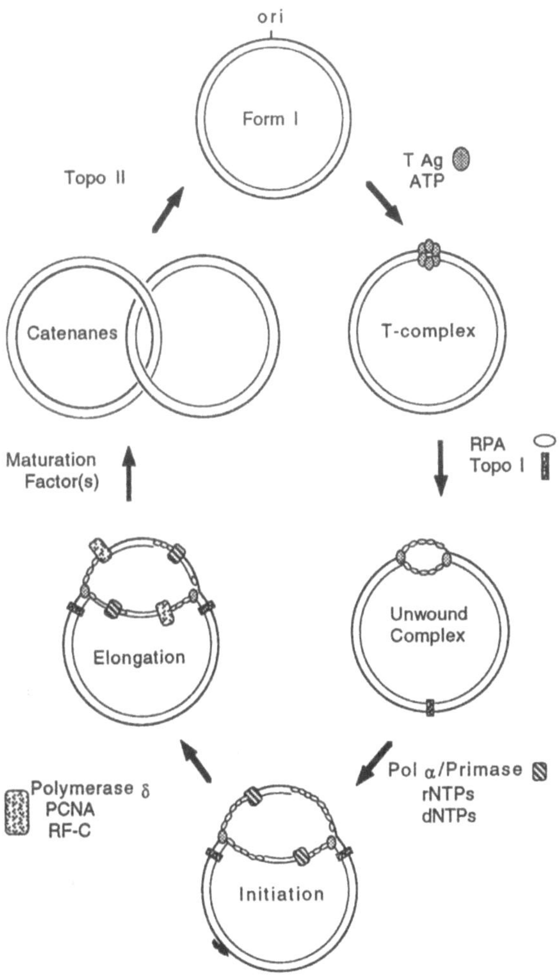

Fig. 3. General scheme for SV40 DNA replication in vitro. See text for details

1990). The products of the lagging strand remain as short Okazaki fragments because of the absence of enzymes that ligate them together in this system. Since omission of either RFC, PCNA or pol δ yield similar reaction products, these results are consistent with the three replication factors acting as a leading strand DNA polymerase complex, with the pol α/primase functioning to replicate the lagging strand template.

These results also imply that the initiation of DNA replication at the origin involves the synthesis of the first Okazaki fragment by a primase mediated RNA priming event, followed by DNA synthesis by pol α (Fig. 3). They also suggest a model for the initiation of DNA replication on the leading strand template in which the 3' end of the first Okazaki fragment is recognized by RFC and PCNA in the presence of ATP and then pol δ binds to this complex and copies the leading strand template (Fig. 3). RPA functions to stimulate both DNA polymerases and to prevent abnormal DNA synthesis on the leading strand template by pol α. Thus, a novel polymerase switching mechanism seems to occur at the replication origin for initiation of both lagging and leading strand DNA synthesis.

4 Initiation of Chromosomal DNA Replication

As indicated above, investigating the replication of eukaryotic virus DNA has contributed greatly to our knowledge of the cellular replication machinery. But these studies rely on the use of a virus encoded initiator protein, the SV40 T antigen, and reveal little about the process of initiation in cellular chromosomes. To address this issue directly, we have focused our attention on DNA replication in *S. cerevisiae*.

4.1 A Cellular Origin of DNA Replication

The *cis*-acting DNA sequences that are required for initiation of DNA replication in yeast chromosomes were first identified because they confer on plasmid DNAs the ability to replicate autonomously in the nucleus of cells and in synchrony with the replication of cellular chromosomes during the S phase of the cell cycle (Stinchcomb et al. 1979; reviewed in Campbell and Newlon 1991). Subsequent studies demonstrated that a subset of these autonomously replicating sequences (ARS) also function directly as origins of DNA replication in the chromosome (Brewer and Fangman 1991). They, therefore, fulfilled the role of 'replicators', as originally defined in the replicon model (Jacob et al. 1963). All ARS sequences contain a close match to a short, 11 bp consensus sequence which is essential for initiation of chromosomal DNA replication (Brewer and Fangman 1991; Campbell and Newlon 1991; Rivier and Rine 1992), but this element cannot function as a replicator element alone. In a recent study, the nature of the *cis*-acting sequences at an origin of DNA replication have been investigated in great detail and the results demonstrate an unexpected complexity (Marahrens and Stillman 1992).

Using a technique called linker-scan mutagenesis, all the DNA sequences within a 185bp region from chromosome IV in *S. cerevisiae*, known as ARS1, were systematically replaced, in eight base-pair blocks, with an eight base pair 'linker' containing the XhoI restriction enzyme recognition site. These mutants were analyzed in two different ways to determine if the mutated DNA could function as an origin of DNA replication. The data demonstrate that the ARS1 origin is composed of four separate DNA sequence elements, called A, B1, B2 and B3 (Fig. 4A). The 15bp A element that contains an 11bp match to the ARS consensus sequence was shown to be essential, but surprisingly, mutagenesis of the B1, B2 and B3 mutations only reduced, but did not eliminate origin function (Fig. 4B). When, however, any two of the B elements were simultaneously mutated, origin function was reduced to below detectable levels. This suggested that the B elements were functionally redundant. Origin function could not be restored by multimerizing any one of the B elements, suggesting that, in spite of the fact that they were partially redundant, the elements could not substitute for each other when present in multiple copies (Fig. 4C). This is in marked contrast to

A

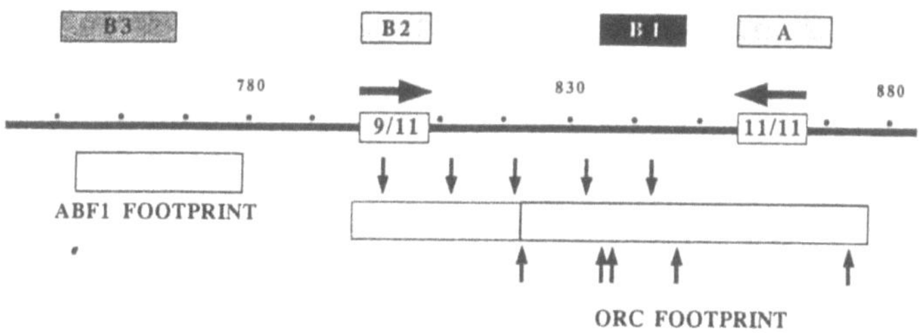

B

B3	B2	B1	A	Origin Function
+	+	+	+	+++
+	+	+	-	-
-	+	+	+	+
+	-	+	+	+
+	+	-	+	+
-	-	+	+	-
-	+	-	+	-
+	-	-	+	-
-	-	-	+	-

C

B1	B1	B1	A	-
B2	B2	B2	A	-
B3	B3	B3	A	-

Fig. 4. A summary of the structure of the chromosomal origin of DNA replication, ARS1. A The schematic shows a 140-bp region of chromosome IV from *S. cerevisiae* (*black line*) and the four functional elements *B3*, *B2*, *B1* and *A*. Matches (*11/11* or *9/11*) to the ARS consensus sequence are shown, together with the 5' to 3' direction of the T-rich strand (*horizontal arrows*). The *white boxes* show the regions of this origin that are protected from DNase 1 digestion by binding of ABF1 or ORC. *Vertical arrows* show DNase 1 hypersensitive sites. B A summary of the function of ARS1 in the presence (+) or absence (–) of combinations of DNA elements. C Reiteration of the individual B elements does not restore origin function

the *cis*-acting DNA sequences that constitute promoters for transcription of eukaryotic genes. In this case, individual promoter elements can be reiterated to create a fully functional promoter in the presence of a TATA-box, depending upon the cell type used to test transcription.

One of the DNA elements at ARS1 was shown to bind the previously identified transcription factor ABF1 (Marahrens and Stillman 1992). The ABF1 protein was originally identified because it bound to ARS1 (Buchman et al. 1988; Diffley and Stillman 1988; Sweder et al. 1988; see Fig. 4A). Subsequently, this protein has been shown to bind to the promoters for a variety of genes and also to the transcriptional silencers at the yeast mating type loci (reviewed in Campbell and Newlon 1991). The ABF1 binding site at ARS1 can be replaced by heterologous transcription factor binding sites, including the bacterial lexA operator sequence, provided that in this latter case, the lexA protein was fused to a eukaryotic transcription activator sequence and was provided in *trans* (Marahrens and Stillman 1992). These observations suggest that a transcription factor binding adjacent to other elements within an origin is able to influence the initiation of DNA replication in cellular chromosomes. This may occur by altering the chromatin structure of the origin or by direct interaction with the DNA replication apparatus, or both.

4.2 Identification of a Putative Cellular Initiator Protein

In the replication model, Jacob et al. (1963) proposed that initiation of DNA replication was mediated by an 'initiator' protein that could bind to the replicator. In bacteria, phage and virus DNA replication, initiator proteins such as the *E. coli* dnaA protein, the phage λ O protein and the SV40 T antigen function as origin recognition proteins and are key factors for the initiation of DNA replication (reviewed by Kornberg and Baker 1991). Despite many years of searching, a cellular initiator protein had not been identified in eukaryotes. Having identified the *cis*-acting DNA sequences in the replicator, we searched for a protein(s) that would bind to the essential A and important B1 and B2 elements of ARS1 with the expectation that the initiator protein would first bind to these essential elements within the origin of DNA replication.

A multi-subunit protein was identified and purified to apparent homogeneity from a nuclear extract of *S. cerevisiae* (Bell and Stillman 1992). This *o*rigin *r*ecognition *c*omplex (ORC) contains polypeptides of apparent molecular weights of 50, 53, 56, 62, 72 and 120 kDa which co-fractionate as a single protein complex. ORC binds in an ATP dependent manner to ARS1 and all other *S. cerevisiae* ARS sequences that have been examined. This mode of nucleotide dependent DNA binding is highly unusual for sequence specific DNA binding proteins.

Evidence suggesting that ORC is the key chromosomal initiator protein derives from the observation that a direct correlation exists between the ability of the protein to bind to mutant ARS1 derivatives and the function of these origin mutants in supporting DNA replication in vivo (Bell and Stillman 1992). ORC primarily recognizes the ARS consensus sequence within the A element, but also protects from nuclease digestion the B1 and B2 elements (Fig. 4A). In fact, the B2 element is a partial, inverted repeat of the ARS consensus sequence in domain A. Point mutations in the essential A element that eliminate ARS activity also eliminate specific ORC binding, whereas other mutations that do not affect origin function do not perturb ORC DNA binding. This correlation between origin function and ORC binding strongly implicates ORC as the chromosomal initiator protein. Furthermore, Diffley and Cocker (1992) have observed nuclease protection and hypersensitive site patterns in genomic footprints of ARS1 that closely resemble the patterns obtained at ARS1 with purified ORC in vitro.

Further biochemical and genetic analysis of the initiation of DNA replication in *S. cerevisiae* should indicate whether or not ORC does indeed initiate DNA replication in cellular chromosomes. If the protein complex is conserved, isolation of functionally homologous proteins from metazoan species should facilitate the analysis of chromosomal origins of DNA replication in these species.

Acknowledgements. I wish to acknowledge the important contributions by current and former members of my laboratory, particularly S. Bell, S. Brill, J. Diffley, S. Din, A. Dutta, M. Fairman, K. Fien, Y. Marahrens, T. Melendy, G. Prelich, and T. Tsurimoto. I thank B. Weinkauff for typing the manuscript. The research in the author's laboratory described in this chapter was supported by grants from the National Institutes of Health (CA13106, AI20460 and GM45436 and the World Business Council).

References

Araki H, Ropp PA, Johnson AL, Johnston LH, Morrison A & Sugino A (1992) DNA polymerase II, the probable homolog of mammalian DNA polymerase ε, replicates chromosomal DNA in the yeast *Saccharomyces cerevisiae*. EMBO J 11:733–740

Bauer GA & Burgers PMJ (1988) The yeast analog of mammalian cyclin/proliferatingn-cell nuclear antigen interacts with mammalian DNA polymerase δ. Proc Natl Acad Sci USA 85:7506–7510

Bauer GA, Heller HM & Burgers PMJ (1988) DNA polymerase III from *Saccharomyces cerevisiae* I purification and characterization. J Biol Chem 263:917–924

Bell SP & Stillman B (1992) ATP-dependent recognition of eukaryotic origins of DNA replication by a multiprotein complex. Nature 357:128–134

Borowiec JA, Dean FB, Bullock PA & Hurwitz J (1990) Binding and unwinding – how T-antigen engages the SV40 origin of DNA replication. Cell 60:181–184

Borowiec JA & Hurwitz J (1988) Localized melting and structural changes in the SV40 origin of replication induced by T-antigen. EMBO J 7:3149–3158

Boulet A, Simon M, Faye G, Bauer GA & Burgers PMJ (1989) Structure and function of the *Saccharomyces cerevisiae* CDC2 gene encoding the large subunit of DNA polymerase III. EMBO J 8:1849–1854

Bramhill D & Kornberg A (1988) A model for initiation at origins of DNA replication. Cell 54:915–918

Brill SJ & Stillman B (1989) Yeast replication factor-A functions in the unwinding of the SV40 origin of DNA replication. Nature 342:92–95

Brill SJ & Stillman B (1991) Replication factor-A from *Saccharomyces cerevisiae* is encoded by three essential genes coordinately expressed at S phase. Genes Dev 5:1589–1600

Buchman AR, Kimmerly WJ, Rine J & Kornberg RD (1988) Two DNA-binding factors recognize specific sequences at silencers, upstream activating sequences, autonomously replicating sequences, and telomeres in Saccharomyces cerevisiae. Mol Cell Biol 8:210–255

Budd M & Campbell JL (1987) Temperature-sensitive mutants of yeast DNA polymerase I. Proc Natl Acad Sci USA 84:2838–2842

Burgers PMJ (1991) *Saccharomyces cerevisiae* replication factor C II. Formation and activity of complexes with the proliferating cell nuclear antigen and with DNA polymerases δ and ε. J Biol Chem 266:22698–22706

Campbell JL & Newlon CS (1991) Chromosomal DNA replication. In: The molecular and cellular biology of the yeast *Saccharomyces*. Cold Spring Harbor, NY, Cold Spring Harbor Laboratory Press

Challberg MD & Kelly TJ (1989) Animal virus DNA replication. Annu Rev Biochem 58:671–717

Chen M, Pan Z-Q & Hurwitz J (1992) Sequence and expression in *Escherichia coli* of the 40-kDa subunit of activator 1 (replication factor C) of HeLa cells. Proc Natl Acad Sci USA 89:2516–2520

Cheng L & Kelly TJ (1989) Transcriptional activator nuclear factor I stimulates the replication of SV40 minichromosomes in vivo and in vitro. Cell 59:541–551

Collins KL & Kelly TJ (1991) Effects of T antigen and replication protein A on the initiation of DNA synthesis by DNA polymerase α-primase. Mol Cell Biol 11:2108–2115

Coverley D, Kenny MK, Munn M, Rupp WD, Lane DP & Wood RD (1991) Requirement for the replication protein SSB in human DNA excision repair. Nature 349:538–541

D'Urso G, Marraccino RL, Marshak DR & Roberts JM (1990) Cell cycle control of DNA replication by a homologue from human cells of the $p34^{cdc2}$ protein kinase. Science 250:786–791

Deb S, DeLucia AL, Baur C-P, Koff A & Tegtmeyer P (1986) Domain structure of the simian virus 40 core origin of replication. Mol Cell Biol 6:1663–1670

DePamphilis ML (1988) Transcriptional elements as components of eukaryotic origins of DNA replication. Cell 52:635–638

DePamphilis ML & Bradley MK (1986) Replication of SV40 and polyoma virus chromosomes. In: Salzman NP (ed) The Papovaviridae, vol I. Plenum, New York, pp 99–246

Diffley JFX & Cocker J (1992) Protein-DNA interactions at a yeast replication origin. Nature 357:169–172

Din S, Brill S, Fairman MP & Stillman B (1990) Cell-cycle-regulated phosphorylation of DNA replication factor A from human and yeast cells. Genes Dev 4:968–977

DiNardo S, Voelkel KA & Sternglanz R (1984) DNA topoisomerase II mutant of *Sacchromyce cerevisiae*: topoisomerase II is required for segregation of daughter molecules at the termination of DNA replication. Proc Natl Acad Sci USA 81:2616–2620

Dodson M, Dean FB, Bullock P, Echols H & Hurwitz J (1987) Unwinding of duplex DNA from the SV40 origin of replication by T antigen. Science 238:964–967

Dornreiter I, Erdile LF, Gilbert IU, von Winkler D, Kelly TJ & Fanning E (1992) Interaction of DNA polymerase α-primase with cellular replication protein A and SV40 T antigen. EMBO J 11:769–776

Dornreiter I, Höss A, Arthur AK & Fanning E (1990) SV40 T antigen binds directly to the large subunit of purified DNA polymerase alpha. EMBO J 9:3329–3336

Dutta A & Stillman B (1992) cdc2 family kinases phosphorylate a human cell DNA replication factor, RPA, and activate DNA replication. EMBO J 11:2189–2199

Erdile LF, Wold MS & Kelly TJ (1990) The primary structure of the 32-kDa subunit of human replication protein. A J Biol Chem 265:3177–3182

Erdile LF, Heyers W-D, Kolodner R & Kelly TJ (1991) Characterization of a cDNA encoding the 70-kDa single-stranded DNA-binding subunit of human replication protein A and the role of the protein in DNA replication. J Biol Chem 266:12090–12098

Fairman MP & Stillman B (1988) Cellular factors required for multiple stages of SV40 replication in vitro. EMBO J 7:1211–1218

Fangman WL & Brewer BJ (1991) Activation of replication origins within yeast chromosomes. Annu Rev Cell Biol 7:375–402

Fanning E (1992) Simian Virus 40 large T antigen: the puzzle, the pieces, and the emerging picture. J Virol 66:1289–1293

Fien K & Stillman B (1992) Identification of replication factor C from *Saccharomyces cerevisiae*: a component of the leading-strand DNA replication complex. Mol Cell Biol 12:155–163

Gannon JV & Lane DP (1987) p53 and DNA polymerase α compete for binding to SV40 T antigen. Nature 329:456–458

Gannon JV & Lane DP (1990) Interactions between SV40 T antigen and DNA polymerase α. New Biol 2:84–92

Goetz GS, Dean FB, Hurwitz J & Matson SW (1988) The unwinding of duplex regions in DNA by the simian virus 40 large tumor antigen-associated DNA helicase activity. J Biol Chem 263:383–392

Heyer W-D, Rao MRS, Erdile LF, Kelly TJ & Kolodner RD (1990) An essential *Saccharomyces cerevisiae* single-stranded DNA binding protein is homologous to the large subunit of human RP-A. EMBO J 9:2321–2329

Holm C, Goto T, Wang JC & Botstein D (1985) DNA topoisomerase II is required at the time of mitosis in yeast. Cell 41:553–563

Ishimi Y, Claude A, Bullock P & Hurwitz J (1988) Complete enzymatic synthesis of DNA containing the SV40 origin of replication. J Biol Chem 263:19723–19733

Jacob F, Brenner S and Cuzin F (1963) On the regulation of DNA replication in bacteria. Cold Spring Harbor Symp Quant. Biol 28:329–348

Kenny MK, Lee S-H & Hurwitz J (1989) Multiple functions of human single-stranded-DNA binding protein in simian virus 40 DNA replication: Single-strand stabilization and stimulation of DNA polymerases α and δ. Proc Natl Acad Sci USA 86:9757–9761

Kenny MK, Schlegel U, Furneaux H & Hurwitz J (1990) The role of human single stranded DNA binding protein and its individual subunits in simian virus 40 DNA replication. J Biol Chem 265:7693–7700

Kimmerly W, Buchman A, Kornberg R & Rine J (1988) Roles of two DNA-binding factors in replication, segregation and transcriptional repression mediated by a yeast silencer. EMBO J 7:2241–2253

Kornberg A & Baker TA (1992) *DNA Replication* 2. ed. Freeman New York

Lanford RE (1988) Expression of simian virus 40 T antigen in insect cells using a baculovirus expression vector. Virology 167:72–81

Lee MYWT, Tan C-K, Downey KM & So AG (1984) Further studies on calf thymus DNA polymerase δ purified to homogeneity by a new procedure. Biochemistry 2:1906–1913

Lee S-H, Eki T & Hurwitz J (1989) Synthesis of DNA containing the simian virus 40 origin of replication by the combined action of DNA polymerases α and δ. Proc Natl Acad Sci USA 86:7361–7365

Lee S-H, Kwong AD, Pan Z-Q & Hurwitz J (1991a) Studies on the activator 1 protein complex, an accessory factor for proliferating cell nuclear antigen-dependent DNA polymerase δ. J Biol Chem 266:594–602

Lee S-H, Zhen-Qiang P, Kwong AD, Burgers PMJ & Hurwitz J (1991b) Synthesis of DNA by DNA polymerase ε in vitro. J Biol Chem 266.22707–22717

Li JJ & Kelly TJ (1984) Simian virus 40 DNA replication in vitro. Proc Natl Acad Sci USA 81:6973–6977

Li JJ & Kelly TJ (1985) Simian virus 40 DNA replication in vitro: Specificity of initiation and evidence for bidirectional replication. Mol Cell Biol 5:1238–1246

Marahrens Y & Stillman B (1992) A yeast chromosomal origin of DNA replication defined by multiple functional elements. Science 255:817–823

Matsumoto T, Eki T & Hurwitz J (1990) Studies on the initiation and elongation reactions in the simian virus 40 DNA replication system. Proc Natl Acad Sci USA 87:9712–9716

Melendy TE & Stillman B (1991) Purification of DNA polymerase δ as an essential SV40 DNA replication factor. J Biol Chem 266:1942–1949

Moore SP, Erdile L, Kelly T & Fishel R (1991) The human homologous pairing protein HPP-1 is specifically stimulated by the cognate single-stranded binding protein hRP-A. Proc Natl Acad Sci USA 88:9067–9071

Morrison A, Araki H, Clark B, Hamatake RK & Sugino A (1990) A third essential DNA polymerase in *S. cerevisiae*. Cell 62:1143–1151
Murakami Y, Wobbe CR, Weissbach L, Dean FB & Hurwitz J (1986) Role of DNA polymerase α and DNA primase in simian virus 40 DNA replication in vitro. Proc Natl Acad Sci USA 8:2869–2873
Pines J & Hunter T (1990) p34^{cdc2}: the S and M kinase. New Biol 2:389–401
Pizzagalli A, Vasasnini P, Plevani P & Lucchini G (1988) DNA polymerase I gene of *Saccharomyces cerevisiae*: nucleotide sequence, mapping of a temperature-sensitive mutation, and protein homology with other DNA polymerases. Proc Natl Acad Sci USA 85:3772–3776
Prelich G, Kostura M, Marshak DR, Mathews MB & Stillman B (1987) The cell-cycle regulated proliferating cell nuclear antigen is required for SV40 DNA replication in vitro. Nature 326:471–475
Prelich G & Stillman B (1988) Coordinated leading and lagging strand synthesis during SV40 DNA replication in vitro requires PCNA. Cell 53:117–126
Prelich G, Tan CK, Kostura M, Mathews MB, So AG, Downey KM & Stillman B (1987) Functional identity of proliferating cell nuclear antigen and a DNA polymerase δ auxiliary protein. Nature 326:517–520
Rivier DH & Rine J (1992) An origin of DNA replication and a transcription silencer require a common element. Science 256:659–663
Roberts JM & D'Urso G (1988) An origin unwinding activity regulates initiation of DNA replication during mammalian cell cycle. Science 241:1486–1489
Simanis V & Lane DP (1985) An immunoaffinity purification procedure for SV40 large T antigen. Virology 144:80–100
Sitney KC, Budd ME & Campbell JL (1989) DNA polymerase III, a second essential DNA polymerase, is encoded by the *S. cerevisiae* CDC2 gene. Cell 56:599–605
Smale ST & Tjian R (1986) T-antigen-DNA polymerase α complex implicated in simian virus 40 DNA replication. Mol Cell Biol 6:4077–4087
Smith S & Stillman B (1989) Purification and characterization of CAF-1, a human cell factor required for chromatin assembly during DNA replication in vitro. Cell 58:15–25
Snapka R, Powelson MA & Strayer JM (1988) Swiveling and decatenation of replicating simian virus 40 genomes in vivo. Mol Cell Biol 8:515–521
Stahl H, Droge P & Knippers R (1986) DNA helicase activity of SV40 large tumor antigen. EMBO J 5:1939–1944
Stillman B (1986) Chromatin assembly during SV40 DNA replication in vitro. Cell 45:555–565
Stillman B (1989) Initiation of Eukaryotic DNA replication in vitro. Ann Rev Cell Biol 5:197–245
Stillman BW & Gluzman Y (1985) Replication and supercoiling of simian virus 40 DNA in cell extracts from human cells. Mol Cell Biol 5:2051–2060
Stinchcomb DT, Struhl K & Davis RW (1979) Isolation and characterization of a chromosomal replicator. Nature 282:39–43
Sundin O & Varshavsky A (1980) Terminal stages of SV40 DNA replication proceed via multiply intertwined catenated dimers. Cell 21:103–114
Sundin O & Varshavsky A (1981) Arrest of segregation leads to accumulation of highly intertwined catenated dimers: dissection of the final stages of SV40 DNA replication. Cell 25:659–669
Sweder K, Rhode PR & Campbell JL (1988) Aerification and characterization of proteins that bind ot yeast ARSs. J Biol Chem 263:17270–17277
Tan CK, Castillo C, So AG & Downey KM (1986) An auxiliary protein for DNA polymerase-d from fetal calf thymus. J Biol Chem 261:12310–12316
Thrash C, Voelkel K, DiNardo S & Sternglanz R (1984) Identification of *Saccharomyces cerevisiae* mutant deficient in DNA topoisomerase I activity. J Biol Chem 259:1375–1377
Tsurimoto T, Fairman MP & Stillman B (1989) Simian virus 40 DNA replication in vitro: identification of multiple stages of initiation. Mol Cell Biol 9:3839–3849

Tsurimoto T, Melendy T & Stillman B (1990) Two DNA polymerase complexes sequentially initiate lagging and leading strand synthesis at the simian virus 40 origin of DNA replication. Nature 346:534–539

Tsurimoto T & Stillman B (1989a) Multiple replication factors augment DNA synthesis by the two eukaryotic DNA polymerases, α and δ. EMBO J 8:3883–3889

Tsurimoto T & Stillman B (1989b) Purification of a cellular replication factor, RF-C, that is required for coordinated synthesis of leading and lagging strands during simian virus 40 DNA replication in vitro. Mol Cell Biol 9:609–619

Tsurimoto T & Stillman B (1990) Functions of replication factor C and proliferating cell nuclear antigen: functional similarity of DNA polymerase accessory proteins from human cells and bacteriophage T4. Proc Natl Acad Sci USA 87:1023–1027

Tsurimoto T & Stillman B (1991a) Replication factors required for SV40 DNA replication in vitro. I. DNA structure specific recognition of a primer-template junction by eukaryotic DNA polymerases and their accessory factors. J Biol Chem 266:1950–1960

Tsurimoto T & Stillman B (1991b) Replication factors required for SV40 DNA replication in vitro. II. Switching of DNA polymerase α and δ during initiation of leading and lagging strand synthesis. J Biol Chem 266:1961–1968

Uemura T & Yanagida M (1984) Isolation of type I and II DNA topoisomerase mutants from fission yeast: single and double mutants show different phenotypes in cell growth and chromatin organization. EMBO J 3:1737–1744

Wang TS-F (1991) Eukaryotic DNA polymerases. Annu Rev Biochem 60:513–552

Wang TS-F, Wong SW & Korn D (1989) Human DNA polymerase α: predicted functional domains and relationships with DNA polymerases. FASEB J 3:14–21

Weaver DT, Fields-Berry SC & DePamphilis ML (1985) The termination region for SV40 DNA replication directs the mode of separation for the two sibling molecules. Cell 41:565–575

Weinberg DH, Collins KL, Simancek P, Russo A, Old MS, Virshup DM & Kelly TJ (1990) Reconstitution of simian virus 40 DNA replication with purified proteins. Proc Natl Acad Sci USA 87:8692–8696

Weinberg DH & Kelly TJ (1989) Requirement for two DNA polymerases in the replication of simian virus 40 DNA in vitro. Proc Natl Acad Sci USA 86:9742–9746

Wessel R, Schweizer J & Stahl H (1992) Simian virus 40 T-antigen DNA helicase is a hexamer which forms a binary complex during bidirectional unwinding from the viral origin of DNA replication. J Virol 66:804–815

Wiekowski M, Droge P & Stahl H (1987) Monoclonal antibodies as probes for a function of large T antigen during the elongation process of simian virus 40 DNA replication. J Virol 61:411–418

Wiekowski M, Schwarz MW & Stahl H (1988) Simian virus 40 large T antigen DNA helicase. Characterization of the ATP-asedependent DNA unwinding activity and its substrate requirements. J Biol Chem 263:436–442

Wobbe CR, Dean F, Weissbach L & Hurwitz J (1985) In vitro replication of duplex circular DNA containing the simian virus 40 DNA origin site. Proc Natl Acad Sci USA 82:5710–5714

Wobbe CR, Weissbach L, Borowiec JA, Dean FB, Murakami Y, Bullock P & Hurwitz J (1987) Replication of simian virus 40 origin-containing DNA in vitro with purified proteins. Proc Natl Acad Sci USA 84:1834–1838

Wold MS & Kelly T (1988) Purification and characterization of replication protein A, a cellular protein required for in vitro replication of simian virus 40 DNA. Proc Natl Acad Sci USA 85:2523–2527

Wold MS, Weinberg DH, Virshup DM, Li JJ & Kelly TJ (1989) Identification of cellular proteins required for simian virus 40 DNA replication. J Biol Chem 264:2801–2809

Yang L, Wold MS, Li JJ, Kelly TJ & Liu LF (1987) Roles of DNA topoisomerases in simian virus 40 DNA replication in vitro. Proc Natl Acad Sci USA 84:950–954

Yoder BL & Burgers PMJ (1991) *Saccharomyces cerevisiae* replication factor C I. Purification and characterization of its ATPase activity. J Biol Chem 266:22689–22697

Protein Kinases and Replication Control

Regulation of the Mitotic CDC2 Protein Kinase

E. A. Nigg[1], W. Krek[1,2], and P. Gallant[1]

1 Introduction

The eukaryotic cell cycle is characterized by two major events, DNA replication (S phase) and mitosis (M phase); these are separated by two so-called "gap"-phases, G1 (prior to S phase) and G2 (prior to M phase). According to the current paradigm of the cell cycle as a "cdc2 cycle", both DNA replication and mitosis are driven by serine-threonine specific protein kinases encoded by functional homologs of the fission yeast cdc2 gene (for review see Nurse 1990; Pines and Hunter 1990; Draetta 1990). In yeasts, a single 34 kD catalytic subunit (called cdc2 in fission yeast, CDC28 in budding yeast) is required prior to the onset of DNA replication for traversing a control point called START, as well as for entry into mitosis. In vertebrates, the mitotic function of $p34^{cdc2}$ is well established, but recent evidence indicates that progression through interphase cell cycle transitions requires the activity of cdc2-related kinases. These kinases were discovered through their association with cyclin subunits, and accordingly, they are referred to as *c*yclin-*d*ependent *k*inases (cdk's; for review see Pines and Hunter 1991a). According to one emerging scenario, $p34^{cdc2}$ may function predominantly to trigger entry into mitosis, while a closely related kinase termed $p33^{cdk2}$ (originally called Eg-1; Paris et al. 1991) may play a role in initiating DNA replication (Fang and Newport 1991; M. Phillipe, Univ. Rennes, pers. comm.).

2 Results and Discussion

2.1 Regulation of cdc2 Kinases by Phosphorylation

The regulation of cdc2 kinases is based on two major mechanisms, i.e. cell cycle-dependent interactions of the 33–34 kD catalytic subunits with other proteins (Booher et al. 1989; Draetta et al. 1989; Giordano et al. 1989; Labbé et al. 1989; Meijer et al. 1989; Moreno et al. 1989; Pines and Hunter 1989; Gautier et al. 1990), and phosphorylation-dephosphorylation reactions (Morla et al. 1989; Gould and Nurse 1989; Krek and Nigg 1991a; Gould et al. 1991; Ducommun et al. 1991; Norbury et al. 1991). Cell cycle-specific phosphorylations occur on both the catalytic 33–34 kD subunits as well

[1] Swiss Institute for Experimental Cancer Research (ISREC), 155, Chemin des Boveresses, CH-1066 Epalinges, Switzerland.
[2] Present Address: Dana-Farber Cancer Institute, 44 Binney Street, Boston, MA 02115, USA.

43. Colloquium Mosbach 1992
DNA Replication and the Cell Cycle

as the cyclin subunits of cdc2 kinases (Pines and Hunter 1989; Meijer et al. 1989; Gautier and Maller 1991; Izumi and Maller 1991). To understand how cdc2 kinases are regulated through the cell cycle, it is important to establish which signalling pathways converge onto cdc2, to identify the kinases and phosphatases involved in these pathways, and to relate the corresponding phosphorylation events to cdc2 kinase activity. This information will be indispensable for understanding the regulatory network that integrates cell growth with cell proliferation.

Comparatively well understood is the role of phosphorylation of $p34^{cdc2}$ on tyrosine 15 (Tyr15). Genetic and biochemical studies in fission yeast have revealed that the phosphorylation state of this residue is under the control of the genes *cdc25*, *wee1* and *mik1*: the *cdc25* gene encodes a protein phosphatase and functions as an activator of $p34^{cdc2}$ (Russell and Nurse 1986; Gould et al. 1990; Strausfeld et al. 1991; Dunphy and Kumagai 1991; Gautier et al. 1991), while both *wee1* and *mik1* encode protein kinases that function as negative regulators of $p34^{cdc2}$ (Russell and Nurse 1987; Featherstone and Russell 1991; Lundgren et al. 1991; Parker et al. 1991). As demonstrated by site-directed mutagenesis, dephosphorylation of Tyr15 is required for activation of $p34^{cdc2}$ and entry of *S. pombe* cells into mitosis (Gould and Nurse 1989). In vertebrates, $p34^{cdc2}$ is phosphorylated not only on Tyr15, but also on the neighbouring residue threonine 14 (Thr14). Both Thr14 and Tyr15 are phosphorylated maximally during G2 phase, and full activation of $p34^{cdc2}$ at the G2/M transition was shown to require dephosphorylation of both residues (Krek and Nigg 1991a, b; Norbury et al. 1991; see Fig. 1).

Based on studies carried out with both *S. pombe* and vertebrates, phosphorylation of Tyr15 has been implicated in establishing the dependency of mitosis on the completion of DNA replication (Enoch and Nurse 1990; Smythe and Newport 1992). In the budding yeast, *S. cerevisiae*, however, the situation appears to be different: although the corresponding residue (tyrosine 19) is also phosphorylated in $p34^{CDC28}$ (the budding yeast homolog of the cdc2 gene product), S-phase feedback controls appear to be independent of Tyr19 phosphorylation (Sorger and Murray 1992; Amon et al. 1992).

Not all phosphorylations occurring on $p34^{cdc2}$ are of an inhibitory type. In fact, phosphorylation of one additional site, identified as Thr167 in *S. pombe* (Gould et al.

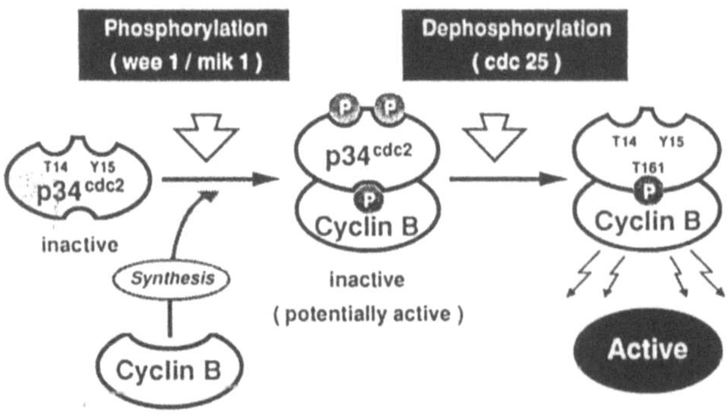

Fig. 1. A schematic model describing the transient inhibition of $p34^{cdc2}$ protein kinase activity by phosphorylation of residues Thr14 and Tyr15 (for detailed information, see Krek and Nigg 1991b)

1991) and Thr161 in vertebrates (Krek and Nigg 1992), appears to be required for p34^{cdc2} protein kinase activity (e.g. Solomon et al. 1990, 1992; Ducommun et al. 1991; Norbury et al. 1991; Gould et al. 1991). The available evidence is interpreted best by a model whereby phosphorylation of Thr161/167 increases the stability of p34^{cdc2}/cyclin complexes (Ducommun et al. 1991; Norbury et al. 1991). In addition, mutational studies carried out with *S. pombe* suggest that dephosphorylation of Thr167 may be important for complete inactivation of p34^{cdc2} at the end of mitosis (Gould et al. 1991; Ducommun et al. 1991). Whether modification of Thr161/167 results from autophosphorylation (Krek and Nigg 1992; Nigg et al. 1992) or from the action of an exogenous kinase (Solomon et al. 1992) remains to be clarified.

Finally, a fourth phosphorylation site was identified in vertebrate p34^{cdc2} as serine 277 (Ser277). In chicken DU249 hepatoma cells, serine phosphorylation was highest in G1 phase cells, but virtually undetectable in G2 cells (Krek and Nigg 1991a). The physiological role of this serine phosphorylation remains to be determined. Recent studies based on the expression of mutant version of avian p34^{cdc2} in fission yeast suggest that serine phosphorylation might be related to monitoring the nutritional state of cells (Krek et al. 1992).

Phosphorylation sites in p34^{cdc2} have been highly conserved during evolution, not only among bonafide cdc2 homologs, but also among p34^{cdc2}-related kinases such as p33^{cdk2} (Nigg et al. 1991; Paris et al. 1991; Elledge and Spottswood 1991; Tsai et al. 1991). Thus, it appears safe to predict that cdk's will also be extensively regulated through phosphorylation.

2.2 Regulation of cdc2 Kinases by Cyclin-Binding

The second major mechanism controlling cdc2 protein kinase activity, besides phosphorylation, is complex formation with other proteins. Prominent among the complex partners of cdc2 kinase catalytic subunits are cyclins. These proteins generally display cyclic patterns of accumulation and destruction during the cell cycle, and they are essential for cdc2 kinase activity. Specifically, they are thought to determine the timing of activation and/or the substrate specificity of the catalytic subunits of cdc2 kinases. According to sequence criteria and their purported functions during the cell cycle, cyclins may be grouped into several classes; the most extensively studied classes include A-type, B-type, and G1 cyclins (for reviews see Hunt 1989; Minshull et al. 1989; Reed 1991; Hunter and Pines 1991; Xiong and Beach 1991; Lew and Reed 1992). While the major mitotic forms of cdc2 kinases consist of complexes between p34^{cdc2} and B-type cyclins (reviewed in Nurse 1990; Draetta 1991; Dorée 1990), A-type cyclins appear to function during S phase (d'Urso et al. 1990; Girard et al. 1991; Walker and Maller 1991; Pagano et al. 1992), as well as the G2/M transition (Lehner and O'Farrell 1990). The complexity of the G1 cyclin family is only beginning to emerge. The first members of this family (the so-called CLNs) were identified in the budding yeast *Saccharomyces cerevisiae* (Nash et al. 1988; Cross 1988; Hadwiger et al. 1989; for review see Reed 1991; Lew and Reed 1992). Subsequently, several novel cyclins (cyclins C, D, and E) with a potential G1 function have been also identified in higher eukaryotes, including vertebrates (Xiong et al. 1991; Koff et al. 1991; Lew et al. 1991; Matsushime et al. 1991; Motokura et al. 1991; Kiyokawa et al. 1992; see also Léopold and O'Farrell 1991). In *S. cerevisiae*, G1 cyclins appear to be major targets of growth-stimulatory and inhibitory agents (e.g. Chang and Herskowitz 1990; Wittenberg et al. 1990; Dirick and Nasmyth 1991), and it is likely that functionally homologous cyclins are targets of mitogenic stimuli also in vertebrates. The study of cyclins may thus be expected to provide important information on the aberrant proliferation behavior of tumor cells (Matsushime et al. 1991; Motokura et al. 1991; for review see Hunter and Pines 1991).

Differences in the subcellular distribution of individual cyclin proteins might play a major role in determining the substrate specificity and/or the regulation of the associated kinase subunits. Whereas both mammalian and avian A-type cyclins localize to the nucleus shortly after their synthesis (Pines and Hunter 1991b; Girard et al. 1991; Pagano et al. 1992; G. Maridor, P. Gallant and E. A. Nigg, unpublished results), B-type cyclins were found to undergo a striking cell cycle-dependent change in subcellular distribution. While chicken cyclin B2 was undetectable in G1 phase cells (due to its destruction during the previous mitosis), newly synthesized protein was mostly cytoplasmic in S and G2 phase cells, but migrated abruptly to the cell nucleus just before the onset of mitosis (Fig. 2). Independently, virtually identical results have been obtained for human cyclin B1 (Pines and Hunter 1991b). These findings are of considerable interest in view of the increasing appreciation that changes in the nucleocytoplasmic distribution of proteins may reflect important regulatory mechanisms (for discussion see Baeuerle and Baltimore 1988; Nigg 1990, 1992).

The observed cell-cycle dependent nuclear translocation of B-type cyclins raises two important questions. First, how is the redistribution of cyclin B2 to the cell nucleus controlled at the onset of mitosis, and second, what is its physiological significance?

With respect to the mechanism of nuclear transport, it is noteworthy that nuclear entry of cyclin B2 occurred very rapidly and prior to any visible alterations of the nuclear envelope (Gallant and Nigg 1992). This suggests that the nuclear translocation of cyclin B2 most probably involves an active transport process. One could argue that B-type cyclins contain an intrinsic nuclear localization signal (NLS), but that this signal is prevented from functioning for most of the interphase of the cell cycle, due to either masking of the NLS or binding of the protein to some cytoplasmic anchoring structure (for discussion, see Nigg 1990; Nigg et al. 1991). Cell cycle-specific modifi-

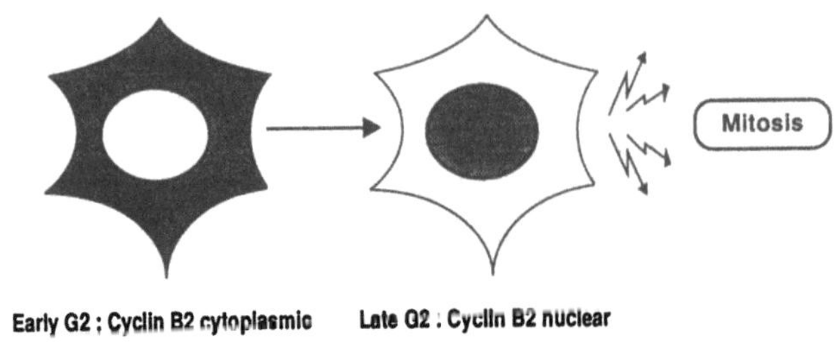

Fig. 2. Illustration of the cell cycle-dependent redistribution of cyclin B2 from the cytoplasm to the nucleus, prior to the onset of mitosis (for discussion, see Gallant and Nigg 1992; Pines and Hunter 1991b)

cation of either cyclin B or the masking/anchoring protein might then trigger activation of the NLS at the G2/M transition. Alternatively, an NLS might be created on cyclin B2 at the G2/M transition, or cyclin B2 might undergo cell cycle-dependent piggy-back transport to the nucleus.

Concerning the physiological role of the nuclear translocation of cyclin B, it is intriguing that $p34^{cdc2}$ is activated through dephosphorylation on Thr14 and Tyr15 at about the time when B-type cyclins redistribute from the cytoplasm to the nucleus (Krek and Nigg 1991a, b). Since cyclin B2 exists in the cytoplasm in G2 phase, and complex formation between $p34^{cdc2}$ and B-type cyclins appears to be required for phosphorylation of $p34^{cdc2}$ on Thr14 and Tyr15 (Solomon et al. 1990; Meijer et al. 1991; Parker et al. 1991), it seems that newly synthesized cyclin B associates with $p34^{cdc2}$ in the cytoplasm, and that this complex is kept inactive through phosphorylation of $p34^{cdc2}$ (on Thr14 and Tyr15) by cytoplasmic kinases. What remains to be determined, is where in the cell dephosphorylation of $p34^{cdc2}$ on Thr14 and Tyr15 occurs (Fig. 3). If dephosphorylation occurs in the nucleus, then nuclear translocation

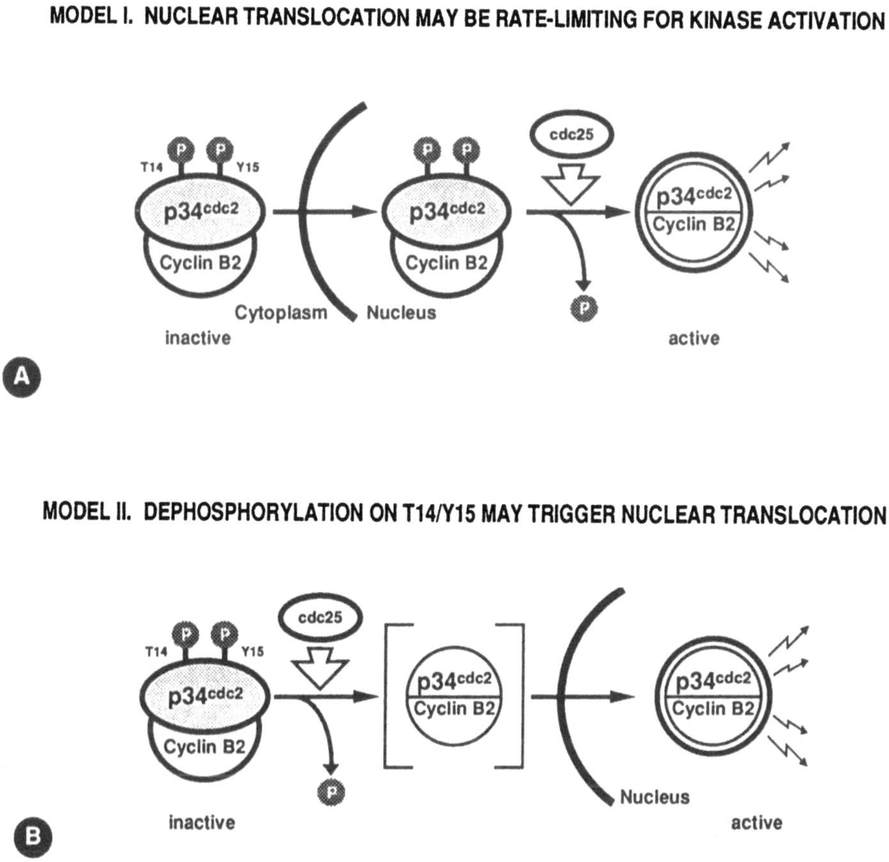

Fig. 3. Two alternative models for explaining the functional significance of cyclin B2 translocation to the nucleus (for explanation, see text)

may represent a prerequisite for activation of the complex, and hence play an important regulatory role (Fig. 3A). Conversely, if dephosphorylation occurs in the cytoplasm, then the concomitant activation of the p34^{cdc2} kinase might play a role in inducing nuclear translocation of the p34^{cdc2}-cyclin B complex (Fig. 3B).

Recently, we have begun to examine the functional consequences of expressing mutant avian cyclins in HeLa cells. The first major finding emerging from these studies relates to the cell cycle-dependent destruction of cyclin proteins. Previous work based on a cell-free cycling *Xenopus* egg extract had lead to the conclusion that destruction of B-type cyclins is required for exit from a mitotic state (Murray et al. 1989; Félix et al. 1990; see also Ghiara et al. 1991). On the other hand, complete destruction of cyclin B does not appear to be required for M phase exit during the rapid early cell cycles in *Drosophila* syncytial embryos (Lehner and O'Farrell 1990; Whitfield et al. 1990), nor during the transition from meiosis I to meiosis II in clams and frogs (Westendorf et al. 1989; Kobayashi et al. 1991). We have examined the consequences of expressing a nondestructible B-type cyclin in somatic vertebrate cells. Since previous data have shown that a conserved arginine in the N-terminus of B-type cyclins is essential for mitotic destruction (Glotzer et al. 1991), the corresponding residue in chicken cyclin B2 (Arg 32) was mutated to serine. When this non-destructible mutant avian cyclin protein was expressed in HeLa cells, more than 60% of the transfected cells became arrested in mitosis within 36 h (Gallant and Nigg 1992). This result demonstrates that the destruction of B-type cyclins is essential for somatic cells to exit from mitosis.

Acknowledgements. This work was supported by grants from the Swiss National Science Foundation (31–33615.92) and the Swiss Cancer League (FOR.205).

References

Amon A, Surana U, Muroff I & Nasmyth K (1992) Regulation of p34^{cdc28} tyrosine phosphorylation is not required for entry into mitosis in *S. cerevisiae*. Nature 355:368–371

Baeuerle PA & Baltimore D (1988) I kappa B: a specific inhibitor of the NF-kappa B transcription factor. Science 242:540–546

Booher RN, Alfa CE, Hyams JS & Beach DH (1989) The fission yeast cdc2/cdc13/suc1 protein kinase: Regulation of catalytic activity and nuclear localization. Cell 58:485–497

Chang F & Herskowitz I (1990) Identification of a gene necessary for cell cycle arrest by a negative growth factor of yeast: FAR1 is an inhibitor of a G1 cyclin, CLN2. Cell 63:999–1011

Cross FR (1988) DAF1, a mutant gene affecting size control, pheromone arrest, and cell cycle kinetics of Saccharomyces cerevisiae. Mol Cell Biol 4675–4684

Dirick L & Nasmyth K (1991) Positive feedback in the activation of G1 cyclins in yeast. Nature 351:754–757

Dorée M (1990) Control of M-phase by maturation-promoting factor. Curr Opinion Cell Biol 2:269–273

Draetta G, Luca F, Westendorf J, Brizuela L, Ruderman J & Beach D (1989) cdc2 protein kinase is complexed with both cyclin A and B: Evidence for proteolytic inactivation of MPF. Cell 56:829–838

Draetta G (1990) Cell cycle control in eukaryotes: Molecular mechanisms of cdc2 activation. Trends Biochem Sci 15:378–383

Draetta G (1991) cdc2-cyclin kinases – master switches for cell cycle regulation. Semin Cell Biol 2 (in press)

Ducommun B, Brambilla P, Félix M-A, Franza BR Jr. Karsenti E & Draetta G (1991) cdc2 phosphorylation is required for its interaction with cyclin. EMBO J 10:3311–3319

Dunphy WG & Kumagai A (1991) The cdc25 protein contains an intrinsic phosphatase activity. Cell 67:189–196

D'Urso G, Marraccino RL, Marshak DR & Roberts JM (1990) Cell cycle control of DNA replication by a homologue from human cells of the $p34^{cdc2}$ protein kinase. Science 250:786–791

Elledge SJ & Spottswood MR (1991) A new human p34 protein kinase, CDK2, identified by complementation of a cdc28 mutation in Saccharomyces cerevisiae, is a homolog of Xenopus Eg1. EMBO J 10:2653–2659

Enoch T & Nurse P (1990) Mutation of fission yeast cell cycle control genes abolishes dependence of mitosis on DNA replication. Cell 60:665–673

Fang F & Newport JW (1991) Evidence that the G1-S and G2-M transitions are controlled by different cdc2 proteins in higher eukaryotes. Cell 66:731–742

Featherstone C & Russell P (1991) Fission yeast p107wee1 mitotic inhibitor is a tyrosine/serine kinase. Nature 349:808–811

Félix M-A, Labbé J-C, Dorée M, Hunt T & Karsenti E (1990) Triggering of cyclin degradation in interphase extracts of amphibian eggs by cdc2 kinase. Nature 346:379–382

Gallant P & Nigg EA (1992) Cyclin B2 undergoes cell cycle dependent nuclear translocation and, when expressed as a non-destructible mutant, causes mitotic arrest in Hela cells. J Biol 117:213–224

Gautier J, Minshull J, Lohka M, Glotzer M, Hunt T & Maller JL (1990) Cyclin is a component of maturation-promoting factor from *Xenopus*. Cell 60:487–494

Gautier J & Maller JL (1991) Cyclin B in *Xenopus* oocytes: implications for the mechanism of pre-MPF activation. EMBO J 10:177–182

Gautier J, Solomon MJ, Booher RN, Bazan JF & Kirschner MW (1991) cdc25 is a specific tyrosine phosphatase that directly activates $p34^{cdc2}$. Cell 67:197–211

Ghiara JB, Richardson HE, Sugimoto K, Henze M, Lew DJ, Wittenberg C & Reed SI (1991) A cyclin B homolog in *S. cerevisiae*: chronic activation of the Cdc28 protein kinase by cyclin prevents exit from mitosis. Cell 65:163–174

Giordano A, Whyte P, Harlow E, Franza BR Jr, Beach D & Draetta G (1989) A 60 kd cdc2-associated polypeptide complexes with the E1A proteins in adenovirus-infected cells. Cell 58:981–990

Girard F, Strausfeld U, Fernandez A & Lamb NJC (1991) Cyclin A is required for the onset of DNA replication in mammalian fibroblasts. Cell 67:1169–1179

Glotzer M, Murray AW & Kirschner MW (1991) Cyclin is degraded by the ubiquitin pathway. Nature 349:132–138

Gould KL & Nurse P (1989) Tyrosine phosphorylation of the fission yeast cdc2+ protein kinase regulates entry into mitosis. Nature 342:39–45

Gould KL, Moreno S, Tonks NK & Nurse P (1990) Complementation of the mitotic activator, $p80^{cdc25}$, by a human protein-tyrosine phosphatase. Science 250:1573–1576

Gould KL, Moreno S, Owen DJ, Sazer S & Nurse P (1991) Phosphorylation at Thr167 is required for Schizosaccharomyces pombe $p34^{cdc2}$ function. EMBO J 10:3297–3309

Hadwiger JA, Wittenberg C, Richardson HE, De Barros Lopes M & Reed SI (1989) A family of cyclin homologs that control the G1 phase in yeast. Proc Natl Acad Sci USA 86:6255–6259

Hunt T (1989) Maturation promoting factor, cyclin and the control of M-phase. Curr Opinion Cell Biol 1:268–274

Hunter T & Pines J (1991) Cyclins and cancer. Cell 66:1071–1074

Izumi T & Maller JL (1991) Phosphorylation of *Xenopus* cyclins B1 and B2 is not required for cell cycle transitions. Mol Cell Biol 3860–3867

Kiyokawa H, Busquets X, Powell CT, Ngo L & Rifkind RA (1992) Cloning of a D-type cyclin from murine erythroleukemia cells. Proc Natl Acad Sci USA 89:2444–2447
Kobayashi H, Minshull J, Ford C, Golsteyn R, Poon R & Hunt T (1991) On the synthesis and destruction of A- and B-type cyclins during oogenesis and meiotic maturation in *Xenopus laevis*. J Cell Biol 114:755–765
Koff A, Cross F, Fisher A, Schumacher J, Leguellec K, Philippe M & Roberts JM (1991) Human cyclin E, a new cyclin that interacts with two members of the CDC2 gene family. Cell 66:1217–1228
Krek W & Nigg EA (1991a) Differential phosphorylation of vertebrate p34cdc2 kinase at the G1/S and G2/M transitions of the cell cycle: Identification of major phosphorylation sites. EMBO J 10:305–316
Krek W & Nigg EA (1991b) Mutations of p34cdc2 phosphorylation sites induce premature mitotic events in HeLa cells: evidence for a double block to p34cdc2 kinase activation in vertebrates. EMBO J 10:3331–3341
Krek W & Nigg EA (1992) Cell cycle regulation of vertebrate p34cdc2 activity: identification of Thr161 as an essential in vivo phosphorylation site. New Biol 4:323–329
Krek W, Marks J, Schmitz N, Nigg EA & Simanis V (1992) Vertebrate p34cdc2 phosphorylation site mutants: effects upon cell cycle progression in the fission yeast Schizosaccharomyces pombe. J Cell Sci 102:43–53
Labbé J-C, Capony J-P, Caput D, Cavadore JC, Derancourt J, Kaghad M, Lelias J-M, Picard A & Dorée M (1989) MPF from starfish oocytes at first meiotic metaphase is a heterodimer containing one molecule of cdc2 and one molecule of cyclin B. EMBO J 8:3053–3058
Lehner CF & O'Farrell PH (1990) The roles of *Drosophila* cyclins A and B in mitotic control. Cell 61:535–547
Lew DJ, Dulic V & Reed SI (1991) Isolation of three novel human cyclins by rescue of G1 cyclin (Cln) function in yeast. Cell 66:1197–1206
Lew DJ & Reed SI (1992) A proliferation of cyclins. Trends Cell Biol 2:77–81
Léopold P & O'Farrell PH (1991) An evolutionary conserved cyclin homolog from *Drosophila* rescues yeast deficient in G1 cyclins. Cell 66:1207–1216
Lundgren K, Walworth N, Booher R, Dembski M, Kirschner M & Beach D (1991) mik1 and wee1 cooperate in the inhibitory tyrosine phosphorylation of cdc2. Cell 64:111–1122
Matsushime H, Roussel MF, Ashmun RA & Sherr CJ (1991) Colony-stimulating factor 1 regulates novel cyclins during the G1 phase of the cell cycle. Cell 65:701–713
Meijer L, Arion D, Golsteyn R, Pines J, Brizuela L, Hunt T & Beach D (1989) Cyclin is a component of the sea urchin egg M-phase specific histone H1 kinase. EMBO J 8:2275–2282
Meijer L, Azzi L & Wang JYJ (1991) Cyclin B targets p34cdc2 for tyrosine phosphorylation. EMBO J 10:1545–1554
Minshull J, Pines J, Golsteyn R, Standart N, Mackie S, Colman A, Blow J, Ruderman JV, Wu M & Hunt T (1989) The role of cyclin synthesis modification and destruction in the control of cell division. J Cell Sci (Suppl) 12:77–97
Moreno S, Hayles J & Nurse P (1989) Regulation of p34cdc2 protein kinase during mitosis. Cell 58:361–372
Morla AO, Draetta G, Beach D & Wang JYJ (1989) Reversible tyrosine phosphorylation of cdc2: dephosphorylation accompanies activation during entry into mitosis. Cell 58:193–203
Motokura T, Bloom T, Kim HG, Jüppner H, Ruderman JV, Kronenberg HM & Arnold A (1991) A novel cyclin encoded by a bcl1-linked candidate oncogene. Nature 350:512–515
Murray AW, Solomon MJ & Kirschner MW (1989) The role of cyclin synthesis and degradation in the control of maturation promoting factor activity. Nature 339:280–286
Nash R, Tokiwa G, Anand S, Erickson K & Futcher AB (1988) The WHI1+ gene of Saccharomyces cerevisiae tethers cell division to cell size and is a cyclin homolog. EMBO J 7:4335–4346
Nigg EA (1990) Mechanisms of signal transduction to the cell nucleus. Adv Cancer Res 55:271–310
Nigg EA (1992) Signal transduction to the cell nucleus. Adv Mol Cell Biol 4:103–131

Nigg EA, Baeuerle PA & Lührmann R (1991) Nuclear import-export: in search of signals and mechanisms. Cell 66:15–22
Nigg EA, Krek W & Peter M (1992) Vertebrate cdc2 kinase: its regulation by phosphorylation and its mitotic targets. Cold Spring Harbor Symp Quant Biol 56:539–547
Norbury C, Blow J & Nurse P (1991) Regulatory phosphorylation of the $p34^{cdc2}$ protein kinase in vertebrates. EMBO J 10:3321–3329
Nurse P (1990) Universal control mechanism regulating onset of M-phase. Nature 344:503–508
Pagano M, Pepperkok R, Verde F, Ansorge W & Draetta G (1992) Cyclin A is required at two points in the human cell cycle. EMBO J 11:961–971
Paris J, Le Guellec R, Couturier A, Le Guellec K, Omilli F, Camonis J, MacNeill S & Philippe M (1991) Cloning by differential screening of a *Xenopus* cDNA coding for a protein highly homologous to cdc2. Proc Natl Acad Sci USA 88:1039–1043
Parker LL, Atherton-Fessler S, Lee MS, Ogg S, Falk JL, Swenson KI & Piwnica-Worms H (1991) Cyclin promotes the tyrosine phosphorylation of $p34^{cdc2}$ in a wee1+ dependent manner. EMBO J 10:1255–1263
Pines J & Hunter T (1989) Isolation of a human cyclin cDNA: Evidence for cyclin mRNA and protein regulation in the cell cycle and for interaction with $p34^{cdc2}$. Cell 58:833–846
Pines J & Hunter T (1990) $p34^{cdc2}$: the S and M kinase? New Biol 2:389–401
Pines J & Hunter T (1991a) Cyclin-dependent kinases: a new cell cycle motif? Trends Cell Biol 1:117–121
Pines J & Hunter T (1991b) Human cyclins A and B1 are differentially located in the cell and undergo cell cycle-dependent nuclear transport. J Cell Biol 115:1–17
Reed SI (1991) G1-specific cyclins: in search of an S-phase-promoting factor. Trends Genet 7:95–99
Russell P & Nurse P (1986) cdc25+ functions as an inducer in the mitotic control of fission yeast. Cell 45:145–153
Russell P & Nurse P (1987) Negative regulation of mitosis by wee1+, a gene encoding a protein kinase homolog. Cell 49:559–567
Smythe C & Newport JW (1992) Coupling of mitosis to the completion of S phase in Xenopus occurs via modulation of the tyrosine kinase that phosphorylates $p34^{cdc2}$. Cell 68:787–797
Solomon MJ, Glotzer M, Lee TH, Philippe M & Kirschner MW (1990) Cyclin activation of $p34^{cdc2}$. Cell 63:1013–1024
Solomon MJ, Lee T & Kirschner MW (1992) Role of phosphorylation of $p34^{cdc2}$ activation: Identification of an activating kinase. Mol Biol Cell 3:13–27
Sorger PK & Murray AW (1992) S-phase feedback control in budding yeast independent of tyrosine phosphorylation of $p34^{cdc28}$. Nature 355:365–368
Strausfeld U, Labbé JC, Fesquet D, Cavadore JC, Picard A, Sadhu C, Russell P & Dorée M (1991) Dephosphorylation and activation of a $p34^{cdc2}$/cyclin B complex in vitro by human CDC25 protein. Nature 351:242–248
Tsai L-H, Harlow E & Meyerson M (1991) Isolation of the human cdk2 gene that encodes the cyclin A- and adenovirus E1A-associated p33 kinase. Nature 353:174–177
Walker DH & Maller JL (1991) Role for cyclin A in the dependence of mitosis on completion of DNA replication. Nature 354:314–317
Westendorf JM, Swenson KI & Ruderman JV (1989) The role of cyclin B in meiosis I. J Cell Biol 108:1431–1444
Whitfield WGF, Gonzales C, Maldonado-Codina G & Glover DM (1990) The A- and B-type cyclins of *Drosophila* are accumulated and destroyed in temporally distinct events that define separable phases of the G2-M transition. EMBO J 9:2563–2572
Wittenberg C, Sugimoto K & Reed SI (1990) G1-specific cyclins of S. cerevisiae: cell cycle periodicity, regulation by mating pheromone, and association with the $p34^{cdc28}$ protein kinase. Cell 62:225–237
Xiong Y & Beach D (1991) Population explosion in the cyclin family. Cur Biol 1:362–364
Xiong Y, Connolly T, Futcher B & Beach D (1991) Human D-type cyclin. Cell 65:691–699

Control of SV40 DNA Replication by Protein Phosphorylation

I. Moarefi[1], C. Schneider[1], K. van Zee[1], A. Höss[1], A. K. Arthur[1] and E. Fanning[1]

1 Introduction

Protein phosphorylation is a major regulatory mechanism in eucaryotic cells. Although much is known about the protein kinases involved, and some information about protein phosphatases is available, a detailed understanding of the role of protein phosphorylation in modulating the function of cellular regulatory proteins has been difficult to achieve. Viral regulatory proteins, such as simian virus 40 (SV40) large tumor antigen (T Ag), represent an experimentally advantageous system to study the modulation of protein function by phosphorylation. T Ag is a complex multifunctional phosphoprotein with sequence-specific DNA binding activity, DNA helicase activity and the ability to associate with a number of cellular proteins (Fig. 1). It is required for the initiation of SV40 DNA replication in infected monkey cells or in a cell-free system in vitro (reviewed recently by Prives 1990; Borowiec et al. 1990; Fanning 1992a; Fanning and Knippers 1992). Indeed, T Ag is the only viral protein

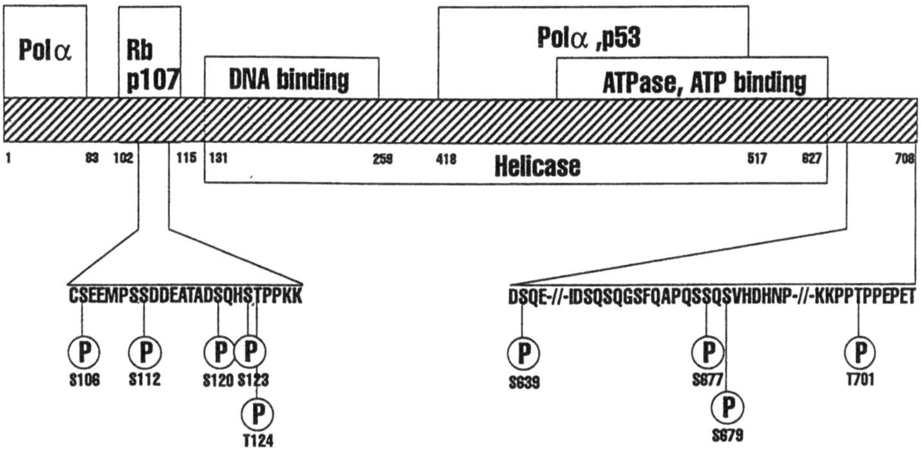

Fig. 1. Functional domains of SV40 T Ag. The regions of the protein that retain DNA binding, ATPase, and DNA helicase activities, and the ability to bind to each of the cellular proteins indicated are diagrammed. *Numbers* below the domains indicate amino acid residue positions. *Asterisks* depict protease-sensitive regions linking postulated structural domains. Residues phosphorylated in T Ag expressed in mammalian cells are circled. (After Fanning 1992a)

[1] Institute for Biochemistry, Karlstraße 23, 8000 Munich 2, FRG.

43. Colloquium Mosbach 1992
DNA Replication and the Cell Cycle

required for viral DNA replication. SV40 DNA replication has therefore been useful in identifying cellular replication proteins and their functions.

Based on biochemical analysis of these replication proteins in vitro, the following sequence of events during the initiation process has been postulated. T Ag binds specifically to a perfect palindrome (DNA binding site II) within the SV40 origin of replication and assembles as a double hexamer on the origin in the presence of ATP, leading to structural distortion of the origin DNA (Borowiec et al. 1990). T Ag then recruits the cellular replication proteins RP-A and DNA polymerase α-primase through direct protein-protein interactions (Dornreiter et al. 1990, 1992) and directs unwinding of the parental DNA strands through its DNA helicase activity (Stahl et al. 1986). DNA primase synthesizes a short RNA primer on each template strand that is then elongated by the polymerase activity of DNA polymerase α-primase. Binding of RF-C and PCNA to the template-primer junction is followed by association of cellular DNA polymerase δ, leading to the assembly of the replication forks (Tsurimoto et al. 1990; Weinberg et al. 1990; Tsurimoto and Stillman 1991a, b). Bidirectional DNA replication ensues, terminating when the replication forks converge (reviewed by Fanning and Knippers 1992; Stillman 1992).

Like cellular DNA synthesis, SV40 DNA replication takes place only in the S phase of the cell cycle. To prepare the cell to support viral replication, T Ag has evolved a set of biochemical activities that induce quiescent host cells to re-enter the cell cycle and to progress through G1 into the S phase. These activities of T Ag are distinct from those involved in initiation of viral DNA replication and include Rb/p107-binding, a C-terminal function most likely to be p53-binding, and unknown activities localized to the DNA binding domain and to the N-terminus (Dobbelstein et al. 1992; A. Dickmanns, unpubl. data) (Fig. 1). Some or all of these activities are also involved in T Ag's ability to immortalize and transform cells in culture, and possibly in its ability to stimulate cellular gene expression (reviewed by Fanning 1992b; Fanning and Knippers 1992).

Early studies of T antigen from infected monkey cells demonstrated that it exists in differentially phosphorylated forms (Fanning et al. 1981; Greenspan and Carroll 1981). Newly synthesized T Ag is initially underphosphorylated, but subsequently matures into a highly phosphorylated oligomer that accumulates and represents the bulk of the T Ag in the cell (Fanning et al. 1981). Interestingly, these two forms of T Ag display different functional properties, in that the underphosphorylated form binds efficiently to the SV40 origin of replication, whereas the highly phosphorylated form has little specific DNA binding activity, particularly on site II (Fanning et al. 1982; Scheidtmann et al. 1984; reviewed in Fanning and Knippers 1992). Phosphopeptide mapping of these forms of T Ag revealed two clusters of phosphoserine and -threonine (Fig. 1); both forms of T Ag carried two phosphothreonine residues, one in each cluster, but the underphosphorylated form contained little phosphoserine (Scheidtmann et al. 1984). These results, together with those from a number of other laboratories, suggested that extensive phosphorylation of T Ag suppressed its origin-specific DNA binding activity and hence, its ability to initiate replication of viral DNA (reviewed in Fanning 1992a; Fanning and Knippers 1992). Moreover, they were also consistent with reports that newly synthesized T Ag was required for viral DNA replication (Yakobson et al. 1977a, b).

The studies summarized in this report were undertaken to further investigate the role of phosphorylation in control of T Ag function. The phosphorylation pattern of T Ag produced in monkey cells was compared with that of T Ag expressed using a baculovirus vector and correlated with functional properties of the two proteins. To assess the importance of each phosphorylation site in the control of viral DNA replication in infected monkey cells and in the cell-free system (Li and Kelly 1984), a series of T Ag mutants altered at each phosphorylation site was characterized. A model for the temporal control of SV40 DNA replication based on differential phosphorylation of T Ag is presented.

2 Results and Discussion

2.1 Baculovirus-Expressed T Ag is Underphosphorylated and Binds to SV40 Origin DNA

The functional characterization of the underphosphorylated form of T Ag from infected monkey cells was difficult due to the small quantities that can be isolated and contamination by the more abundant highly phosphorylated forms. Thus, T Ag expressed in insect cells using a recombinant baculovirus was examined as a better source of T Ag. Preliminary studies had indicated that baculovirus T Ag was phosphorylated, but phosphopeptide mapping was not performed (Lanford 1988). Quantitative comparison of the incorporation of labeled phosphate into T Ag indicated that baculovirus T Ag was likely to be underphosphorylated (Fig. 2A). Phosphopeptide mapping confirmed that although baculovirus T Ag was modified at the same sites as monkey cell T Ag, it carried much less phosphoserine (Höss et al. 1990b). Indeed, the phosphorylation pattern of baculovirus T Ag closely resembled that determined previously (Scheidtmann et al. 1984) for newly synthesized T Ag from SV40-infected monkey cells (Fig. 2B).

This structural resemblance suggested that baculovirus T Ag might also have functional properties different from those of highly phosphorylated monkey T Ag and more like those of the newly synthesized form. To address this question, the DNA binding activities of baculovirus and total Cos1 monkey T Ag were compared using a simple immunoprecipitation DNA binding assay (Höss et al. 1990b). End-labeled DNA fragments were incubated together with T Ag bound in an immune complex. One DNA fragment carried T Ag binding site I, which plays a role in autoregulation of early viral transcription by T Ag, and another fragment carried site II, an essential element of the SV40 origin of DNA replication. Baculovirus T Ag bound well to both sites (Fig. 3A, lanes 1–3), whereas monkey T Ag bound little or no site II DNA fragment (Fig. 3B, lanes 1–3). Treatment of baculovirus T Ag with alkaline phosphatase to dephosphorylate serine residues (Klausing et al. 1988) did not affect its DNA binding activity (Fig. 3A, lanes 4–6). In contrast the dephosphorylation of serines in monkey T Ag increased the DNA binding activity of T Ag markedly, in particular on site II (Fig. 3B, lanes 4–6). These results suggest that those serines that downregulate

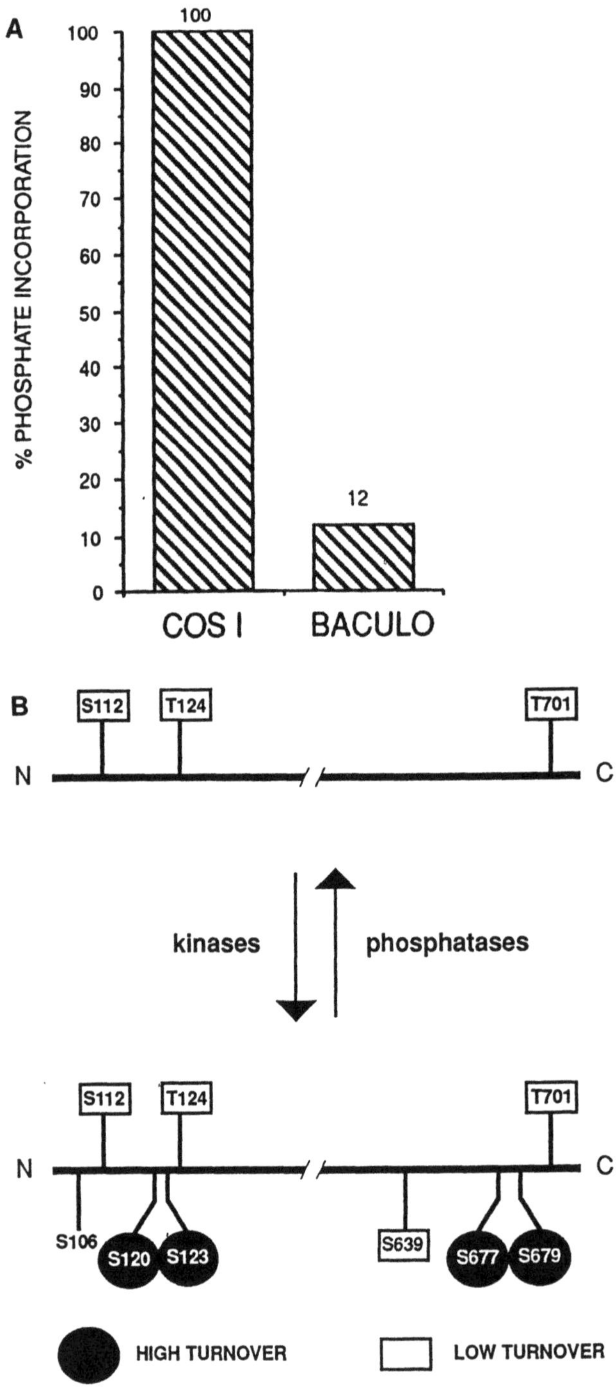
A
% PHOSPHATE INCORPORATION
100
90
80
70
60
50
40
30
20
10
0
100
12
COS I
BACULO
B
S112
T124
T701
N
C
kinases
phosphatases
S106
S120
S123
S639
S677
S679
HIGH TURNOVER
LOW TURNOVER

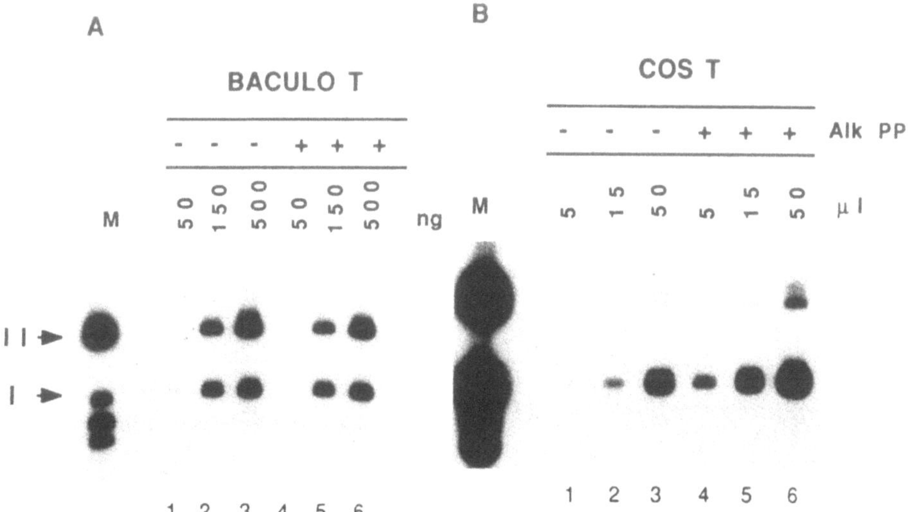

Fig. 3. SV40 DNA binding of baculovirus (A) and Cos1 monkey cell T Ag (B). Immunoprecipitated T Ag treated with (+) or without (–) alkaline phosphatase was incubated in the indicated amounts with end-labeled fragments of SV40 plasmid DNA (M) containing two specific binding sites (*I*, *II*) for T Ag. T Ag-bound fragments were detected by agarose gel electrophoresis and autoradiography (Höss et al. 1990b)

DNA binding on site II are not significantly modified in baculovirus T Ag, but are strongly modified in Cos1 T Ag. Inspection of the phosphopeptide maps of monkey and baculovirus T Ags (Höss et al. 1990b) implies that phosphorylation of serines 120, 123, 677, 679 and perhaps 106 and 639 could be responsible for the down-regulation of site II DNA binding activity.

2.2 Properties of Mammalian T Ag Bearing Mutations in Phosphorylation Sites

The question of the functional role of each mapped phosphorylation site in T Ag was approached by using oligonucleotide-directed mutagenesis to create a matched set of mutants, in which each phosphorylated residue was conservatively substituted by ala-

Fig. 2. Differential phosphorylation of T Ag. **A** SV40-transformed Cos1 monkey cells and Sf9 insect cells infected with a recombinant baculovirus directing the expression of T Ag were labeled with $^{32}PO_4$. T Ag immunoprecipitated from extracts of these cells was analyzed by SDS polyacrylamide gel electrophoresis, silver staining and autoradiography. The diagram depicts the level of phosphate label per arbitrary unit of T Ag, setting Cos-1 T Ag as 100% (After Höss et al. 1990b). **B** Summary of phosphopeptide analysis of baculovirus (*top*) and monkey cell T Ag (*bottom*). These two forms of T Ag closely resemble the newly synthesized monomeric T Ag (*top*) and accumulated T Ag aggregates (*bottom*) in productively infected monkey cells. Phosphate turnover rates at various sites are indicated (Scheidtmann 1986)

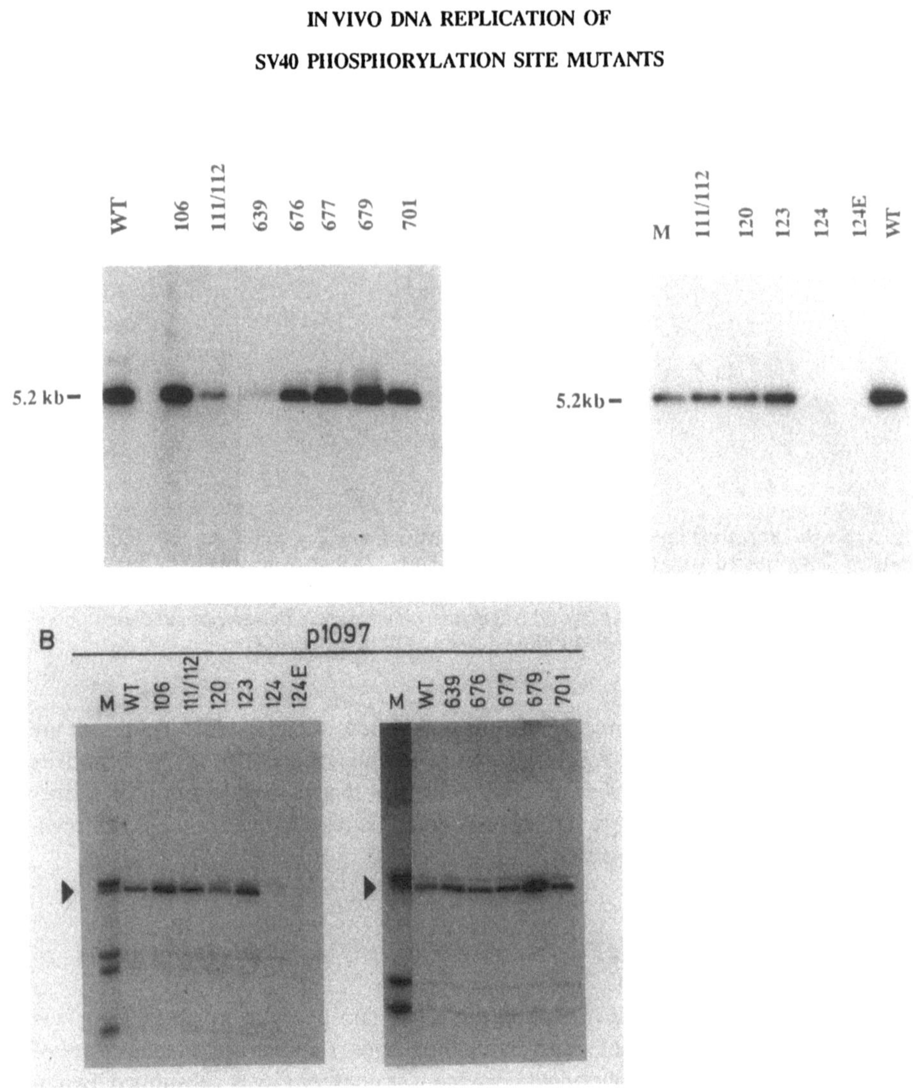

Fig. 4. Characterization of phosphorylation site mutants of T Ag in monkey cells. **A** SV40 plasmid DNAs encoding T Ag with an alanine residue instead of the indicated phosphoserine or -threonine were transfected into TC7 monkey cells (Schneider and Fanning 1988). Two days later, the low molecular weight DNA was isolated, digested with Bam HI and DpnI, and analyzed by agarose gel electrophoresis and blot hybridization. **M** Full-length Bam HI- linearized SV40 DNA as marker. **B** SV40 binding activity of immunoprecipitated mutant T Ag was assayed as in Fig. 2. Site II is indicated by the *arrow*; site I was deleted in this template DNA (Schneider and Fanning 1988)

nine, by cysteine, or to mimic a negative charge, by glutamate (Schneider and Fanning 1988). The biological and biochemical properties of each mutant were characterized after transfection of SV40 plasmid DNA into monkey or rat cells.

Figure 4A depicts the results of a DNA replication assay in TC7 monkey cells. Newly replicated SV40 plasmid DNA was detected by the appearance of a full-length, DpnI-resistant band in Southern blots (arrow). All mutant T Ags replicated to some extent except those at Thr 124, which showed no activity. Mutants 111/112, 120, 123, 639 and 676 were less active than wildtype, mutants 106 and 701 were similar to wildtype, and mutants 677 and 679 replicated as well as, or in some experiments, better than wildtype.

The specific DNA binding activity of the mutant proteins expressed in rat cells was analyzed using an immunoprecipitation assay (Fig. 4B). All of the mutant proteins bound to binding site II in the SV40 origin except those mutated at Thr 124, which showed no activity. This result is consistent with the interpretation that phosphorylation of Thr 124 is required for T Ag binding to site II and for SV40 DNA replication. A mutation at Ser 679 stimulated site II binding activity, correlating with the increased replication activity of this mutant. This result suggests that phosphorylation of Ser 679 contributes to down-regulation of site II DNA binding and replication. The fact that mutations at other sites such as Ser 120 and 123 affect DNA replication, but not site II DNA binding, suggests that in monkey cells, other steps in the replication process subsequent to site II recognition may require phosphorylation of T Ag at these sites.

However, the interpretation of the results obtained with the phosphorylation site mutants is subject to two major limitations:

1. A mutant phenotype may be due either to lack of phosphorylation at the mutant residue, or to the nature of the amino acid replacement. However, a wildtype T Ag expressed in bacteria was not modified at Thr 124 and also lacked site II DNA binding activity (McVey et al. 1989; Höss et al. 1990a). This result implies that lack of phosphorylation at Thr 124 is responsible for the phenotype of Ala 124 T Ag. However, these data cannot be generalized for the other phosphorylation site mutants.

2. A mutation at one phosphorylation site could cause altered phosphorylation at adjacent or remote secondary sites of modification, that would then generate the mutant phenotype. To assess this possibility and to confirm the previously mapped phosphorylation sites, phosphopeptide maps of the mutant proteins were prepared (Scheidtmann et al. 1991a). The results confirmed the phosphorylation sites shown in Fig. 2B. Moreover, we found that a mutation at Ser 677 resulted in underphosphorylation of Ser 120 and 123, and that a mutation at Ser 120 caused underphosphorylation at Ser 123. These results would be consistent with a sequential mode of phosphorylation at these three sites, but this awaits further testing.

2.3 Properties of Baculovirus T Ag Bearing Mutations in Phosphorylation Sites

Characterization of the phosphorylation site mutants of T Ag was extended to mutant proteins expressed in the baculovirus system. The large amounts of purified T Ag obtained with this system facilitate more detailed biochemical analysis of its activi-

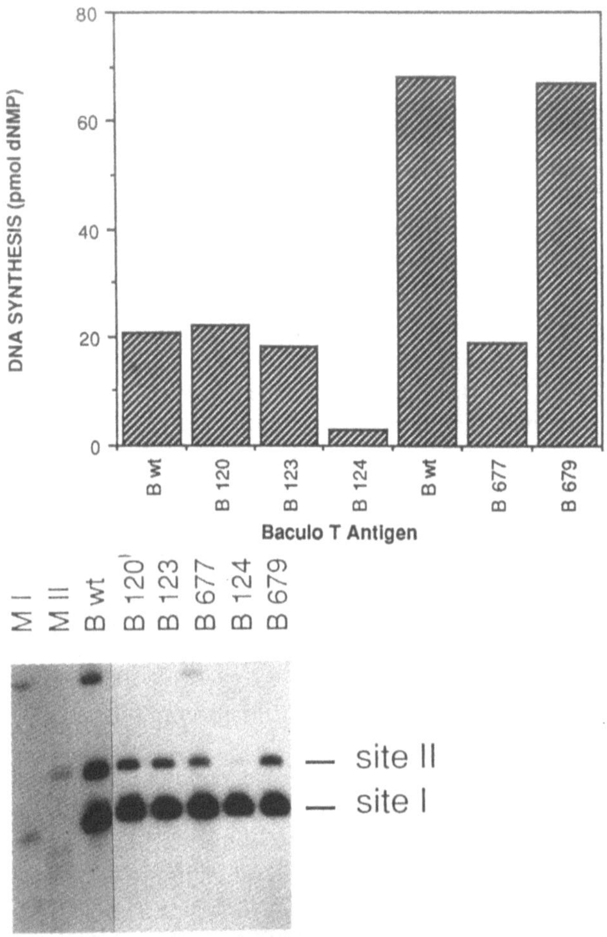

Fig. 5. Characterization of phosphorylation site mutants of T Ag expressed using recombinant baculoviruses. **A** The phosphorylation site mutations indicated were reconstructed into a baculovirus transfer vector pVL941T (Lanford 1988). Recombinant baculoviruses expressing each mutant T Ag were isolated using published methods (Summers and Smith 1987) and verified by PCR sequencing of baculovirus DNA (Moarefi, unpubl. results). Mutant T Ags were purified as described (Höss et al. 1990b) and assayed for replication of SV40 DNA in cell-free human extracts as described (Schneider et al. 1992). The *ordinate* shows the incorporation of labeled dCMP and dTMP (pmol) in replication reactions driven by 600 ng of each T Ag. **B** SV40 DNA binding of 200 ng of each baculovirus T Ag. A mixture of MI and MII DNA fragments was incubated with T Ag immune complexes as in Fig. 3

ties, for example in a cell-free SV40 DNA replication assay. The replication activity of selected phosphorylation site mutants was compared with that of wildtype T Ag (Fig. 5A). A mutant protein carrying Ala rather than Thr 124 was unable to replicate SV40 plasmid DNA in vitro, in agreement with the results of in vivo replication assays (Fig. 4A). The replication activities of Ala 123 and Ala 679 T Ag were indistinguishable from that of wildtype T Ag (Fig. 5A). These results indicate that phosphorylation at Ser 123 and 679 is not necessary for SV40 DNA replication in vitro, although it does appear to play a role in vivo (Fig. 4A).

Mutant T Ags bearing Ala in place of Ser 120 and 677 replicated SV40 DNA in vitro less efficiently than wildtype T Ag (Fig. 4A). This result is in agreement with the in vivo replication data for the mutant Ala 120, but not for Ala 677 T Ag, which replicated at wildtype level in vivo (Fig. 4A). The reason for this discrepancy is not known.

The SV40 DNA binding activity of the baculovirus mutant T Ags was then compared with that of wildtype T Ag (Fig. 5B). All of the mutant proteins bound to site I and site II DNA fragments except Ala 124 T Ag. This mutant T Ag bound to site I, but not to site II in this assay, thereby resembling its behavior when expressed in mammalian cells (Schneider and Fanning 1988; Fig. 4B). This result suggests that phosphorylation of Thr 124 is necessary for T Ag recognition and binding to site II in the SV40 origin of replication. In agreement with this interpretation, bacterially expressed T Ag lacking phosphorylation at Thr 124 was unable to bind to site II DNA or replicate SV40 DNA in vitro; however, after phosphorylation at Thr 124 by a human homolog of *cdc 2* protein kinase, bacterial T Ag bound to site II DNA and directed SV40 DNA replication (McVey et al. 1989).

2.4 Regulation of SV40 Replication by Differential Phosphorylation of T Ag

Differential phosphorylation of T Ag regulates the origin DNA binding and replication activities of the protein (Figs. 2–5; reviewed in Prives 1990; Fanning 1992a). The underphosphorylated form of T Ag, which constitutes a minor fraction of the T Ag in SV40-infected monkey cells and the bulk of baculovirus-expressed T Ag, is highly active in replication. Replication activity, as well as stable binding to site II DNA, requires phosphorylation of Thr 124 (Fig. 6, REPL-ACTIVE). This form of T Ag arises from newly synthesized T Ag, which becomes quickly and nearly quantitatively modified at Thr 124 and Thr 701. These modifications are quite stable, with a phosphate half-life of about 4.5 h (Scheidtmann 1986). Early after infection, this form of T Ag is rapidly converted to a highly phosphorylated form (Fanning et al. 1981). Phosphorylation of serine residues results in marked reduction of site II DNA binding and replication activity (Fig. 6, REPL-INACTIVE).

The highly phosphorylated form of T Ag accumulates to high levels in SV40-infected cells and represents the main form of the protein. Although it lacks replication activity, it appears to have other functions. T Ag associates with the tumor suppressor proteins p53 and Rb, as well as an Rb-related protein p107, thereby abrogating their growth suppressing activities (reviewed by Fanning and Knippers 1992). Since it is the highly phosphorylated form of T Ag that is found in these protein complexes, we speculate that the accumulation of highly phosphorylated T Ag during the early phase of the infection effectively titrates out these growth suppressing proteins, promoting progression into the S phase (Fig. 6). As the cell progresses toward S, Rb becomes more highly phosphorylated and dissociates from T Ag (Ludlow et al. 1990).

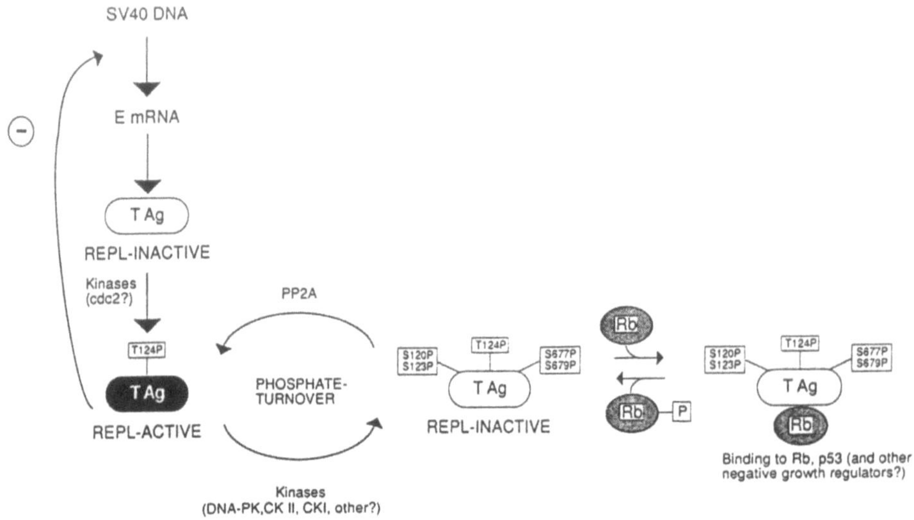

Fig. 6. Speculative model for temporal control of SV40 DNA replication in infected monkey cells (Fanning 1992a)

As the concentration of highly phosphorylated T Ag builds up, it may serve increasingly as a substrate for phosphatases, generating a cycle of phosphate turnover (Fig. 6). In contrast with the stably modified threonine residues, phosphate turnover on several of these serine residues (Ser 120, 123, 677, 679) is rapid (Fig. 2B). A cellular phosphatase catalytic subunit PP2Ac, purified as a stimulatory factor for SV40 DNA replication in vitro, dephosphorylated these serine residues of T Ag preferentially, thereby restoring the replication activity of T Ag (Virshup et al. 1989; Virshup and Kelly 1989; Scheidtmann et al. 1991b). Moreover, addition of okadaic acid, a specific inhibitor of PP2A, to cell-free SV40 replication assays inhibited replication activity (Lawson et al. 1990). Thus, phosphorylation of these four serine residues is likely to be responsible for the poor SV40 replication activity of the highly phosphorylated form of T Ag.

Assuming constant rates of phosphate turnover at these four serine residues of T Ag and continued production of newly synthesized T Ag, the concentration of the replication-active form of T Ag will gradually increase. We speculate that once a threshold concentration of the replication-active T Ag is reached, SV40 DNA replication will begin. The duration of this replication period will be self-limiting, since the replication-active form of T Ag is also active in repression of early transcription, which will eventually result in decreased production of newly synthesized T Ag. Clearly, either a decreased rate of conversion of replication-active T Ag to the highly phosphorylated form, or an increased rate of dephosphorylation of accumulated T Ag would shift the balance between the two forms of T Ag to favor the replication-active form. This could also contribute to the temporal control of initiation of viral replication. Such a change in the equilibrium between the two forms of T Ag could be brought about by either a decrease in protein kinase activities or an increase in protein phosphatase activity, associated for example with cell cycle progression. Indeed, when G1 phase cell extracts, which support SV40 DNA replication in vitro poorly, were supplemented with purified PP2Ac, they became as active as S phase extracts (Virshup and Kelly 1989). These results suggest that PP2Ac activity may increase as the cell enters S phase, at least for some substrates, thereby promoting the activation of SV40 DNA replication.

The scheme proposed in Fig. 6 provides a mechanism for converting a gradual process, the accumulation of highly phosphorylated T Ag inactive in replication, via a measuring device, the turnover of phosphate on certain serine residues of T Ag, into a switch that controls the timing of SV40 DNA replication. Mechanisms such as these would be sensitive to the growth state of the cell and the multiplicity of viral infection, adjusting the timing of the switch from the early phase to the late phase of infection to ensure maximal production of viral DNA.

PP2A is postulated to play a central role in this scheme (Fig. 6). PP2A is also thought to act as a measuring device regulating entry into mitosis (Goris et al. 1989; Felix et al. 1990; Kinoshita et al. 1990; Solomon et al. 1990; Lee et al. 1991). Its role in mitosis, however, is to prevent premature activation of the accumulated mitosis-promoting factor. The apparently reciprocal roles of PP2A in the temporal control of SV40 DNA synthesis and mitosis raise the possibility that SV40 may have borrowed a mechanism governing the timing of cellular DNA replication.

Acknowledgements. We thank H.-P. Nasheuer for thoughtful discussions, Silke Dehde, Ilka Gilbert and Antje Brunahl for expert technical assistance, E. Rohrer for excellent secretarial assistance, and the DFG, BMFT-Genzentrum and Fonds der Chemischen Industrie for financial support.

References

Borowiec JA, Dean FB, Bullock PA & Hurwitz J (1990) Binding and unwinding-how T antigen engages the SV40 origin of DNA replication. Cell 60:181–184

Dobbelstein M, Arthur AK, Dehde S, van Zee K, Dickmanns A & Fanning E (1992) Intracistronic complementation reveals a new function of SV40 T antigen that cooperates with Rb and p53 binding to stimulate DNA synthesis in quiescent cells. Oncogene 7:837–847

Dornreiter I, Höss A, Arthur AK & Fanning E (1990) SV40 T antigen binds directly to the catalytic subunit of DNA polymerase α. EMBO J 9:3329–3336

Dornreiter I, Erdile LF, Gilbert IU, von Winkler D, Kelly TJ & Fanning E (1992) Interaction of DNA polymerase α-primase with cellular replication protein A and SV40 T antigen. EMBO J 11:769–776

Fanning E (1992a) Simian virus 40 large T antigen: the puzzle, the pieces and the emerging picture. J Virol 66:1289–129

Fanning E (1992b) Modulation of cellular growth control by SV40 large T antigen. In: Doerfler W & Böhm P (eds) Malignant transformation by DNA viruses. Molecular mechanisms. Verlag Chemie, Weinheim pp 1–19

Fanning E, Novak B & Burger C (1981) Detection and characterization of multiple forms of simian virus 40 large T antigen. J Virol 37:92–102

Fanning E, Westphal K-H, Brauer D & Cörlin D (1982) Subclasses of simian virus 40 large T antigen: Differential binding of two subclasses of T antigen from productively infected cells to viral DNA. EMBO J 1:1023–1028

Fanning E & Knippers R (1992) Structure and function of simian virus 40 large tumor antigen. Annu Rev Biochem 61 pp 55–85

Felix M-A, Cohen P & Karsenti E (1990) Cdc2 H1 kinase is negatively regulated by a type 2A phosphatase in the *Xenopus* early embryonic cell cycle: evidence from the effects of okadaic acid. EMBO J 9:675–683

Goris J, Hermann J, Hendrix P, Ozon R & Merlevede W (1989) Okadaic acid, a specific protein phosphatase inhibitor, induces maturation and MPF formation in *Xenopus laevis* oocytes. FEBS Lett 245:91–94

Greenspan DS & Carroll RB (1981) Complex of simian virus 40 large tumor antigen and 48 000-dalton tumor antigen. Proc Natl Acad Sci USA 48:105–109

Höss A, Moarefi IF, Fanning E & Arthur AK (1990a) The finger domain of simian virus 40 large T antigen controls DNA-binding specificity. J Virol 64:6291–6296

Höss A, Moarefi I, Scheidtmann K-H, Cisek LJ, Corden JL, Dornreiter I, Arthur AK & Fanning E (1990b) Altered phosphorylation pattern of SV40 T antigen expressed in insect cells using a baculovirus vector. J Virol 64:4799–4807

Kinoshita N, Ohkura H & Yanagida M (1990) Distinct, essential roles of type 1 and 2A protein phosphatases in the control of the fission yeast cell division cycle. Cell 63:405–415

Klausing K, Scheidtmann K-H, Baumann E & Knippers R (1988) Effect of in vitro dephosphorylation on DNA-binding and DNA helicase activities of simian virus 40 large tumor antigen. J Virol 62:1258–126

Lanford RE (1988) Expression of simian virus 40 T antigen in insect cells using a baculovirus expression vector. Virology 167:72–81

Lawson R, Cohen P & Lane DP (1990) Simian virus 40 large T-antigen-dependent DNA replication is activated by protein phosphatase 2A in vitro. J Virol 64:2380–2383

Lee TH, Solomon MJ, Mumby MC & Kirschner MW (1991) INH, a negative regulator of MPF, is a form of protein phosphatase 2A. Cell 64:415–423

Li JJ & Kelly TJ (1984) Simian virus 40 DNA replication in vitro. Proc Natl Acad Sci USA 81:6973–6977

Ludlow JW, Shon J, Pipas JM, Livingston DM & DeCaprio JA (1990) The retinoblastoma susceptibility gene product undergoes cell cycle-dependent dephosphorylation and binding to and release from SV40 large T. Cell 60:387–396

McVey D, Brizuela L, Mohr I, Marshak DR & Gluzman Y (1989) Phosphorylation of large tumor virus antigen by cdc2 stimulates SV40 DNA replication. Nature 341:503–507

Prives C (1990) The replication functions of SV40 T antigen are regulated by phosphorylation. Cell 61:735–738

Scheidtmann K-H (1986) Phosphorylation of simian virus 40 large T antigen: cytoplasmic and nuclear phosphorylation sites differ in their metabolic stability. Virology 150:85–95

Scheidtmann K-H, Buck M, Schneider J, Kalderon D, Fanning E & Smith AE (1991a) Biochemical characterization of phosphorylation site mutants of simian virus 40 large T antigen: evidence for interaction between amino- and carboxyterminal domains. J Virol 65:1479–1490

Scheidtmann K-H, Hardung M, Echle B & Walter G (1984) DNA-binding activity of simian virus 40 large T antigen correlates with a distinct phosphorylation state. J Virol 50:1–12

Scheidtmann K-H, Virshup DM & Kelly TJ (1991b) Protein phosphatase 2A dephosphorylates simian virus 40 large T antigen specifically at residues involved in regulation of DNA-binding activity. J Virol 65:2098–2101

Schneider C, von Winkler D, Dornreiter I, Nasheuer HP, Gilbert I, Dehde S, Arthur AK & Fanning E (1992) Bovine RP-A functions in SV40 DNA replication in vitro, but bovine polymerase α-primase inhibits replication competitively. In: Hughes P, Fanning E & Kohiyama M (eds) Regulatory mechanisms of DNA replication. Springer, Berlin Heidelberg New York, pp 385–398

Schneider J & Fanning E (1988) Mutations in the phosphorylation sites of SV40 T antigen alter its specificity for origin DNA binding sites I or II and affect SV40 DNA replication activity. J Virol 62:1598–1605
Solomon MJ, Glotzer M, Lee TH, Philippe M & Kirschner MW (1990) Cyclin activation of $p34^{cdc2}$. Cell 63:1013–1024
Stahl H, Dröge P & Knippers R (1986) DNA helicase activity of SV40 large tumor antigen. EMBO J 5:1939–1944
Stillman B (1989). Initiation of eucaryotic DNA replication in vitro. Annu Rev Cell Biol 5:197–245
Summers MD & Smith GE (1987) A manual of methods for baculovirus vectors and cloned insect cell culture procedures. Texas Agricultural Experiment Station, College Station, TX
Tsurimoto T & Stillman B (1991a) Replication factors required for SV40 DNA replication in vitro. I. DNA structure-specific recognition of a primer-template junction by eucaryotic DNA polymerases and their accessory proteins. J Biol Chem 266:1950–1960
Tsurimoto T & Stillman B (1991b) Replication factors required for SV40 DNA replication in vitro. II. Switching of DNA polymerase α and δ during initiation of leading and lagging strand synthesis. J Biol Chem 266:1961–1968
Tsurimoto T, Melendy T & Stillman B (1990) Sequential initiation of lagging and leading strand synthesis by two different polymerase complexes at the SV40 DNA replication origin. Nature 346:534–539
Virshup DM & Kelly TJ (1989) Purification of replication protein C, a cellular protein involved in the initial stages of simian virus 40 DNA replication in vitro. Proc Natl Acad Sci USA 86:3584–3588
Virshup DM, Kauffman MG & Kelly TJ (1989) Activation of SV40 DNA replication in vitro by cellular protein phophatase 2A. EMBO J 8:3891–3898
Weinberg DH, Collins KL, Simancek P, Russo A, Wold MS, Virshup DM & Kelly TJ (1990) Reconstitution of simian 40 DNA replication with purified proteins. Proc Natl Acad Sci USA 87:8692–8696
Yakobson E, Prives C, Hartman JR, Winocour E & Revel M (1977a) Inhibition of viral protein synthesis in monkey cells treated with interferon late in simian virus 40 lytic cycle. Cell 12:73–81
Yakobson E, Revel M & Winocour E (1977b) Inhibition of simian virus 40 replication by interferon treatment late in the lytic cycle. Virology 80:225–228

The Mitogen/Oncogene-Activated p70^{s6k}: Its Role in the G_0/G_1 Transition

S. Ferrari[1], and G. Thomas[1]

1 Introduction

The addition of growth factors to quiescent cells in culture triggers a complex array of biochemical processes which culminate in DNA replication and cell division (Pardee 1989). One such process is the activation of protein synthesis which is required not only for the G_0/G_1 transition, but also for cells to progress through G_1 and initiate DNA synthesis (Brooks 1977; Rossow et al. 1979). This event is associated with the phosphorylation of a number of essential translational components, including 40S ribosomal protein S6 (Kozma et al. 1989; Hershey 1989). S6 has been roughly mapped to the anticodon-codon interaction site of the 40S ribosomal subunit (Nygård and Nilsson 1990), and experiments in vivo and in vitro show that multiple phosphorylation of this protein is required for the activation and maintenance of high rates of protein synthesis (Palen and Traugh 1987; Olivier et al. 1988, Šuša et al. 1989).

The major kinase thought to be involved in regulating this response has been identified as a M_r 70 000 polypeptide (Jenö et al. 1988; Jenö et al. 1989; Kozma et al. 1989; Price et al. 1989; Lane and Thomas 1991) and has recently been termed p70^{s6k} (Reinhard et al. 1992; Thomas 1992). The kinase is activated by Ser/Thr phosphorylation (Ballou et al. 1988a), in a biphasic manner (Šuša et al. 1989; Šuša and Thomas 1990; Šuša et al. 1992) and is inactivated by a type 2A phosphatase (Ballou et al. 1988b). Though the sites of phosphorylation exhibit *Ser/Thr*-Pro motifs (Ferrari et al. 1992) it is clear that the p70^{s6k} lies on a distinct signalling pathway from that of the p42^{mapk} (Ballou et al. 1991; Lane et al. 1992; Thomas 1992).

In this article we review the sites of S6 phosphorylation leading to the activation of protein synthesis as well as the recognition motif within S6 which is recognized by the p70^{s6k}. We then discuss the recent discovery of a second isoform of the p70^{s6k}, termed p85^{s6k}, that is derived from the same gene and is apparently targeted to a unique compartment in the cell. Finally, we end by discussing the sites of phosphorylation leading to p70^{s6k} activation and how this information might be exploited in the search for p70^{s6k} kinase.

[1] Friedrich Miescher Institute, P.O. Box 2543, 4002 Basel, Switzerland.

43. Colloquium Mosbach 1992
DNA Replication and the Cell Cycle

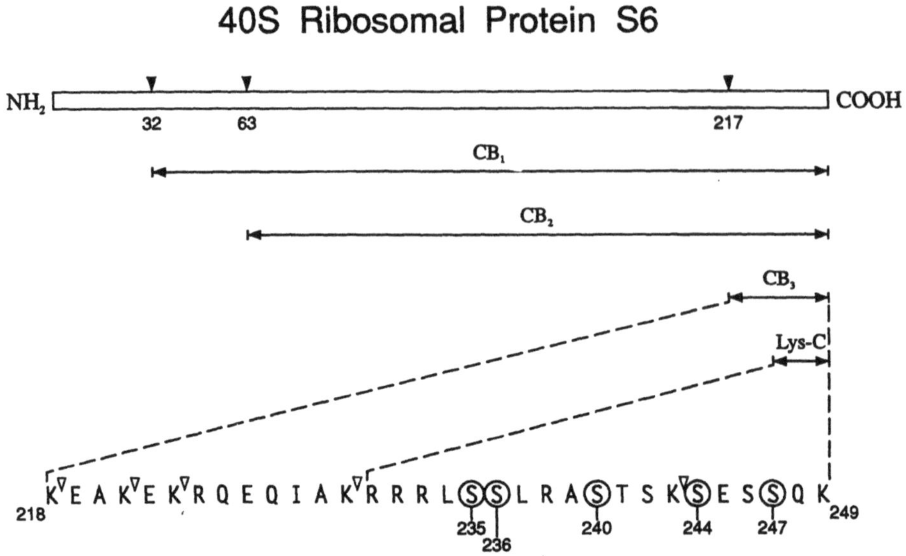

Fig. 1. Sequence of S6. The sites of phosphorylation are indicated by *circles*. *CB* Cyanogen bromide fragment; ∇: endo Lys-C cleavage site. (Bandi 1992a)

2 Results

2.1 Sites of S6 Phosphorylation

Recently we identified the in vivo sites of S6 phosphorylation and showed that they resided in a 32 amino acid CNBr fragment at the carboxy terminus of S6 (Krieg et al. 1988; Bandi et al. 1992b). Prior to determining the ability of the $p70^{s6k}$ to phosphorylate these same sites in vitro, we established the effects of different cations and autophosphorylation on kinase activity (Ferrari et al. 1991). The $p70^{s6k}$ was found to be dependent on Mg^{2+} for activity and this requirement could not be substituted for by Mn^{2+}. Indeed, 50 times lower amounts of Mn^{2+} block the effect of Mg^{2+} on the kinase. In the presence of optimum Mg^{2+} concentrations the enzyme incorporates an average of 1.2 mol of phosphate/mol of $p70^{s6k}$. The autophosphorylation reaction appears to be intramolecular and leads to a 25% reduction in kinase activity towards S6. In the case of S6 all of the sites of phosphorylation appear to reside in a 19-amino acid peptide at the carboxyl end of the protein. Four of these sites have been identified as Ser^{235}, Ser^{236}, Ser^{240} and Ser^{244}, equivalent to four of the five sites previously observed in vivo (Krieg et al. 1988). A fifth mole of phosphate is incorporated at low stoichiometry into the peptide, probably into Ser^{247}, but this site could not be unequivocally assigned (Ferrari et al. 1991). The sites of phosphorylation are depicted in Fig. 1.

2.2 Substrate Recognition Determinants

Having established that the in vivo and in vitro sites of S6 phosphorylation were identical we set out to identify the substrate recognition determinants for the p70^{s6k}. The kinase has an apparent K_m for a synthetic peptide derived from the endoproteinase fragment described above which is only 2.5 times higher than that of the substrate (Flotow and Thomas 1992). This would imply that all the substrate recognition determinants resided within this sequence. A number of shorter peptides revealed that the substrate recognition determinants for the preferred site of phosphorylation, Ser^{236}, reside in a 7-amino acid stretch of S6, residues 231–237 (Fig. 1). Critical to recognition are Arg^{231} and Arg^{233} as well as Leu^{237}. In contrast, replacement of Arg^{232}, Ser^{235} or Ser^{236} with alanine has little effect on substrate recognition. Based on this information the substrate recognition motif would be Arg-X-Arg-(X)-Ser-(X). The two Xs in parentheses have not yet been tested by substitution. A number of kinases known to phosphorylate S6, including cAMP-dependent protein kinase and protein kinase C as well as p90^{rsk}, were also found to use this peptide. However, with the exception of the p90^{rsk}, none of these kinases have very high affinity for the peptide. Surprisingly, in the case of p90^{rsk} the K_m was ten-fold lower for the peptide than for 40S ribosomes. Based on this information as well as a three-dimensional model constructed for p70^{s6k}, synthetic peptides are being generated as potential inhibitors of the kinase.

2.3 Cloning of the p70^{s6k} and p85^{s6k}

Earlier two cDNA clones were isolated for the p70^{s6k} from a rat liver (Kozma et al. 1990) or a hepatoma cDNA library (Banerjee et al. 1990), referred to as clone 1 and clone 2, respectively (Reinhard et al. 1992). Except for a single amino acid change and a 23 amino acid extension in clone 2, the open reading frames of the two clones were identical. Clone 1 was 2.8 kb and contained a 133 nucleotide GC-rich, 5' untranslated region followed by a 502 amino acid residue coding sequence. The nucleotide sequence surrounding the initiation codon conformed to a "strong" consensus start site. The 3' untranslated region was AT-rich and contained no poly(A) tail. Clone 2 was 2.3 kb and contained a short, 21-nucleotide, 5' untranslated region followed by a "weak" consensus translation start site and an open reading frame of 525 amino acids (Banerjee et al. 1990). From a second potential AUG start site, at nucleotide 91, the two clones were identical except for a single amino acid difference (proline 344 in clone 1 and arginine 367 in clone 2). These characteristics are summarized in Fig. 2.

A probe common to both clones revealed four distinct transcripts of 6.0, 4.0, 3.2 and 2.5 kb (Kozma et al. 1990). By using specific probes, it was possible to show that clones 1 and 2 were encoded in the 6.0 transcript, that only clone 1 was present in the 4.0 and 3.2 transcripts, whereas only clone 2 was present in the 2.5 kb transcript (Reinhard et al. 1992). It was also possible by a combination of genomic PCR combined with Southern blot analysis to demonstrate that both clones were derived from a single gene, and that no full-length pseudo-genes existed (Reinhard et al. 1992). Analysis of in vitro translation products of clone 1 and clone 2 using specific anti-

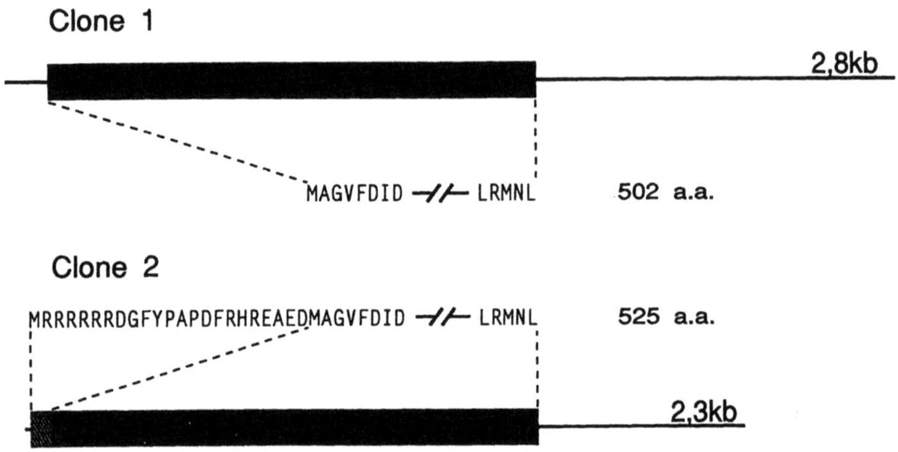

Fig. 2. Rat *s6k* cDNA clones 1 and 2. *Solid thick bar* indicates common coding region; *shaded thick bar* indicates 5' coding region specific for clone 2; *thin bar* indicates untranslated regions. Amino acids are denoted by *single-letter code* (Reinhard et al. 1992)

bodies demonstrates that both clones encode the $p70^{s6k}$, but that the clone harboring the 23 amino acid extension also encodes an additional isoform of the kinase referred to as $p85^{s6k}$ (Reinhard et al. 1992). Employing a specific antibody to the $p85^{s6k}$, it could be further shown that this isoform of the kinase is present in rat liver and is activated following mitogenic stimulation of quiescent Swiss 3T3 cells. Preliminary data further suggests that this isoform is targeted to the nucleus (C. Reinhard, A. Fernandez, N. Lamb and G. Thomas, unpubl. observ.).

2.4 Phosphorylation Sites Associated with $p70^{s6k}$ Activation

During the initial sequencing of the $p70^{s6k}$ 22 tryptic peptides were examined and three were found to contain phosphorylated residues (Ferrari et al. 1992). To ascertain if these sites were associated with $p70^{s6k}$ activation, the kinase was labeled to high specific activity with $^{32}P_i$ in Swiss mouse 3T3 cells (Ferrari et al. 1992). By sequential cleavage with cyanogen bormide, endo proteinase Lys-C and trypsin four phosphopeptides were identified which contained all the sites of phosphorylation. Employing a newly described computer program (Boyle et al. 1991), four tryptic peptides were chosen as the most likely candidates to contain the in vivo sites of phosphorylation. Synthetic peptides based on these sequences were phosphorylated either chemically (Kitas et al. 1991) or enzymatically and found to comigrate by two-dimensional TLE/TLC with the four major in vivo labeled tryptic phosphopeptides. Three of these peptides were equivalent to those sequenced in the rat liver $p70^{s6k}$. Surprisingly, all four sites display the motif *Ser/Thr*-Pro (Fig. 3), typical of cell cycle regulated sites, and are clustered in a putative autoinhibitory domain of the enzyme (Ferrari et al. 1992; Flotow and Thomas 1992).

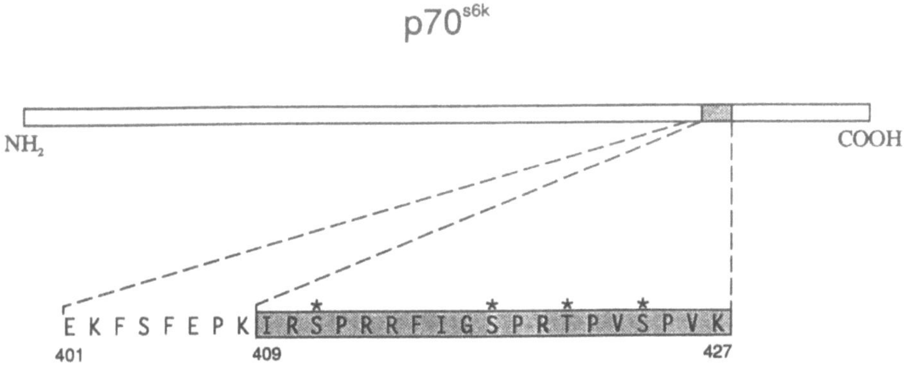

Fig. 3. Domain of p70^{s6k} containing the sites of phosphorylation. The *stippled area* represents the endo Lys-C fragment Ile^{409}-Lys^{427} (Kozma et al. 1990), which contains the four sites of phosphorylation (denoted by *asterisks*) (Ferrari et al. 1992)

3 Future Prospective

At this point in the study of the S6 phosphorylation signalling pathway, the tools are available to test whether the substrate as well as the kinases which modulate this response are essential components of the mitogenic response. By taking advantage of molecular biology and biochemical techniques combined with computer modeling and different biological systems we are presently examining this question in detail.

References

Ballou LM, Siegmann M & Thomas G (1988a) S6 kinase in quiescent Swiss mouse 3T3 cells is activated by phosphorylation in response to serum treatment. Proc Natl Acad Sci USA 85:7154–7158

Ballou LM, Jenö P & Thomas G (1988b) Protein phosphatase 2A inactivates the mitogen-stimulated S6 kinase from Swiss mouse 3T3 cells. J Biol Chem 263:1188–1194

Ballou LM, Luther H & Thomas G (1991) Map2 kinase and 70K S6 kinase lie on distinct signalling pathways. Nature 349:348–350

Bandi HR (1992) Characterization of the sites of phosphorylation of 40S ribosomal protein S6 in mitogen-stimulated Swiss mouse 3T3 fibroblasts. PhD Thesis, University of Basel, Switzerland

Bandi HR, Ferrari S, Krieg J, Meyer HE & Thomas G (1992) Indification of 40S ribosomal protein S6 phosphorylation sites in Swiss mouse 3T3 fibroblasts stimulated with serum. J Biol Chem (in press)

Banerjee P, Ahamad MF, Grove JR, Kozlosky C, Price DJ & Avruch J (1990) Molecular structure of a major insulin/mitogen-activated 70 kDa S6 protein kinase. Proc Natl Acad Sci USA 87:8550–8554

Boyle WJ, van der Geer P & Hunter T (1991) Phosphopeptide mapping and phosphoamino acid analysis by two-dimensional seperation on thin layer cellulose plates. Meth Enzymol 201:110–149

Brooks RF (1977) Continuous protein synthesis is required to maintain the probability of entry into S phase. Cell 12:311–317

Ferrari S, Bandi HR, Hofsteenge J, Bussian BM & Thomas G (1991) Mitogen-activated 70K S6 kinase: identification of in vitro 40S ribosomal S6 phosphorylation sites. J Biol Chem 266:22770–22775
Ferrari S, Bannwarth W, Morley SJ, Totty NF & Thomas G (1992) Activation of p70^{s6k} is associated with phosphorylation of four clustered sites displaying Ser/Thr-Pro motifs. Proc Natl Acad Sci USA 89:7282–7286
Flotow H & Thomas G (1992) Substrate recognition determinats of the mitogen-activated 70K S6 kinase from rat liver. J Biol Chem 267:3074–3078
Hershey JWB (1989) Protein phosphorylation controls translational rates. J Biol Chem 264:20823–20826
Jenö P, Ballou LM, Novak-Hofer I & Thomas G (1988) Identification and characterization of a mitogen-activated S6 kinase. Proc Natl Acad Sci USA 85:406–410
Jenö P, Jäggi N, Luther H, Siegmann M & Thomas G (1989) Purification and characterization of a 40S ribosomal protein S6 kinase from vanadate-stimulated Swiss 3T3 cells. J Biol Chem 264:1293–1297
Kitas EA, Knorr R, Trzeciak A & Bannwarth W (1991) Alternative strategies for the Fmoc solid-phase synthesis of O^4-phospho-L-tyrosine-containing peptides. Helv Chim Acta 74:1314–1328
Kozma SC, Ferrari S, Bassand P, Siegmann M, Totty N & Thomas G (1990) Cloning of the mitogen-activated S6 kinase from rat liver reveals an enzyme of second messenger subfamily. Proc Natl Acad Sci USA 87:7365–7369
Kozma SC, Ferrari S & Thomas G (1989) Unmasking a growth factor/oncogene-activated S6 phosphorylation cascade. Cell Signal 1:219–225
Kozma SC, Lane HA, Ferrari S, Luther H, Siegmann M & Thomas G (1989) A stimulated S6 kinase from rat liver: identity with the mitogen-activated S6 kinase of 3T3 cells. EMBO J 8:4125–4132
Krieg J, Hofsteenge J & Thomas G (1988) Identification of the 40S ribosomal protein S6 phosphorylation sites induced by cycloheximide. J Biol Chem 263:11473–11477
Lane HA, Morley SJ, Doree M, Kozma SC & Thomas G (1992) Identification and early activation of *Xenopus laevis* p70^{s6k} following progesterone-induced meiotic maturation. EMBO J 11:1743–1749
Lane HA & Thomas G (1991) Purification and properties of mitogen-activated S6 kinase from rat liver and 3T3 cells. Meth Enzymol 200:268–291
Nygård O and Nilsson (1990)Translational dynamics. Eur J Biochem 191:1–17
Olivier AR, Ballou LM & Thomas G (1988) Differential regulation of S6 phosphorylation by insulin and epidermal growth factor in Swiss mouse 3T3 cells: insulin activation of type 1 phosphates. Proc Natl Acad Sci USA 85:4720–4724
Palen E & Traugh JA (1987) Phosphorylation of ribosomal protein S6 by cAMP-dependent protein kinase and mitogen-stimulated S6 kinase differentially of globin mRNA. J Biol Chem 262:3518–3523
Pardee AB (1989) G_1 events and regulation of cell proliferation. Science 246:603–608
Price DJ, Nemenoff RA & Avruch J (1989) Purification of a hepatic S6 kinase from cycloheximide treated rats. J Biol Chem 264:13825–13833
Reinhard C, Thomas G & Kozma SC (1992) A single gene encodes two isoforms of the p70 S6 kinase: activation upon mitogenic stimulation. Proc Natl Acad Sci USA 89:4052–4056
Rossow PW, Diddle VGH & Pardee AB (1979) Synthesis of labile,serum-dependent protein in early G_1 controls animals cell growth. Proc Natl Acad Sci USA 76:4446–4450
Šuša M, Olivier AR, Fabbro D & Thomas G (1989) EGF induces biphasic S6 kinase activation: late phase is protein kinase C-dependent and contributes to mitogenicity. Cell 57:817–824
Šuša M & Thomas G (1990) Identical M_r 70,000 S6 kinase is activated biphasically by epidermal growth factor: a phosphopeptide that characterizes the late phase. Proc Natl Acad Sci USA 87:7040–7044
Šuša M, Vulevic D, Lane HA & Thomas G (1992) Inhibition of down-regulation of protein kinase C late-phase p70^{s6k} activation induced by epidermal growth factor but not by platelet-derived growth factor or insulin. J Biol Chem 267:6905–6909
Thomas G (1992) MAP kinase by any other name smells just as sweet. Cell 68:3–6

Ubiquitin-Dependent Protein Degradation

S. Jentsch[1], W. Seufert[1], T. Sommer[1], H.-A. Reins[1], J. Jungmann[1], and H.-P. Hauser[1]

1 Introduction

Intracellular protein levels are controlled by protein synthesis and degradation. One major proteolytic pathway of eukaryotes requires the covalent attachment of ubiquitin, a small (8.5 kDa) and highly conserved protein, to proteolytic substrates prior to degradation (for reviews see Hershko 1991; Finley and Chau 1991; Jentsch et al. 1991, Jentsch 1992a; Jentsch 1992b). This pathway is highly selective and mediates the elimination of abnormal proteins and controls the half-lives of some regulatory proteins. Known targets include transcriptional regulators (Hochstrasser et al. 1991) p53 (Scheffner et al. 1990) and cyclins (Glotzer et al. 1991). In a number of eukaryotes, cyclin degradation is required to exit from mitosis.

The conjugation of ubiquitin to protein substrates is a multistep process. In an initial activating step, a thioester is formed between the C-terminus of ubiquitin and an internal cysteine residue of a ubiquitin-activating (E1) enzyme. Activated ubiquitin is then transferred to a specific cysteine residue of one of several ubiquitin-conjugating (E2) enzymes. Finally, these E2 enzymes donate ubiquitin to protein substrates. This results in branched protein conjugates in which the C-terminus of ubiquitin is linked by an isopeptide bond to specific internal lysine residues of target proteins. Since ubiquitin itself is a substrate for ubiquitination, certain substrates are modified by the attachment of multiubiquitin chains (Chau et al. 1989). Multiubiquitinated substrates are degraded in vivo by the proteasome, a multisubunit protease complex.

Purified enzymes of the *Saccharomyces cerevisiae* ubiquitin-conjugating system have been used to study the enzymology of the ubiquitin system and allowed the cloning of the corresponding genes. So far, ten genes coding for ubiquitin-conjugating enzymes (*UBC* genes) have been isolated from the yeast *S. cerevisiae* (reviewed in Jentsch et al. 1990). These enzymes are located in the cytosol and the nucleus, indicating that both cellular compartments have active ubiquitin-conjugating systems. Here we review our data on the genetic analysis of the ubiquitin system (see Table 1).

[1] Friedrich-Miescher-Laboratorium der Max-Planck-Gesellschaft, Spemannstr. 37–39, 7400 Tübingen, FRG.

43. Colloquium Mosbach 1992
DNA Replication and the Cell Cycle

Table 1. The yeast ubiquitin-conjugation system

Genes	Function	References
UBA1	Ubiquitin-activating enzyme, essential for viability cytoplasmic and nuclear	McGrath et al. (1991)
UBC1	Specifically required after sporulation/germination mediates vital functions together with *UBC4* and *UBC5* G_0 induction	Seufert et al. (1990)
UBC2/RAD6	DNA repair, induced mutagenesis, sporulation, Repression of retrotransposition	Jentsch et al. (1987) Friedberg (1988) Picologlou et al. (1990)
UBC3/CDC34	Essential for viability, G_1-S cell cycle control, DNA replication, spindle pole body separation nuclear protein	Goebl et al. (1988)
UBC4 *UBC5*	Bulk degradation of short-lived and abnormal proteins mediate vital functions together with *UBC1* essential under stress conditions heat shock-inducible, cadmium-inducible	Seufert and Jentsch (1990) Seufert et al. (1990) Seufert and Jentsch (1990)
UBC6	Integral membrane protein of the endoplasmic reticulum function in the secretory pathway	T. Sommer and S. Jentsch (in prep.)
UBC7	Confers cadmium resistance Cadmium-inducible	H.-A. Reins, J. Jungmann, C. Schobert, S. Jentsch (in prep.)
UBC8	No detectable mutant phenotypes	Quin et al. (1991)
UBC9	Essential for viability, cell cycle progression nuclear protein	W. Seufert and S. Jentsch (unpubl.)

2 Results and Discussion

2.1 Functions in Protein Quality Control Pathways

Ubiquitin-dependent proteolysis serves essential functions in probably all eukaryotes (McGrath et al. 1991). The yeast proteins UBC1, UBC4, UBC5 and UBC7 have roles in the degradation of abnormal proteins. *UBC1*, *UBC4* and *UBC5* constitute a gene family essential for viability (the triple mutant is inviable) (Seufert and Jentsch 1990; Seufert et al. 1990). The *ubc1* and *ubc4* single mutants and *ubc4 ubc5* double mutants have a slow growth phenotype and are hypersensitive to amino acid analogs. In vivo data indicate that these mutants are largely deficient in the degradation of otherwise short-lived and abnormal proteins. The *ubc1* mutants show specific defects in growth after germination of ascospores, suggesting that UBC1 has a specific function during this transition period in the yeast life cycle. The enzymes UBC4 and UBC5 mediate essential functions of the stress response: the transcription of *UBC4* and *UBC5* genes is induced by heat and *ubc4 ubc5* double mutants are inviable under stress conditions. Transcription of *UBC7* is induced by cadmium but not by heat shock (H.-A. Reins, J. Jungmann, C. Schobert and S. Jentsch, in prep.). Mutations in *UBC7* do not lead to an altered growth rate or a heat shock sensitive phenotype. Yet *ubc7* mutants and also *ubc4 ubc5* double mutants are cadmium hypersensitive, suggesting that cadmium exposure generates abnormal proteins which are normally subject to UBC4, UBC5 and UBC7-mediated proteolysis. A structural close homolog of UBC4 and UBC5 has been cloned from *Drosophila melanogaster* (Treier et al. 1992). This gene, *UbcD1*, has been shown to be functionally equivalent to yeast *UBC4*, suggesting that UBC4-like enzymes mediate similar functions in probably all eucaryotes.

The UBC6 protein has a C-terminal signal-anchor sequence and is an integral membrane protein of the endoplasmic reticulum, with the active site of the enzyme on the cytoplasmic face of the membrane (T. Sommer and S. Jentsch, in prep.). Recent genetic data suggest that the UBC6 enzyme has a function in the secretory pathway, probably by mediating the degradation of misfolded membrane proteins. Another integral membrane ubiquitin-conjugating enzyme, UbcM1, was recently shown to be expressed in mouse pituitary tumor cells (H.-P. Hauser and S. Jentsch, in prep.). This enzyme is unusually large (500 kDa) and has several potential transmembrane spanning segments.

2.2 Functions in Cellular Regulation

The UBC2 protein (also known as RAD6) plays a prominent role in a central DNA-repair pathway of yeast (Jentsch et al. 1987; Friedberg 1988). Mutants of *UBC2* are highly vulnerable to DNA-damaging agents and deficient in induced mutagenesis. Moreover, *UBC2* is required for sporulation and repression of retrotransposition in yeast (Picologlou et al. 1990). This regulatory role suggests that UBC2 might mediate the degradation of regulatory proteins such as transcriptional repressors for genes encoding proteins for DNA repair, sporulation and transposition functions. Alternatively, since this enzyme conjugates ubiquitin to histones in vitro, UBC2 might modulate chromatin structure (Jentsch et al. 1987).

Only two of the known ten *UBC* genes, *UBC3* and *UBC9*, are essential for viability and both of these encode nuclear enzymes which are required for cell cycle progression (Goebl et al. 1988; W. Seufert and S. Jentsch, unpubl.). Whereas the precise function of *UBC9* in the cell cycle is presently unknown, *UBC3* (also known as *CDC34*) is required for the transition from the G_1 into the S-phase of the cell cycle (Goebl et al. 1988). Temperature-sensitive *ubc3* mutants arrest at the restrictive temperature in G_1 with multiple elongated buds and have unreplicated DNA. The spindle pole bodies duplicate but fail to undergo separation. The in vivo substrates of the UBC3 enzyme are presently unknown but may include the yeast G_1-cyclins.

Acknowledgements. Research in this laboratory is supported by grants to S. J. from the Deutsche Forschungsgemeinschaft, The German-Israeli Foundation for Scientific Research and Development and The Human Frontier Science Program.

References

Chau V, Tobias JW, Bachmair A, Marriott D, Ecker DJ, Gonda DK & Varshavsky A (1989) A multiubiqquitin chain is confined to specific lysine in a targeted short-lived protein. Science 243:1576–1583

Friedberg EC (1988) Deoxyribonucleic acid repair in the yeast *Saccharomyces cerevisiae*. Microbiol Rev 52:70–102

Finley D & Chau V (1991) Ubiquitination. Annu Rev Cell Biol 7:25–69

Glotzer M, Murray AW & Kirschner MW (1991) Cyclin is degraded by the ubiquitin pathway. Nature 349:132–138

Goebl MG, Yochem J, Jentsch S, McGrath JP, Varshavsky A & Byers B (1988) The yeast cell cycle gene *CDC34* encodes a ubiquitin-conjugating enzyme. Science 241:1331–1335

Hershko A (1991) The ubiquitin pathway for protein degradation. Trends Biochem Sci 16:265–268

Hochstrasser M, Ellison MJ, Chau V & Varshavsky A (1991) The short-lived MATα2 transcriptional regulator is ubiquitinated in vivo. Proc Natl Acad Sci USA 88:4606–4610

Jentsch S (1992a) Ubiquitin-dependent protein degradation: a cellular perspective. Trends Cell Biol 2:98–103

Jentsch S (1992b) The ubiquitin-conjugation system. Annu Rev Genet 26:177–205

Jentsch S, McGrath JP & Varshavsky A (1987) The DNA repair gene *RAD6* encodes a ubiquitin-conjugating enzyme. Nature 329:131–134

Jentsch S, Seufert W & Hauser H-P (1991) Genetic analysis of the ubiquitin system. Biochim Biophys Acta 1089:127–139

Jentsch S, Seufert W, Sommer T & Reins H-A (1990) Ubiquitin-conjugating enzymes: novel regulators of eukaryotic cells. Trends Biochem Sci 15:195–198

McGrath JP, Jentsch S & Varshavsky A (1991) *UBA1*: an essential yeast gene encoding ubiquitin-activating enzyme. EMBO J 10:227–236

Picologlou S, Brown N & Liebman SW (1990) Mutations in *RAD6*, a yeast gene encoding a ubiquitin-conjugating enzyme, stimulate retrotransposition. Mol Cell Biol 10:1017–1022

Qin S, Nakajima B, Nomura M & Arfin SM (1991) Cloning and characterization of a *Saccharomyces cerevisiae* gene encoding a new member of the ubiquitin-conjugating protein family. J Biol Chem 266:15549–15554

Scheffner M, Werness BA, Huibregtse JM, Levine AJ & Howley PM (1990) The E6 oncoprotein encoded by human papillomavirus types 16 and 18 promotes the degradation of p53. Cell 63:1129–1136
Seufert W, McGrath JP & Jentsch S (1990) *UBC1* encodes a novel member of an essential subfamily of yeast ubiquitin-conjugating enzymes involved in protein degradation. EMBO J 9:4535–4541
Seufert W & Jentsch S (1990) Ubiquitin-conjugating enzymes UBC4 and UBC5 mediate selective degradation of short-lived and abnormal proteins. EMBO J 9:543–550
Treier M, Seufert W & Jentsch S (1992) *Drosophila UbcD1* encodes a highly conserved ubiquitin-conjugating enzyme involved in selective protein degradation. EMBO J 11:367–372

Transcriptional Control of Cell Proliferation

Ternary Complex Formation at the Human c-*fos* Serum Response Element

R. A. Hipskind[1], R. Janknecht[1], C. G. F. Mueller[1], and A. Nordheim[1]

1 Introduction

Quiescent cells can be stimulated to active growth by the addition of serum growth factors or mitogens. These proliferative signals activate an intracellular signalling pathway that leads to the rapid and transient transcriptional activation of a large family of genes, the so-called "immediate-early" genes (Almendral et al. 1988; Bravo 1990). To attain a comprehensive understanding of cellular growth control will require the elucidation of the mechanisms by which the signal transduction cascade reaches and targets regulatory elements in individual "immediate-early" gene promoters.

Among these genes, the c-*fos* proto-oncogene is the best studied and serves as a paradigm for their transcriptional regulation (Cohen and Curran 1989). A multitude of stimuli (e.g. serum, individual growth factors, mitogens, increased intracellular cAMP, insulin, phorbol esters, Ca^{2+} ionophores) efficiently activate the c-*fos* promoter within minutes, and this rapid stimulation is followed quickly by a FOS-dependent transcriptional repression (for review: Rivera and Greenberg 1990). Several promoter elements, depicted in Fig. 1, have been identified to contribute to this complex pattern of transcriptional control (Treisman 1985; Gilman et al. 1986; Fisch et al. 1987; Runkel et al. 1991; Lucibello et al. 1991). Among them, a central role is assumed by the serum response element (SRE), originally identified by Treisman (1985) and Gilman et al. (1986). The SRE has been shown to mediate the transcriptional activation of c-*fos* by many stimuli, including serum, individual growth factors (e.g. NGF, EGF), UV light, cAMP, insulin, phorbol esters, and oncogenic ras, and SREs can be found in the promoters of other immediate-early genes (for review: Treisman 1990). In addition, the SRE is also required for repression of c-*fos* transcription (Shaw et al. 1989a; König et al. 1989; Leung and Miyamoto 1989; Subramaniam et al. 1989; Rivera et al. 1990).

The recognition of the SRE by specific transcriptional regulatory proteins is emphasized herein. Since the SRE is a target for a variety of signals, it is not surprising that several different proteins have been reported to bind to it (Metz and Ziff 1991; and references therein). We will concentrate on the specific interaction of the serum response factor (SRF) and SRF-dependent ternary complex factors (TCFs) with the SRE. The SRF binds the SRE with high affinity (Treisman 1986), and the SRE-SRF binary complex additionally recruits the accessory protein $p62^{TCF}$ to form a ternary

[1] Institut für Molekularbiologie, Medizinische Hochschule Hannover, Konstanty-Gutschow-Straße 8, 3000 Hannover 61, FRG.

43. Colloquium Mosbach 1992
DNA Replication and the Cell Cycle

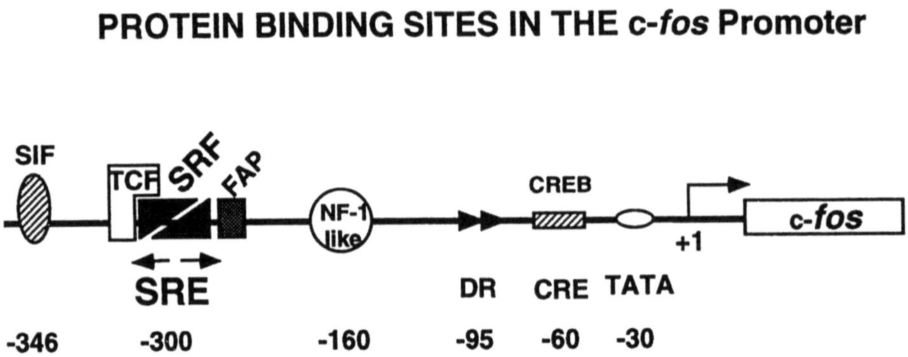

Fig. 1. Protein binding sites within the human c-*fos* promoter. Indicated are sites of protein interaction within the c-*fos* promoter that play important roles in the regulation of either basal level transcription or the induced and repressed states of promoter activity. Abbreviations for DNA elements: *TATA* TATA box; *CRE* cAMP response element; *DR* direct repeats, *SRE* serum response element; and proteins: *CREB* CRE binding protein; *NF-1* nuclear factor 1; *FAP fos* AP1-like binding protein; *SRF* serum response factor; *TCF* ternary complex factor; *SIF* v-sis inducible factor. This schematic display of *cis*-elements and *trans*-factors is described in more detail in Runkel et al. (1991) and Metz and Ziff (1992)

complex (Shaw et al. 1989b; Schröter et al. 1990). Such ternary complexes are likely to constitute the main target in the c-*fos* promoter for intracellularly transmitted proliferative signals.

2 Results and Discussion

2.1 The Serum Response Factor (SRF)

The ability of the SRE to confer serum inducibility directly correlates with the binding of SRF to the SRE (Gilman et al. 1986; Treisman 1986). SRF has been purified biochemically (Prywes and Roeder 1987; Schröter et al. 1987; Treisman 1987) and the corresponding cDNA has been cloned (Norman et al. 1988). It encodes a 508 amino acid protein with an apparent molecular weight of 62–67 kDa. SRF binds the dyad-symmetric SRE as a dimer (Norman et al. 1988; Schröter et al. 1990) thereby establishing close contacts to the two pairs of G residues within the central $CC(A/T)_6GG$ sequence (also called CArG-box, Fig. 2). SRF purified from nuclear cell extracts contains the post-translational modifications of O-glycosylation (Schröter et al. 1990) and phosphorylation (Prywes et al. 1988). Whereas the role of glycosylation is unknown at present, phosphorylation was shown to influence SRE binding (Prywes et al. 1988; Manak and Prywes 1991; Manak et al. 1990; Schalasta and Doppler 1990; Misra et al. 1991). More specifically, phosphorylation at individual serines within a potential casein kinase II target region (Fig. 3) drastically influences the kinetics of SRE–SRF interaction while the affinity for the SRE is only weakly affected (Janknecht et al. 1992; Marais et al. 1992). SRF contains a central region of

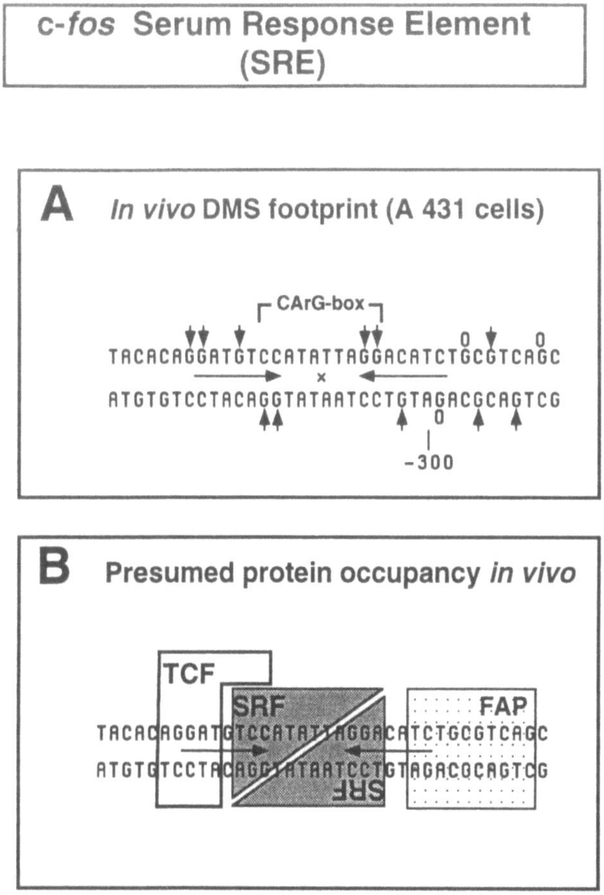

Fig. 2. In vivo DNA-protein complexes at the c-*fos* SRE. The data shown are based on the work of Herrera et al. (1989) and König (1991). At least three protein components are presumed to be bound to the *SRE in vivo*: *SRF, TCF* and the yet undefined *FAP* (*fos* AP1-site binding protein) (Velcich and Ziff 1990). These binding sites are occupied before, during, and after transcriptional stimulation and repression of the c-*fos* promoter. *Arrowheads* point to G-residues that are protected from *in vivo* DMS reactivity, whereas *open ovals* indicate those G's displaying increased reactivity

90 amino acids which displays significant degrees of homology to other eukaryotic transcription factors, namely the yeast regulatory proteins MCM1 and ARG80 and the two plant homeotic proteins AG and DEF A (Sommer et al. 1990; Yanofsky et al. 1990). The region of homology has therefore been termed a MADS box (Sommer et al. 1990) (Fig. 3). A truncated SRF, or SRF-core ($SRF_{132\text{-}222}$), that encompasses the MADS box has at least three functions, namely SRF dimerization, specific SRE binding, and ternary complex formation (Norman et al. 1988; Schröter et al. 1990; Mueller and Nordheim 1991).

The mechanisms by which SRF contributes to the transcriptional regulation of c-*fos* are unclear at present. However, genomic footprinting of the SRE in several cell

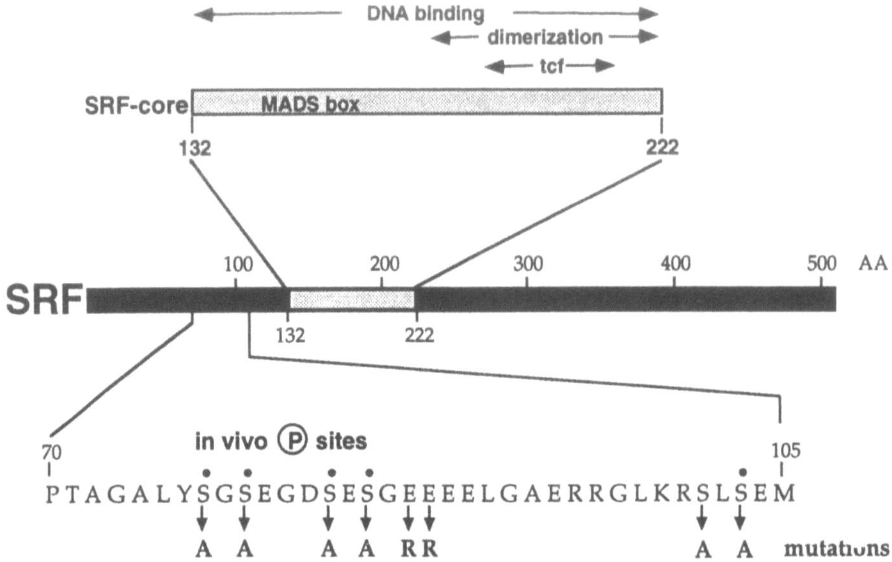

Fig. 3. Functional elements within the serum response factor (SRF). The centrally positioned core element, spanning amino acid coordinates 132 to 222, encodes at least three functions, namely DNA binding to the SRE, dimerization and interaction with a ternary complex factor. This region is highly conserved among eukaryotic regulatory proteins and contains the MADS box (Sommer et al. 1990) in the N-terminal half of core. SRF protein produced in HeLa cells using the vaccinia virus expression system permitted the direct identification of phosphorylated serine residues within amino acid positions 77–85 and at Ser103 (Janknecht et al. 1992). These sites represent substrate sequences for the kinase CKII and Ser*103* could potentially also be a target for PKA or the Calmodulin/Ca^{2+}-dependent kinase. The mutations shown were introduced singly or in various combinations and were found to strongly affect the kinetics of interaction with the SRE while only weakly influencing the overall affinities to the SRE (Janknecht et al. 1992; see also Marais et al. 1992)

types suggests the presence of SRF at this site before, during and after c-*fos* induction (Herrera et al. 1989; König 1991). This supports the notion that the SRF fulfills essential functions in both activation and repression of the c-*fos* promoter. The recent identification of several SRF-related cDNAs suggests that a number of similar proteins may be able to mediate these functions (Pollock and Treisman 1991).

2.2 Ternary Complex Factors (TCFs)

During the course of our purification of SRF from chloroquine-extracted HeLa nuclei (Schröter et al. 1987) we noticed a novel DNA binding protein activity, called $p62^{TCF}$ (*T*ernary *C*omplex *F*actor), which specifically interacted with the SRE-SRF complex (Shaw et al. 1989b). This protein could not interact with the SRE alone, but rather required the SRE to be complexed by SRF in order to stably bind. The resulting higher order complex, or 'ternary complex', was generated by the binding of a dimer of SRF and a monomer of $p62^{TCF}$ (Schröter et al. 1990). The latter factor is bound asymmetrically to the 5' boundary of the SRE. Methylation interference and protection studies identified two G residues (–318, –319; Fig. 2) within a CAGGAT stretch that are in close contact with the protein (Shaw et al. 1989b). The triple transversion mutant EL (CAGGAT into ACTGAT) strongly reduced ternary complex formation without affecting the binding of SRF to the SRE. This indicated that p62^{TCF} directly contacted the DNA double helix in the ternary complex, a notion confirmed subsequently by UV crosslinking studies (P. E. Shaw 1992). Protein-protein interactions between SRF and p62^{TCF} appear to be important within the ternary complex, since domains required for protein contacts can be identified in both proteins (Mueller and Nordheim 1991; Dalton and Treisman 1992; Janknecht and Nordheim 1992) (Figs. 4 and 5). In the absence of DNA, however, there is no evidence for protein-protein interactions between SRF and p62^{TCF} (Shaw et al. 1989b; V. Pingoud and A. Nordheim, unpubl.). In summary, these characteristics help define a ternary complex factor (TCF): 1) it interacts selectively with a DNA-protein complex, in this case SRE-SRF, establishing specific contacts with both the DNA and the DNA-bound protein, and 2) no high affinity complexes can be formed with either the specific DNA (SRE) or the free protein (SRF) alone. This does not preclude the possibility of the TCF interacting with other DNA sequences directly (see below).

The biochemical identification of p62^{TCF} provided the first example of an SRE/SRF-dependent ternary complex factor. As will be discussed below (see also Table 1), cDNAs encoding three SRE/SRF-dependent ternary complex factors have recently been identified (Hipskind et al. 1991; Dalton and Treisman 1992). In addition, ternary complex factors selective for other DNA-protein complexes are known, e.g. for P-Box/MCM1 or CGGAAR/GABPα (Table 1; for review: Shaw 1990). Genomic footprinting indicated that, in vivo, p62^{TCF} was also bound to the SRE

Table 1. Ternary complex factors (TCFs) in eukaryotic transcriptional control

	TCF	Reference
1. SRE/SRF-dependent TCFs:	p62^{TCF}	Shaw et al. (1989b)
	Elk-1	Hipskind et al. (1991)
	SAP-1a, SAP-1b	Dalton and Treisman (1992)
	SAP-2	Dalton and Treisman (1992)
2. P-Box/MCM1-dependent TCFs:	α1	Bender and Sprague (1987)
	α2	Kelleher et al. (1988)
	STE12	Errede and Ammerer (1989)
	SFF	Lydall et al. (1991)
3. DNA-protein complexes recruiting other TCFs:		
EBS/E2F	E4	Hardy et al. (1989)
Octamer/Oct1	VP16	O'Hare and Goding (1988)
CGGAAR/GABPα	GABPβ	Thompson et al. (1991)
Pu/PU.1	NF-EM5	Pongubala et al. (1992)

(Herrera et al. 1989; Fig. 2). Interestingly, the occupation of this site was unchanged before, during and after c-*fos* induction, implicating p62^{TCF}, together with SRF, in both the induction and the subsequent repression of c-*fos* transcription. Transient transfection studies also indicated an involvement of p62^{TCF} in a subset of signal-mediated c-*fos* induction events in some cells (Shaw et al. 1989b; Graham and Gilman 1991; Malik et al. 1991) but not in others (König 1991). These results, together with the interesting observation by Malik et al. (1991) of inducible SRE/SRF-specific ternary complex formation in vitro, would support the idea that p62^{TCF} represents a target for incoming transcriptional stimulatory signals.

2.3 The Ets-Related Proteins Elk-1 and SAP-1 Are Ternary Complex Factors

The sequence CAGGAT contacted by p62^{TCF} in the ternary complex resembles the direct recognition site for members of the Ets-family of DNA binding proteins (Karim et al. 1990). Indeed, proteins that interact with the PEA3 element, an Ets protein binding site, can also bind to this CAGGAT sequence (Gutman and Wasylyk 1990). Therefore it seemed reasonable to screen various Ets protein for TCF activity. None of those initially tested scored positive (R.A. Hipskind and A. Nordheim, unpubl.). However, after R. Treisman and S. Dalton communicated the identification of TCF cDNA clones with their ETS domain at the amino terminus (see below), we found that the previously uncharacterized Elk-1 protein (Rao et al. 1989) displayed biochemical characteristics indistinguishable from p62^{TCF} (Hipskind et al. 1991). Treisman and Dalton (1992) used genetic selection in yeast to screen a human cDNA library for clones encoding SRF-associated proteins. This approach identified the Ets-protein SAP-1a, its close relative SAP-1b, and the yet uncharacterized SAP-2. These proteins have the ETS domain at the N-terminus (Fig. 4). As is shown graphically in Fig. 4, distinct homologies exist between Elk-1 and SAP-1 in three regions, namely

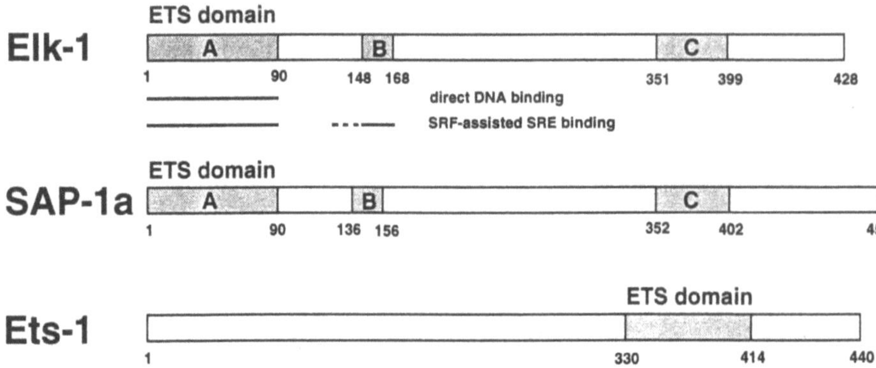

Fig. 4. Comparison of the ETS domain-containing proteins Elk-1, SAP-1 and Ets-1. The cDNA cloning of the SRF-associated protein SAP-1 (Dalton and Treisman 1992), a SRE/SRF-dependent ternary complex factor, revealed three regions of homology (namely *A*, *B* and *C*) to Elk-1. Region *A* encodes their N-terminally positioned ETS domain, which is required both for direct DNA binding and SRF-assisted SRE binding. The latter also requires the presence of sequences encompassing region *B* (Dalton and Treisman 1992; Janknecht and Nordheim 1992)

A, B, and C, with lengths of 90, 21, and 48 amino acids, respectively. Region A contains the ETS domain. SAP-1 displays the characteristics of a ternary complex factor similar to $p62^{TCF}$ (Dalton and Treisman 1992).

We showed that Elk-1 strongly resembles $p62^{TCF}$ by several criteria. They have the same DNA sequence requirements for ternary complex formation and antibodies against Elk-1 block the binding of both proteins. Like $p62^{TCF}$, Elk-1 forms complexes with the yeast SRF-homologue MCM1, but not with yeast ARG80 (Hipskind et al. 1991). Although Elk-1 is unable to bind directly to the SRE on its own (Fig. 5), it can nevertheless interact independently and specifically with related DNA sequences, such as the E74 binding site (E74-BS) (Janknecht and Nordheim 1992; Rao and Reddy 1992). This demonstrates that Elk-1 displays two types of DNA sequence recognition, namely SRF-dependent binding to the SRE and direct binding to E74-BS. Both types of DNA binding require the entire ETS domain spanning amino acid positions 3–86. SRF-dependent SRE binding additionally requires sequences encompassing the B-region (Janknecht and Nordheim 1992; Fig. 5). This was determined with truncated Elk-1 proteins produced either in transfected HeLa cells (Janknecht and Nordheim 1992) or by coupled in vitro transcription/translation (Fig. 5). It therefore appears that the subset of Ets-proteins with the ETS domain at the N-terminus contain additional domains enabling protein-assisted DNA recognition for ternary complex formation. The relationship between these proteins and the original HeLa-derived $p62^{TCF}$ activity (Shaw et al. 1989b) remains to be established.

2.4 Ternary Complexes in Transcriptional Control: Evolutionarily Conserved Modules

The homology in the core domain between mammalian SRF and transcription factors in yeast and plants (Norman et al. 1988; for review, Treisman and Ammerer 1992) suggested that mechanisms for DNA recognition and dimerization are phylogenetically conserved, as might be the concept of ternary complex formation associated with transcriptional control. Indeed, there are striking similarities between SRF-$p62^{TCF}$ interactions and those of yeast MCM1 with a variety of transcriptional regulatory proteins, including α2, α1, STE12 and SFF (Dolan and Fields 1991; Treisman and Ammerer 1992). We therefore investigated the functional validity of this apparent similarity, as well as its structural basis (Hipskind et al. 1991; Mueller and Nordheim 1991; see also Primig et al. 1991). These findings can be summarized as follows: MCM1 and SRF not only bind to the same DNA sequence but also form ternary complexes. MCM1 is able to recruit STE12, $p62^{TCF}$ and Elk-1 into complexes, while SRF can interact with $p62^{TCF}$ and Elk-1, but not with STE12. In contrast, the other yeast MADS box protein ARG80 is unable to form ternary complexes with any of the abovementioned proteins. However, by exchanging three residues of ARG80-core with those of corresponding positions in SRF (residues 198, 200, and 203), the ARG80-core protein acquires the ability to recruit $p62^{TCF}$ or Elk-1 into a ternary complex. Similarly, the replacement of four SRF amino acids by the corresponding MCM1-derived residues (amino acids 73, 75, 77 and 78) now allows SRF-core to interact with STE12. This identified specific amino acids in MCM1 and SRF, critical

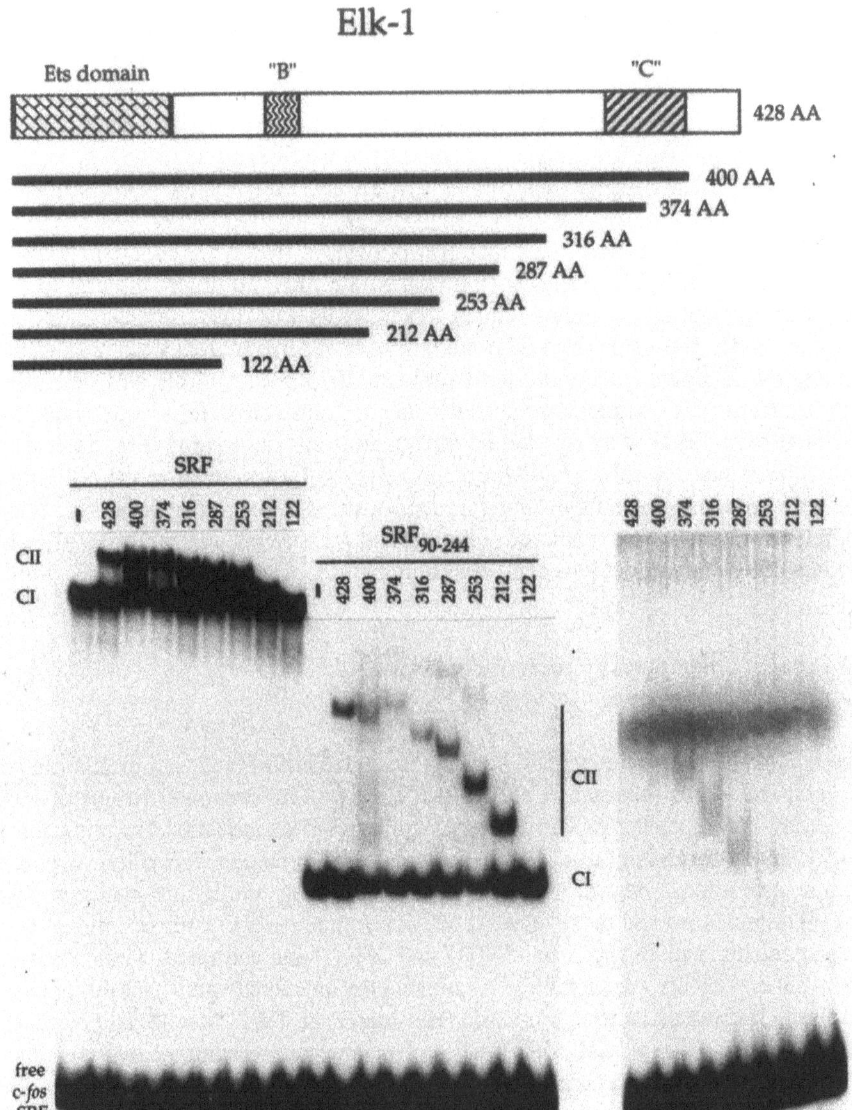
Elk-1
Ets domain
"B"
"C"
428 AA
400 AA
374 AA
316 AA
287 AA
253 AA
212 AA
122 AA
SRF
- 428 400 374 316 287 253 212 122
SRF90-244
- 428 400 374 316 287 253 212 122
428 400 374 316 287 253 212 122
CII
CI
CII
CI
free
c-fos
SRE

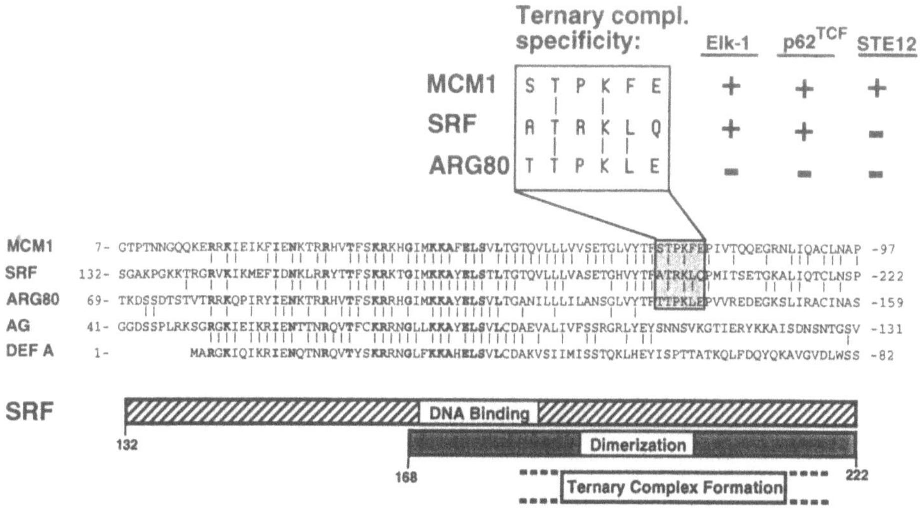

Fig. 6. Alignment of MADS box proteins and identification of sequences critical for ternary complex formation. The various proteins shown are identified in the text. The *vertical lines* indicate common amino acids, and positions shown in *boldface* are identical in all 5 proteins. The protein regions in *MCM1, SRF*, and *ARG80* responsible for ternary complex formation with $p62^{TCF}$, Elk-1 or STE12 could be mapped to the C-terminal half of the core element (see also Fig. 3) and a short segment of 6 amino acids, shown in the *box*, was identified as critical in determining the specificity for ternary complex formation (Mueller and Nordheim 1991)

for ternary complex formation, that map to equivalent positions within the C-terminal half of the shared domains (Fig. 6). It therefore appears that the structural basis for specific protein-protein interactions has been conserved in evolution within a class of transcription factors that exert control of gene activity by utilizing ternary complex structures (Mueller and Nordheim 1991).

This phenomenon is also observed in viral systems. The herpes simplex virus HSV, for example, regulates early events of its life cycle by employing two types of ternary complexes. First, the octamer sequence TAATGARAT interacts with the cellular Oct1 and the viral VP16 proteins (O'Hare and Goding 1988) and, second, the purine-rich CGGAAR sequence is recognized by the Ets-protein GABPα, which

Fig. 5. Mapping of functional domains within Elk-1. The diagram at the *top* represents the Elk-1 protein and indicates the 3 regions of homology to SAP-1. The *lines below* this represent the truncated versions of Elk-1 synthesized by coupled in vitro transcription/translation from Elk-1 templates linearized with the appropriate restriction enzymes. The efficiency of translation was confirmed by analysis on protein gels. The Elk proteins were diluted and incubated together with a ^{32}P-labeled c-*fos* SRE probe and either SRF or a deleted version (SRF_{90-244}) spanning the core domain (Hipskind et al. 1991). Alternatively, they were tested for direct binding with the SRE probe alone using 30-fold more Elk protein than in the other reactions. The protein-DNA complexes were separated by polyacrylamide gel electrophoresis and visualized by autoradiography of the dried gel. *CI* indicates the binary complex of SRF or SRF_{90-244} with the SRE probe, and *CII* denotes the ternary complex including the corresponding Elk protein

forms a complex together with the GABPβ protein. This complex is dependent upon the DNA-binding ETS domain of GABPα plus an adjacent C-terminal segment of 37 amino acids (Thompson et al. 1991). Ternary complexes are also implicated in adenovirus early gene expression in that the E2F and E4 proteins interact at the E2F binding site within the E2 gene promoter (Hardy et al. 1989; Table 1).

3 Perspectives

The recent identification of cDNAs encoding SRE/SRF-dependent ternary complex factors (Hipskind et al. 1991; Dalton and Treisman 1992) should lead to rapid progress in our understanding of the functional involvement of these SRF accessory proteins in c-*fos* transcriptional control. More generally, this knowledge should provide insights into the mechanisms by which proliferative signals activate the immediate-early class of genes. The potential involvement of signal-activated kinases or phosphatases can now be tested directly on the transcription factors presumed to be targets for the signal. The identification of a family of SRE/SRF-dependent TCFs suggests the presence of additional such proteins yet undetected, that may participate in c-*fos* regulation in different cells and under different conditions. The fact that a variety of Ets proteins mediate their effect through forming ternary complexes (Elk-1, SAPs, GABPα, PU.1; Table 1) implies that protein-protein interaction is an important aspect of Ets protein function.

Acknowledgements. We appreciate the dedicated technical support from P. Delany and U. Wiedemann and gratefully acknowledge secretarial assistance by A. Borchert. Our work described in this article was funded by the DFG (No 120/7-1) and the Fonds der Chemischen Industrie.

References

Almendal JM, Sommer D, MacDonald-Bravo H, Burkhardt J, Perera J & Bravo R (1988) Complexity of the early genetic response to growth factors on mouse fibroblasts. Mol Cell Biol 8:2140–2148

Bender A & Sprague GJ (1987) MAT alpha 1 protein, a yeast transcription activator, binds synergistically with a second protein to a set of cell-type-specific genes. Cell 50:681–691

Bravo R (1990) Genes induced during G0/G1 transition in mouse fibroblasts. Semin Cancer Biol 1:37–46

Cohen DR & Curran T (1989) The structure and function of the fos proto-oncogene. Crit Rev Oncogenesis 1:65–88

Dalton S & Treisman R (1992) Characterization of SAP-1, a protein recruited by serum response factor to the c-*fos* serum response element. Cell 68:597–612

Dolan JW & Fields S (1991) Cell-type-specific transcription in yeast. Biochim Biophys Acta 1088:155–169

Errede B & Ammerer G (1989) STE12, a protein involved in cell-type-specific transcription and signal transduction in yeast, is part of protein-DNA complexes. Genes & Dev 3:1349–1361

Fisch TM, Prywes R & Roeder RG (1987) c-*fos* sequences necessary for basal expression and induction by epidermal growth factor, 12-0-tetradecanoyl phorbol-13-acetate, and the calcium ionophore. Mol Cell Biol 7:3490–3502

Gilman MZ, Wilson RN & Weinberg RA (1986) Multiple protein-binding sites in the 5' flanking region regulate c-*fos* expression. Mol Cell Biol 6:4305–4306

Graham R & Gilman M (1991) Distinct protein targets for signals acting at the c-*fos* serum response element. Science 251:189–192

Gutman A & Wasylyk B (1990) The collagenase gene promoter contains a TPA and oncogene-responsive unit encompassing the PEA2 and AP-1 binding sites. EMBO J 9:2241–2246

Hardy S, Engel DA & Shenk T (1989) An adenovirus early region 4 gene product is required for induction of the infection-specific form of cellular E2F activity. Genes Dev 3:1062–1074

Herrera RE, Shaw PE & Nordheim A (1989) Occupation of the c-*fos* serum response element in vivo by a multi-protein complex is unaltered by growth factor induction. Nature 340:68–70

Hipskind RA, Rao VN, Mueller CGF, Reddy EP & Nordheim A (1991) Ets-related protein Elk-1 is homologous to the c-*fos* regulatory factor p62TCF. Nature 354:531–534

Janknecht R, Hipskind RA, Houthaeve T, Nordheim A & Stunnenberg HG (1992) Identification of multiple SRF N-terminal phosphorylation sites affecting DNA binding properties. EMBO J 11:1045–1054

Janknecht R & Nordheim A (1992) Elk-1 protein domains required for direct and SRF-assisted DNA-binding. Nucl Acids Res 20:3317–3324

Karim FD, Urness LD, Thummel CS, Klemsz MJ, McKercher SR, Celada A, Van Beveren C, Maki RA, Gunther CV, Nye JA & Graves BJ (1990) The ETS-domain: a new DNA-binding motif that recognizes a purine-rich core DNA sequence. Genes Dev 4:1451–1453

Kelleher CA, Gontte C & Johnson AD (1988) The yeast cell type-specific repressor α2 acts co-operatively with a non-cell-type-specific protein. Cell 53:927–936

König H (1991) Cell-type specific multiprotein complex formation over the c-*fos* serum response element in vivo: ternary complex formation is not required for the induction of c-fos. Nucl Acids Res 19:3607–3611

König H, Ponta H, Rahmsdorf U, Büscher M, Schönthal A, Rahmsdorf HJ & Herrlich P (1989) Autoregulation of *fos*: the dyad symmetry element as the major target of repression. EMBO J 8:2559–2566

Leung S & Miyamoto NG (1989) Point mutational analysis of the human c-*fos* serum response factor binding site. Nucleic Acids Res 17:1177–1195

Lucibello FC, Ehlert F & Müller R (1991) Multiple interdependent regulatory sites in the mouse c-*fos* promoter determine basal level transcription: cell-type-specific effects. Nucleic Acids Res 19:3583–3591

Lydall D, Ammerer G & Nasmyth K (1991) A new role for MCM1 in yeast cell cycle regulation of SW15 transcription. Genes Dev 5:2405–2419

Malik RK, Roe MW & Blackshear PJ (1991) Epidermal growth factor and other mitogens induce binding of a protein complex to the c-*fos* serum response element in human astrocytoma and other cells. J Biol Chem 266:8576–8582

Manak JR, De BN, Kris RM & Prywes R (1990) Casein kinase II enhances the DNA-binding activity of serum response factor. Genes Dev 4:955–967

Manak JR & Prywes R (1991) Mutation of serum response factor phosphorylation sites and the mechanism by which its DNA-binding activity is increased by casein kinase II. Mol Cell Biol 11:3652–3659

Marais RM, Hsuan JJ, McGuigan CM, Wynne J & Treisman R (1992) Casein kinase II phosphorylation increases the rate of serum response factor-binding site exchange. EMBO J 11:97–105

Metz R & Ziff E (1992) The helix-loop-helix protein rE12 and the C/EBP-related factor rNFIL-6 bind to neighboring sites within the c-*fos* serum response element. Oncogene 6:2165–2178

Misra RP, Rivera VM, Wang JM, Fan PD & Greenberg ME (1991) The serum response factor is extensively modified by phosphorylation following its synthesis in serum-stimulated fibroblasts. Mol Cell Biol 11:4545–4554

Mueller CGF & Nordheim A (1991) A protein domain conserved between yeast MCM1 and human SRF directs ternary complex formation. EMBO J 10:4219–4229

Norman C, Runswick M, Pollock R & Treisman R (1988) Isolation and properties of cDNA clones encoding SRF, a transcription factor that binds to the c-*fos* serum response element. Cell 55:989–1003

O'Hare P & Goding CR (1988) Herpes simplex virus regulatory elements and the immunoglobulin octamer domain bind a common factor and are both targets for virion transactivation. Cell 52:435–445

Pollock R & Treisman R (1991) Human SRF-related proteins: DNA-binding properties and potential regulatory targets. Genes Dev 5:2327–2341

Pongubala JMR, Nagulapalli S, Klemsz MJ, McKercher SR, Maki RA & Atchison ML (1992) PU.1 recruits a second nuclear factor to a site important for immunoglobulin κ 3' enhancer activity. Mol Cell Biol 12:368–378

Primig M, Winkler H & Ammerer G (1991) The DNA-binding and oligomerisation domain of MCM1 is sufficient for its interaction with other regulatory proteins. EMBO J 10:4209–4218

Prywes R & Roeder RG (1987) Purification of the c-*fos* enhancer-binding protein. Mol Cell Biol 7:3482–3489

Prywes R, Dutta A, Cromlish JA & Roeder RG (1988) Phosphorylation of serum response factor, a factor that binds to the serum response element of the c-FOS enhancer. Proc Natl Acad Sci USA 85:7206–7210

Rao VN & Reddy ESP (1992) A divergent ets-related protein, Elk-1, recognizes similar c-ets-1 proto-oncogene target sequences and acts as a transcriptional activator. Oncogen 7:65–70

Rao VN, Huebner K, Isobe M, Ar-Rushdi A, Croce CM & Reddy ESP (1989) *elk*, Tissue-specific *ets*-related genes on chromosomes X and 14 near translocation breakpoints. Science 244:66–70

Rivera VM & Greenberg ME (1990) Growth factor-induced gene expression: the ups and downs of c-*fos* regulation. New Biol 2:751–758

Rivera VM, Sheng M & Greenberg ME (1990) The inner core of the serum response element mediates both the rapid induction and subsequent repression of c-*fos* transcription following serum stimulation. Genes Dev 4:255–268

Runkel L, Shaw PE, Herrera RE, Hipskind RA & Nordheim A (1991) Multiple basal promoter elements determine the level of human c-*fos* transcription. Mol Cell Biol 11:1270–1280

Schalasta G & Doppler C (1990) Inhibition of c-*fos* transcription and phosphorylation of the serum response factor by an inhibitor of phospholipase C-type reactions. Mol Cell Biol 10:5558–5561

Schröter H, Shaw PE & Nordheim A (1987) Purification of intercalator-released p67, a polypeptide that interacts specifically with the c-*fos* serum response element. Nucl Acids Res 15:10145–10158

Schröter H, Mueller CG, Meese K & Nordheim A (1990) Synergism in ternary complex formation between the dimeric glycoprotein p67SRF, polypeptide p62TCF and the c-*fos* serum response element. EMBO J 9:1123–1130

Shaw PE (1990) Multicomponent transcription factor complexes: the exception or the rule? New Biol 2:111–118

Shaw PE (1992) Ternary complex formation over the c-fos serum reponse element: p62^{TCF} exhibits dual component specificity with contacts to DNA and an extended structure in the DNA-binding domain of p67^{SRF}. EMBO J 11:3011–3019

Shaw PE, Frasch S & Nordheim A (1989a) Repression of c-fos transcription is mediated through p67^{SRF} bound to the SRE. EMBO J 8:2567–2574

Shaw PE, Schröter H & Nordheim A (1989b) The ability of a ternary complex to form over the serum response element correlated with serum inducibility of the human c-*fos* promoter. Cell 56:563–572

Sommer H, Beltrán J-P, Huijser P, Pape H, Lönnig W-E, Saedler H & Schwarz-Sommer Z (1990) Deficiens, a homeotic gene involved in the control of flower morphogenesis in *Antirrhinum majus*: the protein shows homology to transcription factors. EMBO J 9:605–613

Subramaniam M, Schmidt LJ, Crutchfield III, CE & Getz MJ (1989) Negative regulation of serum-responsive enhancer elements. Nature 340:64–66

Thompson CC, Brown TA & McKnight SL (1991) Convergence of Ets- and notch-related structural motifs in a heteromeric DNA binding complex. Science 253:762–768

Treisman R (1985) Transient accumulation of c-fos RNA following serum stimulation requires a conserved 5' element and c-*fos* 3' sequences. Cell 42:889–902

Treisman R (1986) Identification of a protein-binding site that mediates transcriptional response of the c-*fos* gene to serum factors. Cell 46:567–574

Treisman R (1987) Identification and purification of a polypeptide that binds to the c-*fos* serum response element. EMBO J 6:2711–2717

Treisman R (1990) The SRE: a growth-factor responsive transcriptional regulator. Semin Cancer Biol 1:47–58

Treisman R & Ammerer G (1992) The SRF and MCM1 transcription factors. Curr Opinion Genet Dev 2:221–226

Velcich A & Ziff EB (1990) Functional analysis of an isolated fos promoter element with AP-1 site homology reveals cell type-specific transcriptional properties. Mol Cell Biol 10:6273–6282

Yanofsky MF, Ma H, Bowman JL, Drews GN, Feldmann KA & Meyerowitz EM (1990) The protein encoded by the *Arabidopsis* homeotic gene agamous resembles transcription factors. Nature 346:35–39

Function and Regulation of DNA Replication Genes in Budding Yeast

P. Plevani[1] and G. Lucchini[1,2]

1 Introduction

Eukaryotic DNA replication is confined to the S phase of the cell cycle and is highly regulated. A prerequisite to address questions related to the control of DNA replication in eukaryotic organisms rests on the comprehensive knowledge of the proteins involved in this process and the isolation of the corresponding genes. This goal has been partially achieved in the yeast *Saccharomyces cerevisiae* by a combination of biochemical and genetic approaches. Several enzymes related to the replication of the yeast genome have been purified and their corresponding genes have been isolated by complementation of known temperature-sensitive mutants defective in DNA replication or by reverse genetic strategies. Cloning of DNA replication genes offered the opportunity to characterize their translational products and to verify whether their expression is under cell cycle control.

2 Results and Discussion

2.1 Genes and Proteins Required for Yeast DNA Replication

The potential for genetic manipulation is certainly the most attractive feature of *S. cerevisiae* as a model for studying eukaryotic DNA replication. However, because of this general assumption, it has been often over-looked that a number of replication proteins, analogous to those found by using the SV40 in vitro DNA replication system (Stillman 1989), have been purified and characterized from yeast cells in several laboratories (for a review see Campbell and Newlon 1991).

As it is summarized in Fig. 1, three essential DNA polymerases (called polα, polδ and polε, according to the revised nomenclature proposed by Burgers et al. 1991) are likely to be involved in replication of the yeast nuclear genome, and the catalytic DNA polymerase polypeptides are found associated in multi-protein complexes (Plevani et al. 1988; Boulet et al. 1989; Morrison et al. 1990). However, with the exception of the two DNA primase subunits (p48 and p58), which are found tightly associated with polα (Francesconi et al. 1991), the function of the other polypeptides is

[1] Dipartimento di Genetica e di Biologia dei Microrganismi, Università degli Studi di Milano, Via Celoria 26, 20133 Milano, Italy.
[2] Istituto di Genetica, Università di Sassari, Via Mancini 5, 07100 Sassari, Italy.

43. Colloquium Mosbach 1992
DNA Replication and the Cell Cycle

DNA replication proteins in *Saccharomyces cerevisiae*

	SUBUNIT (kDa)	CLONED GENE (deduced M.W.)	ESSENTIALITY	CELL CYCLE CONTROL	FUNCTION
DNA polymerase α-primase	180	*POL 1* (167)	+	+	Discontinuous DNA synthesis on the lagging strand
	86	*POL 12* (79)	+	+	
	58	*PRI 2* (62)	+	+	
	48	*PRI 1* (48)	+	+	
DNA polymerase δ	125	*POL 3* (124)	+	+	Leading strand DNA synthesis ?
	55				
DNA polymerase ε	200	*POL 2* (256)	+	+	Leading strand DNA synthesis ?
	80	*DPB 2* (79)	+	+	
	34	*DPB 3* (23)	-	+	
	29				
PCNA	26	*POL 30* (29)	+	+	Auxiliary factor in pol δ elongation
RF-A	69	*RFA 1* (70)	+	+	Single-stranded DNA binding protein
	36	*RFA 2* (30)	+	+	
	13	*RFA 3* (14)	+	+	
RF-C	130				Auxiliary factor in pol δ (pol ε) elongation on natural DNA template
	86				
	41				
	40				
	37				
	27				
DNA ligase	87	*CDC 9* (85)	+	+	Joining of Okazaki fragments
DNA topoisomerase I	90	*TOP 1* (90)	-	NT	Relieves positive and negative superhelicity
DNA topoisomerase II	150	*TOP 2* (164)	+	+	Relieves positive and negative superhelicity

Fig. 1. DNA replication proteins in *Saccharomyces cerevisiae*. The listed proteins and their corresponding genes are directly involved in budding yeast DNA replication (for complete references, see Campbell and Newlon 1991). Essentiality of the genes has been tested by analyzing spore viability of heterozygous strains containing one deleted copy of the gene. Cell cycle control is related to the observation that mRNA level of the indicated genes fluctuates during cell cycle, reaching a peak at the G1/S phase boundary; *NT* not tested

still unknown. While the role of polα in discontinuous DNA synthesis on the lagging strand is well documented, the biochemical properties of both polδ and polε are well suited for a role in leading strand replication (Wang 1991). To clarify their reciprocal role in DNA replication and in other aspects of DNA metabolism, both the detailed biochemical and physiological characterization of *pol2* and *pol3* temperature-sensitive mutants (Morrison et al. 1991; Araki et al. 1992) and careful in vitro reconstitution studies with purified components (Tsurimoto et al. 1990) will be essential.

The Proliferating Cell Nuclear Antigen (PCNA) and the Replication Factor-C (RF-C) are accessory proteins that interact with polδ, and possibly with polε, increasing their processivity and activity on certain DNA templates (So and Downey 1992; Burgers 1991). While it has been shown that PCNA is essential for yeast cell viability (Bauer and Burgers 1990), the in vivo role of RF-C in yeast DNA replication is not defined since none of the genes encoding the RF-C subunits have yet been cloned. Replication Factor-A (RF-A) is a three-subunit, single stranded DNA binding protein which, by comparison with its role in SV40 DNA replication, is inferred to be required for DNA unwinding at yeast DNA replication origins (Brill and Stillman 1989; 1991).

With the exception of the *CDC9* and *POL3* genes, that have been isolated by complementation of available mutants defective in DNA replication and cell cycle progression at the non-permissive temperature (Hartwell et al. 1973), all the other yeast replication genes listed in Fig. 1 have been cloned by reverse genetic strategies. This experimental approach requires a purified protein that can be used to make specific antibodies, or whose known partial amino acid sequence allows the preparation of oligonucleotide probes specific for the gene. These reagents can then be used to clone the gene from an appropriate library. After in vitro mutagenesis or disruption of the gene, the modified copy can be reintroduced into the yeast genome to verify whether its corresponding gene product is a bona fide constituent of the replication apparatus and is essential for yeast cell viability (Guthrie and Fink 1991).

It is quite remarkable that the general polypeptide structure and the biological function of the DNA replication proteins listed in Fig. 1 have been conserved from yeast to mammalian cells (Burgers 1989). Moreover, when the genes encoding the same replication protein in unicellular or multicellular eukaryotes have been isolated and sequenced, comparison of the deduced primary sequence of the corresponding gene product revealed a level of amino acid identity ranging from 29 to 50% (polα = 33%; large primase subunit = 40%; small primase subunit = 35%; polδ = 45%; PCNA = 35%; p36 RF-A subunit = 29%; DNA ligase = 50%; DNA topoisomerase I = 42%; DNA topoisomerase II = 41%). In spite of these structural and functional similarities, unexpectedly, human polα and p34 RF-A subunit or mouse primase subunits can not substitute for the homologous yeast proteins in vivo (Brill and Stillman 1991; Santocanale et al. 1992; Francesconi et al. 1992). On the contrary, expression of human DNA ligase or topoisomerase I and II genes can complement the defect of mutations in the corresponding yeast genes (Wang 1987; Barnes et al. 1990). This discrepancy can be interpreted by assuming that the polα-primase complex and RF-A play an essential role in mediating species specific protein-protein interactions with other components involved in DNA replication, or they might also interact with some specific initiation factor(s) that function prior to the onset of DNA synthesis. On the other hand, DNA ligase is certainly involved in the maturation of the replication products and DNA topoisomerases relieve the torsional stress on the template DNA molecule, but their in vivo function seems not to be constrained by species specific interactions with the replication machinery.

2.2 Cell Cycle-Dependent Expression of Yeast DNA Synthesis Genes

The temporal pattern of expression of the yeast DNA synthesis genes listed in Fig. 1 has been tested by examining the level of the corresponding mRNAs in yeast cultures undergoing synchronous mitotic cell cycles. From this extensive analysis, the genes encoding proteins working at the replication fork appear to be preferentially expressed at the G1/S phase boundary, clearly preceeding the peak of histone gene transcription (reviewed in Campbell and Newlon 1991). Other genes, encoding enzymes involved in dNTPs biosynthesis (such as *CDC8*, thymidylate kinase, *CDC21*, thymidylate synthase, and *RNR1* and *RNR2*, ribonucleotide reductase subunits) are also transiently expressed with identical kinetics, suggesting that many, if not all, of

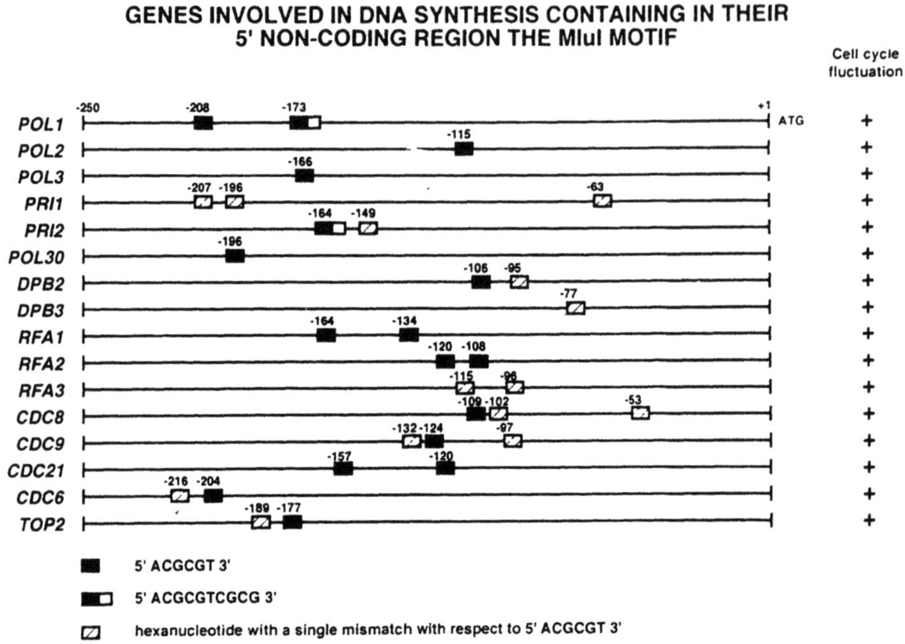

Fig. 2. Genes involved in DNA synthesis containing in their 5' non-coding region the MluI motif. The transcripts of all the listed genes fluctuate with the same kinetics during the mitotic cell cycle (for complete references, see Campbell and Newlon 1991). The positions of the MluI motifs (5'ACGCGT3') or of a hexanucleotide containing a single mismatch with respect to the MluI sequence are indicated

the genes required for DNA synthesis might be coordinately expressed in the same defined interval of the yeast cell cycle.

As it is shown in Fig. 2, a DNA sequence survey of the sequenced 5' non-coding regions of this gene family revealed, in almost all cases, the presence of at least one copy of the MluI motif (5'ACGCGT3') in the regions between –250 and –50 with respect to the translational initiation codon, and this finding led to the initial proposal that the MluI sequence might, at least in part, define a cell cycle-specific upstream activating sequence (McIntosh et al. 1988; Pizzagalli et al. 1988). This assumption has been recently substantiated by promoter deletion analysis of the *CDC21* and *POL1* genes, and by the observation that oligonucleotides corresponding to *CDC21*, *CDC9* and *POL1* promoter regions containing the MluI motif are sufficient to convey periodic fluctuation of a *CYC1-lacZ* reporter gene (Gordon and Campbell 1991; Lowndes et al. 1991; McIntosh et al. 1991; Pizzagalli et al. 1992). Interestingly, the MluI cell cycle box (MCB) is the target of specific protein factor(s), called DSC1, whose binding to the MCB appears to fluctuate during cell cycle as is expected for a *trans*-acting factor required for the periodic expression of the DNA synthesis genes (Lowndes et al. 1991).

By using a genetic approach, we have searched for *trans*-acting mutations affecting *POL1* gene expression. After mutagenesis of a haploid yeast strain carrying, on a multi-copy plasmid, a *POL1-lacZ* fusion, whose transcript fluctuates during cell cycle with the same kinetic of *POL1* mRNA, we have screened for mutants showing increased or decreased β-galactosidase activity. Clones carrying chromosomal mutations responsible for this phenotype were identified by curing the original mutant of the multi-copy plasmid, followed by integration of the same *POL1-lacZ* fusion in single copy at the *URA3* locus in each putative mutant, and by verifying the effect of the mutation on the expression of the integrated fusion. Genetic analysis has, so far, allowed the characterization of six independent, recessive, single-gene, chromosomal mutations, two of which confer a decreased level of *POL1-lacZ* expression, while the remaining four cause an increase in β-galactosidase production. Since none of them belong to the same complementation group, they should identify, respectively, two positive and four negative regulators. Furthermore, all the mutations are associated with a temperature-sensitive phenotype, suggesting that they might affect functions essential for cell viability. Two of the mutations identifying putative negative regulators seem to be highly specific, because they affect the expression of a *CYC1-lacZ* fusion only when the *CYC1* UAS is substituted with *POL1* promoter regions containing the MluI motif. Since this sequence has been shown to act as a positive *cis*-acting regulatory element, these negative regulators might function upstream of the putative positive *trans*-acting factor(s) discussed above. Therefore, *POL1* cell cycle control might be mediated through a cascade of regulatory events.

2.3 DNA Replication and the Yeast Cell Cycle

Progression through the cell division cycle requires a number of highly coordinated controls and, in recent years, much progress has been made in identifying and characterizing components of the molecular machinery that drive the cell cycle in apparently all eukaryotes (for recent reviews see Hartwell and Weinert 1989; Nurse 1990; Enoch and Nurse 1991; Wittenberg and Reed 1991). In this respect, DNA replication is certainly a central event of the cell cycle and must be properly activated and controlled. For example, the firing of replication origins is temporally regulated, but it must be executed only once per cycle (Laskey et al. 1989; Fangman and Brewer 1991); moreover, the onset of mitosis is tightly coupled to completion of S phase, since any attempt to separate incompletely replicated chromosomes would be deleterious for the cell (Enoch and Nurse 1991).

The genetic, biochemical and physiological characterization of yeast mutants defective in the progression through the cell cycle (*cdc*) has been essential to reveal a fundamental aspect of cell cycle control, namely that the execution of late events is dependent on the completion of early events. It was proposed that some control mechanisms should monitor S phase and inhibit M phase until S has been completed (Hartwell and Weinert 1989; Enoch and Nurse 1991). This model predicted that mutations abolishing the dependence of M phase on S phase should identify components of such control mechanisms. Indeed, such mutations have been identified and the corresponding gene products appear to be essential to "check" the completion of DNA replication and to drive the entry of the cells into mitosis.

Perhaps one of the most important issues in the comprehension of the regulation of cell cycle is to determine how many such "checkpoints" there are in the eukaryotic cell cycle. Their presence can be resolved by genetic analysis evaluating if a loss-of-function mutation in one gene can relieve the dependence of certain cell cycle events. The existance of such "feedback control" is taken as indicative of regulatory mechanisms that make passage through checkpoints dependent on the successful completion of previous events. By using this approach, the existence of a feedback control that makes the exit from mitosis dependent on the completion of spindle assembly has been recently found in *S. cerevisiae* (Li and Murray 1991).

However, it is well known that the life cycle of the budding yeast is principally regulated in G1 (Pringle and Hartwell 1981). Once a yeast cell has passed through a point in G1 called START, it becomes committed to complete the mitotic cell cycle. Among the genes required for START, the most studied is *CDC28* which, like its counterpart (*CDC2*) in *Schizosaccharomyces pombe* and multicellular eukaryotes, encodes a protein kinase required not only for the passage from G1 to S, but also for the execution of mitosis. This dual role of the CDC28-kinase is likely to be achieved by differential association with regulatory subunits, known, respectively, as G1 and G2 cyclins (Fig. 3; Wittenberg and Reed 1991; Surana et al. 1991). Several groups have simultaneously discovered that the budding yeast G1 cyclins cln1 and cln2 are engaged in a positive feedback loop operating at the level of transcription of their own genes (reviewed in North 1991). As we have discussed in the previous section, discontinuous gene expression seems to be a general feature of the yeast cell cycle. At least four classes of genes are periodically transcribed with different kinetics: the *CLN1* and *CLN2* cyclin genes are maximally expressed in G1, the DNA replication gene family at the G1/S phase transition, the histone genes in middle S and the G2 cyclin genes in S/G2 (Fig. 3).

The cell cycle-dependent expression of these classes of genes can be correlated to the requirement for protein synthesis. Limited protein synthesis extends the length of G1 phase, and this observation has been connected to the accumulation of labile protein(s) (Shilo et al. 1979). The finding that the accumulation of G1 cyclins likely controls the execution of START indicates that the positive feedback loop controlling

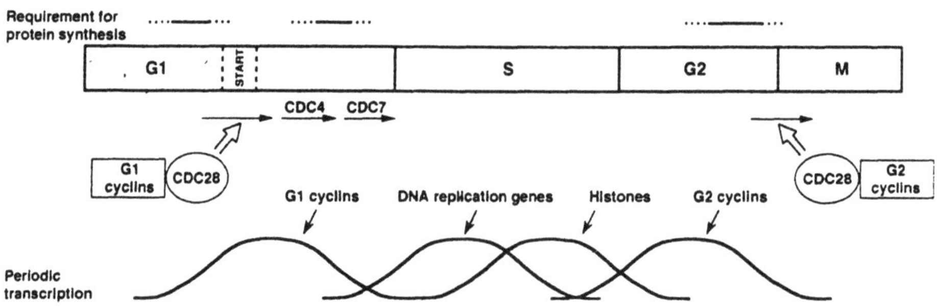

Fig. 3. A cell cycle scheme for periodically expressed genes. The kinetics of fluctuation of the four major classes of transcripts showing periodic cell cycle accumulation and the intervals in which protein synthesis is required for further cell cycle progression are superimposed to the four intervals generally used to characterize the eukaryotic cell cycle (see text for details)

the level of G1 cyclin mRNAs is crucial for determining the cell cycle transition. The temporal requirement for protein synthesis has been mapped relative to a number of *cdc* mutants, and it was found that there are two requirements for protein synthesis in G1: one for the completion of START, and the other after the *CDC4*-dependent step, but before the *CDC7*-dependent step (Burke and Church 1991). Interestingly, transcript fluctuation of the MluI gene family is dependent upon completion of START, being blocked in cdc28 mutants, while it is independent of the *CDC4*-mediated event (Johnston et al. 1987; Andrews and Herskowitz 1990). In contrast, the expression of the histone genes requires the activity of the *CDC4* gene product. The last requirement for protein synthesis during cell cycle is to progress from the end of DNA synthesis to nuclear division and cytokinesis, and probably it reflects the necessity to accumulate G2 cyclins for the onset of mitosis.

From these observations, a link between cell cycle-dependent transcription and the requirement for protein synthesis can be easily imagined to explain the temporal expression of genes encoding rate-limiting regulators of cell cycle progression, such as G1 or G2 cyclins. Transient transcription at a specific interval, in conjunction with a burst of protein synthesis followed by rapid protein turnover, will provide the molecular signal to drive a specific cell cycle transition event. A functional role for cell cycle regulation of genes whose products perform house-keeping functions and are not labile proteins, such as histones or DNA replication enzymes, is more difficult to interpret. Since many genes appear to be co-regulated by common transcription factor(s), this might provide a simple target to control cell cycle mechanisms, such as DNA replication, that are carried out by the regulated genes. However, since the level and activity of the replication enzymes, as well as their posttranslational modifications, have not been yet carefully analyzed during the cell cycle, regulation at the post-transcriptional level might also play a critical role. Constitutive overexpression of individual DNA synthesis genes does not appear to affect cell viability or growth rate to a significant extent (Santocanale et al. 1992). However, as previously discussed, we have identified *trans*-acting mutations affecting *POL1* expression and they all show an associate temperature-sensitive phenotype, as is usually observed for mutations in genes essential for cell viability. It will be interesting to verify whether this phenotype is related to the simultaneous impairment of cell cycle control on a number of genes involved in DNA replication. Similarly, it will be important to evaluate whether the effect of overexpression of the putative regulator DSC1 will affect the timing of entry into S phase. The final role for transient gene expression in yeast might be related to the inherent physiology of the yeast cell. This unicellular, totipotent eukaryotic organism might have evolved this sophisticated temporal control as a means to efficiently utilize cellular resources by limiting RNA and subsequent protein synthesis to the cell cycle phase(s) during which the corresponding gene product is effectively utilized.

Acknowledgments. Work in our laboratory has been partially supported by grants from Target Projects Biotechnology and Bioinstrumentation and Genetic Engineering C.N.R., Italy and by grant N° SC1-0479-C (A) from the European Economic Community. We wish to thank all the members of our laboratory for experimental work and constructive discussions.

References

Andrews BJ & Herskowitz I (1990) Regulation of cell cycle dependent gene expression in yeast. J Biol Chem 265:14057–14060

Araki H, Ropp P, Johnson AL, Johnston LH, Morrison A & Sugino A (1992) DNA polymerase II, the probable homolog of mammalian DNA pole, replicates chromosomal DNA in the yeast *Saccharomyces cerevisiae*. EMBO J 11:733–740

Barnes DE, Johnston LH, Kodama KI, Tomkinson AE, Lasko DD & Lindahl T (1990) Human DNA ligase I cDNA: Cloning and functional expression in *Saccharomyces cerevisiae*. Proc Natl Acad Sci USA 87:6679–6683

Bauer GA & Burgers PMJ (1990) Molecular cloning, structure and expression of the yeast proliferating cell nuclear antigen. Nucl Acids Res 18:261–265

Boulet A, Simon M, Faye G, Bauer GA & Burgers PMJ (1989) Structure and function of the *Saccharomyces cerevisiae CDC2* gene encoding the large subunit of DNA polymerase III. EMBO J 8:1849–1854

Brill SJ & Stillman B (1989) Yeast replication factor A functions in the unwinding of the SV40 origin of DNA replication. Nature 342:92–95

Brill SJ & Stillman B (1991) Replication factor-A from *Saccharomyces cerevisiae* is encoded by three essential genes coordinately expressed at S phase. Genes Dev 5:1589–1600

Burgers PMJ (1989) Eukaryotic DNA polymerases α and δ: Conserved properties and interactions from yeast to mammalian cells. Progress in Nucl Acids Res 37:235–280

Burgers PMJ (1991) *Saccharomyces cerevisiae* replication factor C. J Biol Chem 266:22698–22706

Burgers PMJ, Bambara RA, Campbell JL, Chang LMS, Downey KM, Hübscher U, Lee MYWT, Linn SM, So AG & Spadari S (1990) Revised nomenclature for eukaryotic DNA polymerases. Eur J Biochem 191:617–618

Burke DJ & Church D (1991) Protein synthesis requirements for nuclear division, cytokinesis and cell separation in *Saccharomyces cerevisiae*. Mol Cell Biol 11:3691–3698

Campbell JL & Newlon CS (1991) Chromosomal DNA replication. In: The molecular biology of the yeast *Saccharomyces cerevisiae:* genome dynamics, protein synthesis and energetics. Cold Spring Harbor Laboratory Press, New York, pp 41–146

Enoch T & Nurse P (1991) Coupling M phase and S phase: controls maintaining the dependence of mitosis on DNA replication. Cell 60:921–923

Fangman WL & Brewer BJ (1991) Activation of replication origins within yeast chromosomes Annu Rev Cell Biol 7:375–402

Francesconi S, Copeland W & Wang TSF (1992) The homologous DNA polymerase α from human and yeast displays species specificity in vivo. J Cell Biochem Supplement 16B pp 32

Francesconi S, Longhese MP, Piseri A, Santocanale C, Lucchini G & Plevani P (1991) Mutations in conserved yeast DNA primase domains impair DNA replication in vivo. Proc Natl Acad Sci USA 88:3877–3881

Gordon CB & Campbell JL (1991) A cell cycle-responsive transcriptional control element and a negative control element in the gene encoding DNA polymerase α in *Saccharomyces cerevisiae*. Proc Natl Acad Sci USA 88:6058–6062

Guthrie C & Fink GR (eds) (1991) Guide to yeast genetics and molecular biology. Methods in enzymology, vol 194. Academic Press, New York

Hartwell LH & Weinert TA (1989) Checkpoints: controls that ensure the order of cell cycle events. Science 246:629–634

Hartwell LH, Mortimer RK, Culotti J & Culotti M (1973) Genetic control of the cell division cycle in yeast. V. Genetic analysis of mutants. Genetics 74:267–286

Johnston LH, White JHM, Johnson AL, Lucchini G & Plevani P (1987) The yeast DNA polymerase I transcript is regulated in both the mitotic cell cycle and in meiosis and is also induced after DNA damage. Nucl Acids Res 15:5017–5030

Laskey RA, Fairman MP & Blow JJ (1989) S phase of the cell cycle. Science 246:609–614

Li R & Murray AW (1991) Feedback control of mitosis in budding yeast. Cell 66:519–531

Lowndes NF, Johnson AL & Johnston LH (1991) Coordination of expression of DNA synthesis genes in budding yeast by a cell cycle regulated *trans* factor. Nature 350:247–250

McIntosh EM, Akinson T, Storms RK & Smith M (1991) Characterization of a short, *cis*-acting DNA sequence which conveys cell cycle stage-dependent transcription in *Saccharomyces cerevisiae*. Mol Cell Biol 11:329–337

McIntosh EM, Ord RW & Storms RK (1988) Transcriptional regulation of the cell cycle-dependent thymidylate synthase gene of *Saccharomyces cerevisiae*. Mol Cell Biol 8:4616–4624

Morrison A, Araki H, Clark AB, Hamatake RK & Sugino A (1990) A third essential DNA polymerase in *Saccharomyces cerevisiae*. Cell 62: 1143–1151

Morrison A, Bell JB, Kunkel TA & Sugino A (1991) Eukaryotic DNA polymerase amino acid sequence required for 3' to 5' exonuclease activity. Proc Natl Acad Sci USA 88:9473–9477

North G (1991) Starting and stopping. Nature 351:604–605

Nurse P (1990) Universal control mechanism regulating onset of M-phase. Nature 344:503–508

Pizzagalli A, Piatti S, Derossi D, Gander I, Plevani P & Lucchini G (1992) Positive *cis*-acting regulatory sequences mediate proper control of *POL1* transcription in *Saccharomyces cerevisiae*. Curr Genet 21:183–189

Pizzagalli A, Valsasnini P, Plevani P & Lucchini G (1988) DNA polymerase I gene of *Saccharomyces cerevisiae:* nucleotide sequence, mapping of a temperature-sensitive mutation and protein homology with other DNA polymerases. Proc Natl Acad Sci USA 85:3772–3776

Plevani P, Foiani M, Muzi Falconi M, Pizzagalli A, Santocanale C, Francesconi S, Valsasnini P, Comedini A, Piatti S & Lucchini G (1988) The yeast DNA polymerase-primase complex: genes and proteins. Biochem Biophys Acta 951:268–273

Pringle JR & Hartwell LH (1981) The *Saccharomyces cerevisiae* cell cycle. In: Strathern SN, Jones EW & Broach JR (eds) The molecular biology of the yeast *Saccharomyces* – life cycle and inheritance. Cold Spring Harbor Laboratory Press, New York pp 97–142

Santocanale C, Locati F, Muzi Falconi M, Piseri A, Tseng BY, Lucchini G & Plevani P (1992) Overproduction and functional analysis of DNA primase subunits from yeast and mouse. Gene 113:199–205

Shilo B, Riddle GH & Pardee AB (1979) Protein turnover and cell cycle initiation in yeast. Exp Cell Res 123:221–227

So AG & Downey KM (1992) Eukaryotic DNA replication. Crit Rev Biochem Mol Biol 27:129–155

Stillman B (1989) Initiation of eukaryotic DNA replication in vitro. Annu Rev Cell Biol 5:197–245

Surana U, Robitsch H, Price C, Schuster T, Fitch I, Futcher AB & Nasmyth K (1991) The role of *CDC28* and cyclins during mitosis in the budding yeast *S. cerevisiae*. Cell 65:145–161

Tsurimoto T, Melendy T & Stillman B (1990) Sequential initiation of lagging and leading strand synthesis by two different polymerase complexes at the SV40 DNA replication origin. Nature 346:534–539

Wang JC (1987) Recent studies of DNA topoisomerases. Biochem Biophys Acta 909:1–9

Wang TSF (1991) Eukaryotic DNA polymerases. Annu Rev Biochem 60:513–552

Wittenberg C & Reed SI (1991) Control of gene expression and the yeast cell cycle. Crit Rev Eukaryotic Gene Expression 1:189–205

Functional Analysis of Recombinant mUBF, a Murine rDNA Transcription Initiation Factor Which is Phosphorylated by Casein Kinase II

R. Voit[1], H.-M. Jantzen[2], C. Fieger[1], A. Schnapp[1], A. Kuhn[1], and I. Grummt[1]

1 Introduction

Promoter recognition by any of the three classes of cellular DNA-dependent RNA polymerases requires the interaction of multiple proteins, both with the DNA and with one another, to generate the specificity and regulation of transcription initiation. Similar to transcription initiation by RNA polymerases II and III, initiation by RNA polymerase I (pol I) from the ribosomal gene promoter is a multistage process which requires the action of at least four initiation factors which in the mouse system have been called TIF-IA, TIF-IB, TIF-IC, and mUBF, respectively. These factors assemble in a sequential order at the rDNA promoter together with pol I to form a productive preinitiation complex (Schnapp and Grummt 1991). Both the growth-dependent factor TIF-IA (Buttgereit et al. 1985; Gokal et al. 1990; Schnapp et al. 1990a) and factor TIF-IC associate with the promoter via specific interaction with pol I. Promoter recognition is brought about by TIF-IB (Clos et al. 1986a, b; Schnapp et al. 1990b), a DNA binding protein which in different systems has been given different names, i. e. TFID (Tower et al. 1986; Mishima et al. 1982; Tanaka et al. 1990), SL1 (Learned et al. 1985; Smith et al. 1990), or Rib1 (McStay et al. 1991).

This selectivity factor forms a strong cooperative complex at the rDNA promoter together with another DNA-binding protein, designated UBF (Bell et al. 1988). Mammalian UBF consists of two polypeptides with molecular weights of 97 and 94 kDa (Jantzen et al. 1990). Recently, cDNAs for this factor have been cloned from human (Jantzen et al. 1990), rat (O'Mahony and Rothblum 1991) and *Xenopus laevis* (Bachvarov and Moss 1991; McStay et al. 1991).

In this communication we report the purification of mouse UBF (mUBF) to apparent homogeneity as well as the isolation and the characterization of cDNA clones encoding mUBF. The deduced primary structure reveals high homology to the human, rat and frog factor indicating the maintenance of essential structural elements during evolution. Furthermore, we present data demonstrating that UBF is a phosphoprotein, and that phosphorylation of UBF is accomplished by a cellular protein kinase which by several criteria closely resembles casein kinase II (CKII). Comparison of the phosphorylation pattern of full-length UBF and C-terminally truncated deletion mutant indicates that the hyperacidic tail of UBF represents the main target for

[1] Institute of Cell and Tumor Biology, German Cancer Research Center, Im Neuenheimer Feld 280, D-6900 Heidelberg, FRG.
[2] Howard Hughes Medical Institute, Department of Molecular and Cell Biology, 401 Barker Hall, Berkeley, CA 94720, USA.

43. Colloquium Mosbach 1992
DNA Replication and the Cell Cycle

modification. Both the degree of phosphorylation and the *trans*-activating properties of UBF correlate with cell growth (O'Mahony et al. 1992; Voit et al. 1992). Therefore, it is suggested that post-translational modification of this pol I transcription factor may play a central role in the chain of events by which extracellular signals are transferred into the nucleolus and link rRNA synthesis with cell growth.

2 Results and Discussion

2.1 Purification of UBF

Both the mammalian and the frog UBF have been shown to play a key role in basic rDNA transcription initiation (Bell et al. 1989, 1990; Jantzen et al. 1990; Pikaard et al. 1990a, b). As a necessary step to elucidate the functional role of UBF in ribosomal gene transcription, mouse UBF was purified to apparent homogeneity from extracts derived from Ehrlich ascites cells including chromatography on DEAE-Sepharose, Heparin-Ultrogel, MonoQ FPLC and a sequence-specific DNA affinity column. DNA binding activity was monitored by DNase footprint analysis. Since UBF alone has been shown to bind poorly to the mouse ribosomal gene promoter, and even saturating amounts of the factor relatively weakly protect the region between –88 and –108 (Bell et al. 1990; our own results), the footprinting studies shown in this paper have been performed with a spacer fragment containing three of the repetitive 140 bp enhancer elements (Grummt and Gross 1980; Kuhn et al. 1990).

As shown in Fig. 1A, DNase footprint analysis of individual fractions from the affinity column demonstrated that fractions 16 through 19 interact with the enhancer repeats. In agreement with previous studies (Pikaard et al. 1990), the most remarkable feature of the UBF footprints is the appearance of enhanced cleavage sites within the repeats (marked by arrows in Fig. 1A) which are flanked by protected regions. This pattern of interaction is very similar to that observed with xUBF, the analogous factor from *X. laevis* (Pikaard et al. 1989). Also the doublet of 97 and 94 kD polypeptides present in the active fractions (Fig. 1B) corresponds to the size of mammalian UBF polypeptides reported before (Bell et al. 1988, 1990; Pikaard et al. 1990a, b).

Next we tested the ability of the affinity column fractions to reconstitute transcriptional activity in the presence of the other three auxiliary factors (Fig. 1C). For this, the individual column fractions were incubated in transcription reactions containing the mouse rDNA template, pol I, TIF-IA, TIF-IB, and TIF-IC (Heilgenthal and Grummt 1991; Schnapp and Grummt 1991). The combination of those four protein fractions is not sufficient to support transcription initiation from the murine rDNA template (lane 1). Specific transcription requires addition of UBF which is contained either in the pool of MonoQ column fractions that have been applied to the affinity column (lane 2), or in the affinity column fractions 16 through 19 (lane 5–8). These are the same fractions which contained both the DNA binding activity and the 97 and 94 kDa polypeptides. Based on the co-purification of these proteins with specific

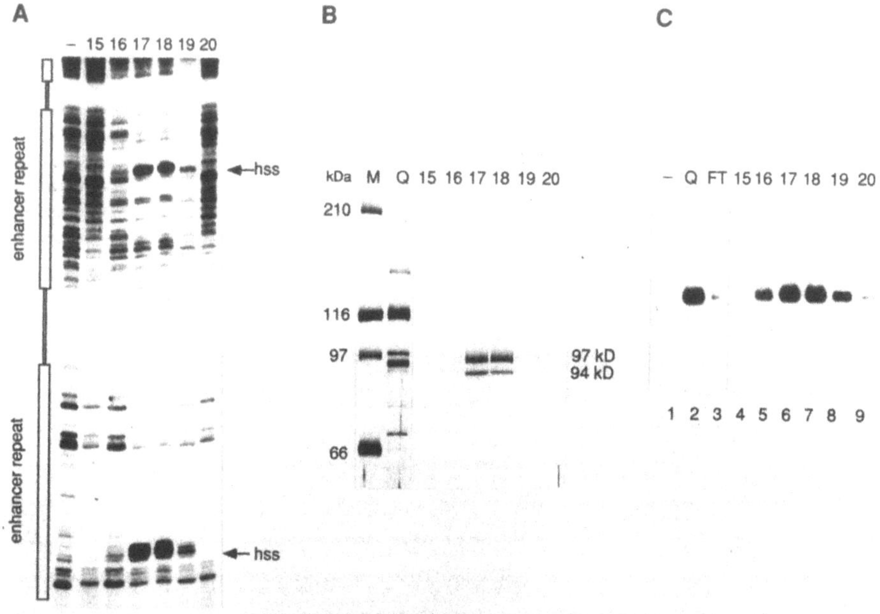

Fig. 1. Purification of mUBF. UBF was purified by chromatography on DEAE-Sepharose, Heparin-Ultrogel, Mono Q (FPLC) and a sequence-specific DNA affinity column as described before (Voit et al. 1992). Individual fractions eluted from the second DNA affinity column were analyzed. **A** DNase I footprinting. UBF-DNA interaction was monitored at a rDNA enhancer fragment extending from –640 to –168. All reactions contained 0.1–0.5 μg of poly(dAT) competitor and 3 fmoles of DNA probe. 12 μl fractions were preincubated for 5 min at 30 °C with the end-labeled probe, digested with 0.3–3 units of DNase I for 1 min and analyzed on denaturing 6% polyacrylamide sequencing gels. The footprint of UBF present in individual fractions from the affinity column is shown (*lanes 15–20*) and compared with the digestion pattern of the naked DNA (*lane*–). The cluster of T residues which flanks the individual repetitive elements are indicated by a ***thin line***, the enhancer repeats are represented by *boxes*, the hypersensitive sites (hss) caused by UBF binding are marked by *arrows*. **B** Silver-stained SDS-polyacrylamide gel of DNA affinity column fractions. 40 μl of the DNA affinity column fractions (*lanes 15–20*) were loaded onto a 7% polyacrylamide gel along with molecular weight size markers (*M*) and a pool of less purified UBF eluted from the MonoQ column (*Q*). The sizes of the molecular weight markers are indicated at the *left*. **C** Transcriptional activity of the fractions eluted from the DNA affinity column. The reconstituted transcription assay contained pol I, TIF-IA, TIF-IB, and TIF-IC at the amounts described before (Schnapp and Grummt 1991) and either no UBF (*lane–*), the pool of active UBF MonoQ fractions that have been applied to the DNA affinity column (*lane Q*), the flow through fraction of the affinity column (*FT*), or the individual fractions from the affinity column (*lanes 15–20*)

A

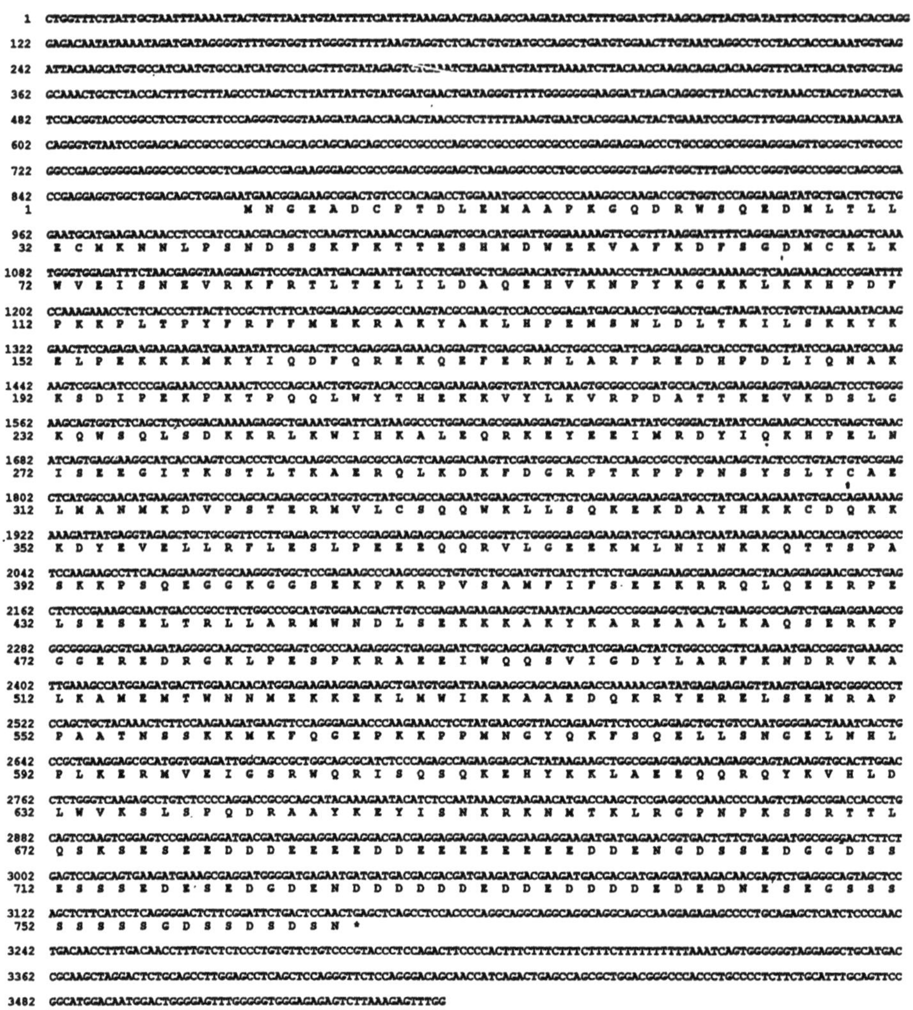

Fig. 2. Nucleotide and amino acid sequence of mUBF. **A** Nucleotide sequence of mUBF cDNA derived from two overlapping cDNA clones. cDNA from hUBF (Jantzen et al. 1990) was used to screen a mouse cDNA library. Two overlapping clones were isolated containing cDNA inserts of 1920 bp and 2090 bp, respectively. The amino acid sequence is shown below the nucleotide sequence in single letter code. **B** Comparison of the deduced amino acid sequences of UBF from mouse (mUBF), rat (O'Mahony and Rothblum 1991), human (Jantzen et al. 1990), and frog (Bachvarov and Moss 1991; McStay et al. 1991). The gap in the *X. laevis* (xUBF) sequence is indicated by *dashes*. Amino acid identities are marked by *dots*

binding and transcriptional activity, we conclude that we have purified functionally active mouse UBF (mUBF) to apparent homogeneity.

2.2 Cloning and Sequencing of mUBF cDNA

To clone cDNA encoding mUBF, a mouse cDNA library was screened with the human UBF cDNA probe. Two overlapping clones were isolated, subcloned into plasmid vectors, and sequenced (Fig. 2A). The overlapping cDNAs contain an open reading frame coding for 765 amino acids which is preceded by a 878 nt 5'-untranslated region. Neither the common polyadenylation signals nor a poly(A)-stretch are present at the 3' end suggesting that the 3' end of the mRNA is not contained within this cDNA. A sequence comparison of UBF from different species is shown in Fig. 2B. There is a remarkable homology of mUBF to its human and rat counterpart (Jantzen et al. 1990; O'Mahony and Rothblum 1991) differing only in 2% of the amino acid residues. Thus, 77% of the sequence is identical between UBF from mammals and amphibians (Bachvarov and Moss 1991; McStay et al. 1991). Furthermore, the amino acid exchanges found are for the most part conservative indicating that the structural organization of UBF is essentially the same in the species analyzed.

B

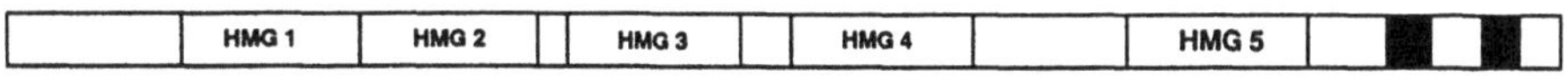


```
1   MNGEADCPTD LEMAAPKGQD RWSQEDMLTL LECMKNNLPS NDSSKFKTTE SHMDWEKVAF KDFSGDMCKL KWVEISNEVR KFRTLTELIL DAQEHVKNPY  mUBF
    .......... .......... .......... .......... .......... .......... .......... .......... .......... ..........  rUBF
    .......... .......... .......... .......... .......... .......... .......... .......... .......... ..........  hUBF
    ...A.GGD.Q GK.T...D.. Q......... .QT..TL..G Q.N....... ..L...HY.. S..RQ..... ..M....... .......... ..D...RH..  xUBF

101 KGKKLKKHPD FPKKPLTPYF RFFMEKRAKY AKLHPEMSNL DLTKILSKKY KELPEKKKMK YIQDFQREKQ EFERNLARFR EDHPDLIQNA KKSDIPEKPK  mUBF
    .......... .......... .......... .......... .......... .......... .......... .......... .......... ..........  rUBF
    .......... .......... .......... .......... .......... .......... .......... .......... .......... ..........  hUBF
    .........E .......... .......... .......... .......... .......... .......... D...M.K... .E....M..P ....V.....  xUBF

201 TPQQLWYTHE KKVYLKVRPD ATTKEVKDSL GKQWSQLSDK KRLKWIHKAL EQRKEYEEIM RDYIQKHPEL NISEEGITKS TLTKAERQLK DKFDGRPTKP  mUBF
    .......... .......... .......... .......... .......... .......... .......... .......... .......... ..........  rUBF
    .......... .......... .......... .......... .......... .......... .......... .......... .......... ..........  hUBF
    .......N.. R.....LHA. .S..DI..A. .......P.. .......... ....Q..GV. .E.M...... ..T....R.. .......... ..........  xUBF

301 PPNSYSLYCA ELMANMKDVP STERMVLCSQ QWKLLSQKEK DAYHKKCDQK KKDYEVELLR FLESLPEEEQ QRVLGEEKML NINKKQTTSP ASKKPSQEGG  mUBF
    .......... .......... .......... .......... .......... .......... .......... .......... .......... ..........  rUBF
    .......... .......... .......... .......... .......... .......... .......... .......... ......A... .....A....  hUBF
    ......M... .......... .......... R......... .......E.R ........M. ...N...... ....A....V GMKR.R.NT. ...MATEDAA  xUBF

401 KGGSEKPKRP VSAMFIFSEE KRRQLQEERP ELSESELTRL LARMWNDLSE KKKAKYKARE AALKAQSERK PGGEREDRGK LPESPKRAEE IWQQSVIGDY  mUBF
    .......... .......... .......... .......... .......... .......... .......... .......... .......... ..........  rUBF
    .......... .......... .......... .......... .......... .......... .......... ......E... .......... ..........  hUBF
    .VK.RS---- ---------- ---------- ---------- ---------- ---------- ------GQAD KKKAA.E.A. ...T..T... ..........  xUBF

501 LARFKNDRVK ALKAMEMTWN NMEKKEKLMW IKKAAEDQKR YERELSEMRA PPAATNSSKK MKFQGEPKKP PMNGYQKFSQ ELLSNGELNH LPLKERMVEI  mUBF
    .......... .......... .......... .......... .......... .......... .......... .......... .......... ..........  rUBF
    .......... .......... .......... .......... .......... .......... .......... .......... .......... ..........  hUBF
    ........A. ...V..A..L .......I.. .......... ......D..S T..P.TAG.. V..L.....A .......... .......... ..........  xUBF

601 GSRWQRISQS QKEHYKKLAE EQQRQYKVHL DLWVKSLSPQ DRAAYKEYIS NKRKNMTKLR GPNPKSSRTT LQSKSESEED DDEEEEDDEE EEEEEEDDEN  mUBF
    .......... .......... .......... .......... .......... .......... .......... .......... ...-.D..DD D.........  rUBF
    .......... .......... .......... .......... .......... .......... .......... .......... .E.-D....D .D........  hUBF
    ....H...PT ..DY...... D...L.RTQF .T.M.G..T. .......QNT ....ST..IA Q.SS..KLVI QSKSDDD.D. E.D-.DEEDD DDDDD..K.D  xUBF

701 GDSSEDGGDS SESSSEDESE DGDENDDDDD DEDDEDDDDE DEDNESEGSS SSSSSSGDSS DSDSN*                                        mUBF
    .......... .......... .....E.... ...DE..... .......... .......... ......                                        rUBF
    .......... .......... .....EE..E ....DE.... .......... .......... ......                                        hUBF
    SSEDG.SS.. .SDEDSE.G. ENEDEE.EE. .DE.NEE..D .NESG.SS.. ...AD.S..D SN*---                                        xUBF
```

Fig. 2 B.

Previously it was shown that hUBF contains several repeats which are structurally homologous to a domain present in the chromosomal proteins HMG1 and HMG2 (Jantzen et al. 1990). These "HMG-boxes" represent the DNA binding domains of UBF. mUBF and hUBF contain five "HMG-boxes", whereas four HMG repeats are present in xUBF. The mammalian HMG domain 4 is absent from xUBF, and this deletion of 60 amino acids accounts for the apparent different molecular weights between mammalian (97/94 kDa) and amphibian UBF (85/82 kDa). Furthermore, the sequence alignment reveals a high degree of homology between individual HMG boxes of different species, suggesting that each box exerts a distinct functional role.

Another striking feature of UBF is the primary structure of its C-terminus. Like HMG 1 and 2, UBF has a very acidic C-terminus. Of the terminal 89 amino acids, 64% are acidic including two uninterrupted stretches of glutamic and aspartic acid residues of 21 and 18 amino acids, respectively. Furthermore, the high percentage (25%) of serine residues within this acidic domain, which represents ideal target sites for casein kinase II, suggests that this region may be post-translationally modified.

2.3 UBF Is a Phosphoprotein

To investigate whether the potential casein kinase II (CKII) sites present in the acidic tail of UBF are functionally important, we tested whether or not UBF is a phosphoprotein. The presence or absence of phosphate groups often alters the mobility of a protein on denaturing SDS-polyacrylamide gels. Therefore, the electrophoretic mobility of UBF was analyzed before and after treatment with calf intestine alkaline phosphatase (CIP). Having the cDNA for the 97 kDa polypeptide of mUBF allowed us to produce recombinant UBF in the vaccinia virus system (UBF_{vac}) and to analyze the electrophoretic mobility of the purified protein before or after phosphatase (CIP) treatment. As shown in Fig. 3A, after phosphatase treatment the mobility of UBF increased (lane 2), whereas in the presence of the phosphatase inhibitor sodium orthovanadate, the mobility of CIP-treated UBF compared to that of the untreated control (lanes 1 and 3) was the same. This change of electrophoretic mobility after phosphatase treatment suggests that UBF is a phosphoprotein.

In an attempt to prove directly that UBF is modified posttranslationally by phosphorylation, we tried to phosphorylate the protein in vitro with γ-(^{32}P) ATP. Surprisingly, a significant amount of phosphate was transferred to purified UBF in the absence of any exogenous protein kinase (Fig. 3C, lane 1). After phosphatase treatment the label was quantitatively removed (lane 2). Thus, the ability to label UBF in vitro with radioactive ATP and to enzymatically remove the modification by treatment with phosphatase demonstrates that UBF is a phosphoprotein. To determine the nature of the phosphoamino acid linkages and to characterize the site of phosphorylation, in vitro phosphorylated UBF was excised from an SDS-polyacrylamide gel and phosphoamino acids were identified after acidic hydrolysis. Serine was the only amino acid labeled, irrespective whether UBF was phosphorylated by the endogenous kinase or by purified CKII (Fig. 3D).

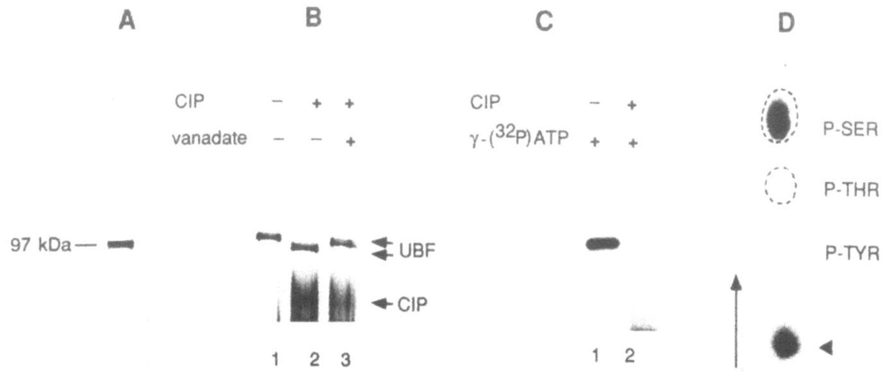

Fig. 3. UBF is a phosphoprotein. **A** SDS-PAGE of recombinant UBF from vaccinia virus infected HeLa cells. UBF was purified from vaccinia virus infected HeLa cells as described (Voit et al. 1992), analyzed on a 8% SDS-polyacrylamide gel, and visualized by silver staining. **B** Phosphatase treatment of UBF. About 3–5 ng of purified UBF_{vac} (*lane 1*) was treated with 0.2 U of calf intestine alkaline phosphatase for 15 min at 30 °C in the absence (*lane 2*) or presence (*lane 3*) of 250 μM sodium orthovanadate following electrophoresis on 8% SDS-polyacrylamide gels and silver staining. UBF molecules corresponding to the phosphorylated and dephosphorylated form are indicated by *arrows*. **C** Purified UBF is phosphorylated in vitro. About 3 ng of purified UBF was incubated with γ-(^{32}P)ATP in kinase buffer (80 mM KCl, 5 mM $MgCl_2$, 12% v/v glycerol, 20 mM Tris HCl, pH 7.9, 0.5 mM DTE, 0.5 mM PMSF) for 30 min at 30 °C. The reactions were either stopped by addition of SDS-sample buffer (*lane 1*) or carried on further 15 min at 30 °C in the presence of alkaline phosphatase (*lane 2*). The products were run on 8% SDS-polyacrylamide gels, and detected by autoradiography. **D** Phosphoamino acid analysis of mUBF. UBF was labeled with γ-(^{32}P)ATP and the band containing the labeled protein was excised from the gel. The labeled protein was concentrated by lyophilization, and subsequently hydrolyzed by incubation in 6 N HCl for 60 min at 110 °C. The recovered phosphoamino acids were analyzed by the one-dimensional thin-layer system according to Hunter and Sefton (1980)

2.4 UBF Is Phosphorylated by Casein Kinase II in Vitro

Next we tested whether the copurifying protein kinase is casein kinase II. CKII is known to use GTP as well as ATP as substrate (Rose et al. 1981). When the phosphorylation was performed in the presence of increasing amounts of nonradioactive ATP and GTP, both nucleoside triphosphates competed with almost the same efficiency for UBF phosphorylation (Fig. 4A), indicating that both are used by the endogenous kinase. Another hallmark of CKII is its inhibition by low concentrations of heparin (Hathaway et al. 1980; Rose et al. 1981). In Fig. 4B the heparin sensitivity of the endogenous UBF kinase is shown. The result demonstrates that UBF phosphorylation is inhibited at identical concentrations of heparin as CKII (50% inhibition at 0.1–0.5 μg per ml) supporting the assumption that all or most UBF phosphorylation is brought about by CKII. Furthermore, the ATP analog DRB (5,6-dichloro-1-β-D-ribofuranosylbenzimidazole), an inhibitor of CKII (Zandomeni et al. 1986), eliminates label-

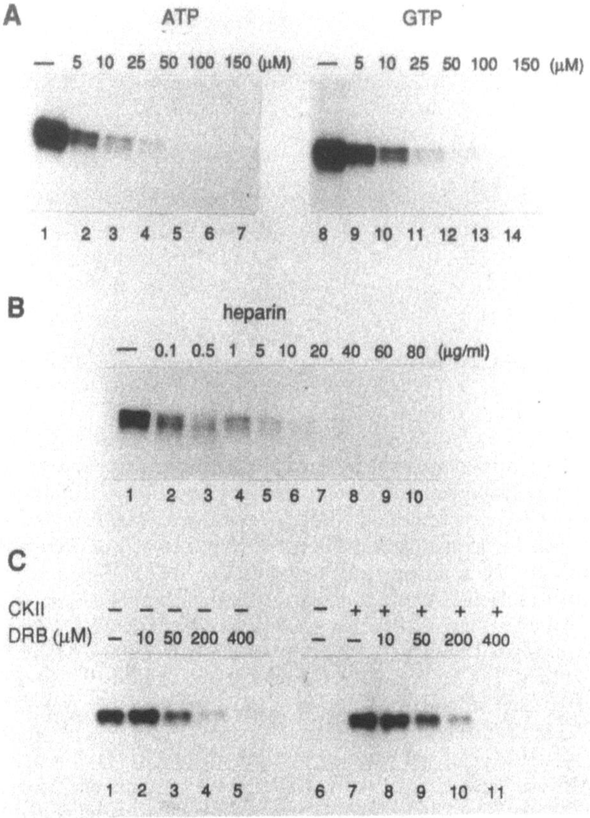

Fig. 4. UBF is phosphorylated by caseine kinase II. **A** The endogenous UBF kinase uses both ATP and GTP as substrates. The phosphorylation assays contained about 3 ng of affinity-purified, dephosphorylated UBF, 5 μCi of γ-(^{32}P)ATP and either increasing amounts of cold ATP (*lanes 2–7*) or GTP (*lanes 9–14*) as indicated above the lanes. **B** The endogenous UBF kinase is sensitive to heparin. Phosphorylation experiments were performed in the presence of 5 μM of ATP, 5 μCi of γ-(^{32}P)ATP and increasing amounts of heparin as indicated above the lanes. **C** Inhibition of UBF phosphorylation by DRB. The phosphorylation reaction was performed either by the endogenous protein kinase (*lanes 1–5*) or (after heat inactivation of the endogenous kinase) by exogenous CKII (*lanes 7–11*) in the presence of increasing concentrations of DRB as indicated above the lanes

ing of UBF by the endogenous kinase at similar concentrations as authentic CKII (Fig. 4C). These inhibitor studies, together with previous findings that of several protein kinases tested only casein kinase II efficiently labeled UBF (O'Mahony et al. 1992; Voit et al. 1992), strongly supports the idea that CKII is the enzyme that modifies UBF.

2.5 UBF Phosphorylation Resides Within the C-Terminal Tail

To investigate whether most or all phosphates are incorporated into the C-terminal region, the phosphorylation of intact UBF and a C-terminal deletion mutant was compared. In this experiment UBF was translated in vitro, and assayed for both autophosphorylation and phosphorylation by exogenous CKII, respectively. As shown in Fig. 5C, in vitro translated full-length UBF is phosphorylated both by an endogenous protein kinase and by exogenous CKII. For efficient labeling, the protein had to be treated with phosphatase before incubation with γ-(^{32}P)ATP and the kinase. There-

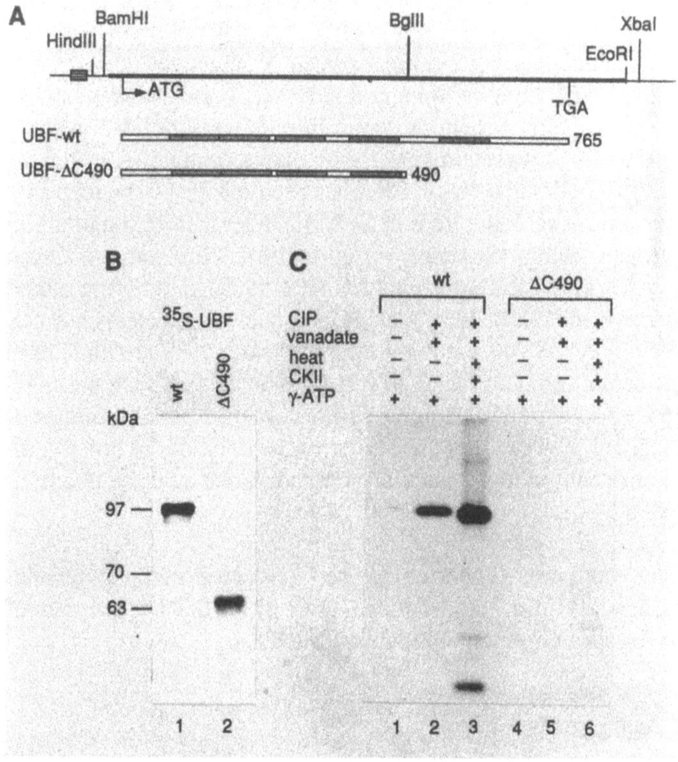

Fig. 5. In vitro phosphorylation of wild-type UBF and a carboxy-terminal deletion mutant by casein kinase II. **A** Diagram of recombinant UBF in the expression vector pBAT (Annweiler et al. 1991). The *hatched regions* indicate the position of the HMG boxes. **B** SDS-PAGE and fluorography of affinity-purified ^{35}S-labeled wild-type UBF and mutant Δ C490 synthesized in reticulocyte lysates. The coding region of mUBF cDNA was cloned between the HindIII and EcoRI sites of pBAT adjacent to the T3 RNA polymerase promoter and β-globin leader sequence. After in vitro transcription and translation, both the wild-type protein (*lane 1* and mutant Δ C490 (*lane 2*) were further purified on specific DNA-affinity columns as described by Jantzen et al. (1990). The purified proteins were analyzed on 8% SDS-polyacrylamide gels and visualized by fluorography. **C** In vitro phosphorylation of recombinant UBF. Wild-type UBF and mutant Δ C490 were expressed in vitro, purified, dephosphorylated by CIP treatment and then labeled with γ-(^{32}P)ATP by either the endogenous protein kinase or by purified CKII. Following SDS-PAGE, the labeled proteins were visualized by autoradiography

fore, UBF translated in the reticulocyte lysate system appears to be highly phosphorylated, too.

We then truncated the carboxy-terminal region of UBF by cleaving the template with Bgl II. The in vitro translation product, a 63 kD polypeptide (Fig. 5B), encodes the first 490 amino acids and lacks the fifth HMG box as well as the acidic tail. When this deletion mutant was assayed for phosphorylation, no significant labeling by either the endogenous kinase nor by exogenous CKII was observed (Fig. 5C, lanes 4–6). A very weak signal in the lanes containing the mutant protein, perhaps nonspecific, could be observed upon longer exposure of the gel. This failure of the C-terminally truncated UBF to be phosphorylated both by the endogenous enzyme and by purified CKII suggests that the potential CKII recognition sequences present in the acidic tail are the sites of UBF phosphorylation.

Although we cannot exclude that other cellular protein kinases may contribute to phosphorylation of UBF as well, both the finding that (I) CKII phosphorylates UBF and (II) UBF phosphorylation is important in controlling its transcriptional activity (Voit et al. 1992) suggests that this protein kinase may play a central role in linking cell growth and rDNA transcriptional activity. CKII has been reported to be increased in transformed cells (Prowald et al. 1984; Rose et al. 1981), after administration of insulin and epidermal growth factors (Sommercorn et al. 1987; Klarlund and Czech 1988; Ackerman and Osheroff 1989; Carroll and Marshak 1989), and during mouse embryogenesis (Schneider et al. 1987). Moreover, several nucleolar proteins and cellular oncoproteins as myc, myb, SV40 T-antigen (Lüscher et al. 1989, 1990; Rihs et al. 1991) as well as the serum response factor SRF (Manak et al. 1990) are phosphorylated by CKII. At present we are determining whether specific serine residues of UBF are differently modified in vivo under different growth conditions in order to ascertain the significance of changes in the state of phosphorylation and the transactivating properties of UBF.

Acknowledgements. This word was supported by the Deutsche Forschungsgemeinschaft, the Fonds der Chemischen Industrie and by a grant of the Science Program of the Commission of the European Community (SCI*–0259–C).

References

Ackerman P & Osheroff N (1989) Regulation of casein kinase II activity by epidermal growth factor in human A-431 carcinoma cells. J Biol Chem 264:11958–11965

Annweiler A, Hipskind RA & Wirth T (1991) A strategy for efficient in vitro translation of cDNAs using the rabbit β-globin leader sequence. Nucl Acids Res 19:3750

Bachvarov D & Moss T (1991) The RNA polymerase I transcription factor xUBF contains 5 tandemly repeated HMG homology boxes. Nucl Acids Res 19:2331–2335

Bell SP, Learned RM, Jantzen H-M & Tjian R (1988) Functional cooperativity between transcription factors UBF1 and SL1 mediates human ribosomal RNA synthesis. Science 241:1192–1197

Bell SP, Pikaard CS, Reeder RH & Tjian R (1989) Molecular mechanisms governing species-specific transcription of ribosomal RNA. Cell 59:489–497

Bell SP, Jantzen H-M & Tjian R (1990) Assembly of alternative multiprotein complexes directs rRNA promoter selectivity. Genes Dev 4:943–954

Buttgereit D, Pflugfelder G & Grummt I (1985) Growth-dependent regulation of rRNA synthesis is mediated by a transcription initiation factor (TIF-IA). Nucl Acids Res 13:8165–8180

Carroll D & Marshak DR (1989) Serum-stimulated cell growth causes oscillations in casein kinase II activity. J Biol Chem 264:7345–7348

Clos J, Buttgereit D & Grummt I (1986a) A purified transcription factor (TIF-IB) binds to essential sequences of the mouse rDNA promoter. Proc Natl Acad Sci USA 83:604–608

Clos J, Normann A, Öhrlein A & Grummt I (1986b) The core promoter of mouse rDNA consists of two functionally distinct domains. Nucl Acids Res 14:7581–7595

Gokal PK, Mahajan PB & Thompson A (1990) Hormonal regulation of transcription of rDNA. J Biol Chem 265:16234–16243

Grummt I & Gross HJ (1980) Structural organization of mouse rDNA: comparison of transcribed and non-transcribed regions. Mol Gen Genet 177:223–229

Hathaway GM, Lubbeb TH & Traugh JA (1980) Inhibition of casein kinase II by heparin. J Biol Chem 255:8038–8041

Heilgenthal G & Grummt I (1991) Isolation of multiple protein factors involved in ribosomal DNA transcription. J Chromatogr 587:25–32

Hunter T & Sefton BM (1980) Transforming gene product of Rous sarkoma virus phosphorylates tyrosine. Proc Natl Acad Sci USA 77:13311–1315

Jantzen H-M, Admon A, Bell SP & Tjian R (1990) Nucleolar transcription factor hUBF contains a DNA-binding motif with homology to HMG proteins. Nature 344:830–836

Klarlund JK & Czech MP (1988) Insulin-like growth factor I and insulin rapidly increase casein kinase II activity in BALB/c 3T3 fibroblasts. J Biol Chem 263:15872–15875

Kuhn A, Deppert U & Grummt I (1990) A 140-base-pair repetitive sequence element in the mouse spacer enhances transcription by RNA polymerase I in a cell-free system. Proc Natl Acad Sci USA 87:7527–7531

Learned RM, Cordes S & Tjian R (1985) Purification and characterization of a transcription factor that confers promoter specifity to human RNA polymerase I. Mol Cell Biol 5:1358–1369

Learned RM, Learned TK, Haltiner MM & Tjian R (1986) Human rRNA transcription is modulated by the coordinate binding of two factors to an upstream control element. Cell 45:847–857

Lüscher B, Kuenzel EA, Krebs EG & Eisenman RN (1989) Myc oncoproteins are phosphorylated by casein kinase II. EMBO J 8:1111–1119

Lüscher B, Christenson E, Litchfield DW, Krebs EG & Eisenman RN (1990) Myb DNA binding inhibited by phosphorylation at a site deleted during oncogenic activation. Nature 344:517–522

Manak JR, de Bisshop N, Kris RM & Prywes R (1990) Casein kinase II enhances binding activity of serum response factor. Genes Dev 4:955–967

McStay B, Hu CH, Pikaard CS & Reeder RH (1991) xUBF and Rib1 are both required for formation of a stable polymerase I promoter complex in *X. laevis*. EMBO J 10:2297–2303

Mishima Y, Finanscek I, Kominami R & Muramatsu M (1982) Fractionation and reconstitution of factors required for accurate transcription of mammalien ribosomal genes: identification of a species-dependent factor. Nucl Acids Res 10:6659–6670

O'Mahony DJ & Rothblum LI (1991) Identification of two forms of the RNA polymerase I transcription factor UBF. Proc Natl Acad Sci USA 88:3180–3184

O'Mahony DJ, Xie WQ, Smith SD, Singer HA, Rothblum LI (1992) Differential phosphorylation and localization of the transcription factor UBF in vivo in response to serum deprivation. J Biol Chem 267:35–38

Pikaard CS, McStay B, Schultz MC, Bell SP & Reeder RH (1989) The Xenopus ribosomal gene enhancers bind an essential polymerase I transcription factor, xUBF. Genes Dev 3:1779–1788

Pikaard CS, Smith SD, Reeder RH & Rothblum L (1990a) rUBF, a RNA polymerase I transcription factor from rats, produces DNAse I footprints identical to those produced by xUBF, its homolog from frog. Mol Cell Biol 10:3810–3812
Pikaard CS, Pape LK, Henderson SL, Ryan K, Paalman MH, Lopota MA, Reeder RH & Sollner-Webb B (1990b) Enhancers for RNA polymerase I in mouse ribosomal DNA. Mol Cell Biol 10:4816–4825
Prowald K, Fischer H & Issinger O-G (1984) Enhanced casein kinase II activity in human tumour cell cultures. FEBS Lett 176:479
Rihs H-P, Jans DA, Fan H & Peters R (1991) The rate of nucleolar cytoplasmic protein transport is determined by the casein kinase II site flanking the nuclear localization sequence of the SV40 T-antigen. EMBO J 10:633–639
Rose KM, Bell LE, Siefken DA & Jacob ST (1981) A heparin-sensitive nuclear protein kinase. J Biol Chem 256:7468–7477
Schnapp A & Grummt I (1991) Transcription complex formation at the mouse rDNA promoter involves the stepwise association of four transcription factors and RNA polymerase I. J Biol Chem 266:24588–24595
Schnapp A, Clos J, Hädelt W, Schreck R, Cvekl A & Grummt I (1990b) Isolation and functional characterization of TIF-IB, a factor that confers promoter specifity to mouse RNA polymerase I. Nucl Acids Res 18:1385-1393
Schnapp A, Pfleiderer C, Rosenbauer H & Grummt I (1990a) A growth-dependent transcription initiation factor (TIF-IA) interacting with RNA polymerase I regulates mouse ribosomal RNA synthesis. EMBO J 9:2857–2863
Schneider HR, Reichert GH, Issinger OG (1987) Enhanced casein kinase II activity during mouse embryogenesis. Eur J Biochem 161:733–739
Smith DD, Oriahi E, Lowe D, Yang-Yeng H-F, O'Mahony D, Rose K & Rothblum LI (1990) Characterization of factors that direct transcription of rat ribosomal DNA. Mol Cell Biol 10:3105–3116
Sommercorn J, Mulligan JA, Lozeman FJ, & Krebs EG (1987) Activation of casein kinase II in response to insulin and epidermal growth factor. Proc Natl Acad Sci USA 84:8834–8838
Tanaka N, Kato H, Ishikawa Y, Hisitake Y, Tashiro K, Kominami R & Muramatsu M (1990) Sequence-specific binding of a transcription factor TFID to the promoter region of mouse ribosomal RNA gene. J Biol Chem 265:13836–13842
Tower J, Culotta V & Sollner-Webb B (1986) The factors and nucleotide sequences that direct rDNA transcription and their relationship to the stable transcription complex. Mol Cell Biol 6:3451–3462
Voit R, Schnapp A, Kuhn A, Rosenbauer H, Hirschmann P, Stunnenberg HG & Grummt I (1992) The nucleolar transcription factor mUBF is phosphorylated by casein kinase II in the C-terminal hyperacidic tail which is essential for transactivation. EMBO J 11:2211–2218
Zandomeni R, Zandomeni MC, Shugar D & Weinmann R (1986) Casein kinase type II is involved in the inhibition by 5,6-dichloro-1-β-D-ribofuranosylbenzimidazole of specific RNA polymerase. J Biol Chem 261:3414–3419

Antisense Strategies for Modulating the Repression of Cell Division

M. Strauss[1], J. Hamann[1], H. Müller[1], A. Lieber[1], V. Sandig[1], D. Bauer[1], and S. Bähring[1]

1 Introduction

Investigations on the regulation of cell division require both genetic studies to identify crucial steps in the process and biochemical studies to assign a particular function to the identified protein. As genetic studies are difficult to carry out in the mammalian system most of the hitherto known regulatory proteins were first identified in yeast and their presence and function was subsequently confirmed in mammalian cells. However, human cancers provide a source for naturally occuring mutations in genes which are related to the regulation of cell division. The products of protooncogenes which are changed in their function or activity in particular cancers are mainly known as positive regulators or modulators of cell division (Hunter 1991). Tumor suppressor genes were identified by their loss in certain cancers and are supposed to be negative regulators of cell division in normal cells (Weinberg 1991). The current view of cancer development, however, implies two or more subsequent genetic changes (Fearon and Vogelstein 1990, Marshall 1991) making the cancer cell itself an unsuitable system for studying the direct consequences of the respective changes. Thus, alternative ways for the inactivation of gene functions are required.

Whereas the knockout of genes by homologous recombination results in stable mutations, the inactivation of a mRNA by basepairing to an antisense nucleic acid offers the general advantage of exerting a transient effect and allows for modulation of protein synthesis. Biological effects can be studied immediately. Several protooncogenes, like c-myc (Bacon and Wickstrom 1991), N-myc (Rosolen et al. 1990), c-myb (Venturelli et al. 1990) have been inactivated by antisense oligonucleotides or by antisense RNA in the case of c-fos (Holt et al. 1986) with the consequence of growth inhibition, confirming their role as positive regulators of cell division. We have used for the first time different antisense strategies to study the role of a presumptive negative growth regulator, the retinoblastoma gene product pRb, in the regulation of cell division and in the process of malignant transformation.

[1] Max-Planck-Gesellschaft, Arbeitsgruppe Zellteilungsregulation & Gensubstitution, Max-Delbrück-Centrum für molekulare Medizin, Robert-Rössle-Str. 10, 1115 Berlin-Buch, FRG.

43. Colloquium Mosbach 1992
DNA Replication and the Cell Cycle

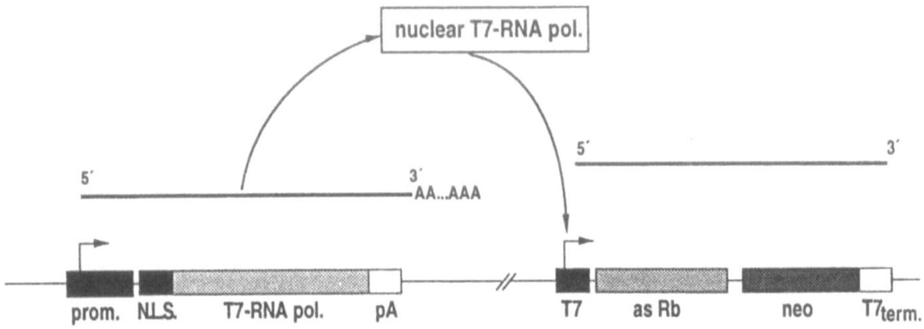

Fig. 1. Schematic representation of the T7 polymerase/promoter system used for the expression of antisense RNA (Lieber et al. 1992)

2 Antisense Nucleic Acids for the Inhibition of Protein Synthesis

Antisense oligonucleotides and RNA sequences capable of specific base-pairing with mRNA molecules have already proved to be powerful tools for the inhibition of specific gene expression. Inhibition of mRNA expression can theoretically be exerted at various stages between transcription and translation (Helene and Toulme 1991). Normal oligodeoxynucleotides are rapidly degraded within culture medium and within the cells. Therefore, several modifications have been introduced rendering the molecules resistant to nucleases. Phosphothioate derivatives are nuclease-resistant and are still negatively charged allowing them to be introduced into cells by conventional transfection techniques or even without additives. With regard to specificity their length should be between 15 and 21 nucleotides. 18mers seem to work fine in most cases (Helene and Toulme 1991).

In order to achieve long-term effects stable expression of antisense RNA from respective constructs is the preferred strategy. Data obtained with different target genes suggest that the amount of antisense RNA required to inactivate translation may vary considerably (Ch'ng et al. 1989) ranging from 1:1 to 1000:1 ratio. We have developed an expression system based on transcription of T7 promoters by a nuclear T7-RNA polymerase (Fig. 1) which allows for very high levels of RNA synthesis (Lieber et al. 1989, 1992). In addition to the synthesis of classical antisense RNA the system is also suited for the expression of ribozymes having a viroid-derived hammerhead sequence in between the flanking antisense sequences (Lieber et al. 1992). In the following we demonstrate the successful use of both antisense oligonucleotides and RNA to inhibit the synthesis of the tumor suppressor gene product pRb.

3 Inhibition of pRb Synthesis by Antisense Oligodeoxynucleotides Stimulates Mitotic Activity

In an initial study published recently (Strauss et al. 1992) we used phosphothioate derivatives of antisense oligodeoxynucleotides directed to different regions of the mRNA for pRb with a length of 18 nucleotides. First, we demonstrated by radioactive labeling that oligonucleotides applied as naked molecules for 24 hours in the culture medium were stable within the cells for at least two weeks. After two weeks RNA was isolated and analyzed by the RNase protection technique using an in vitro transcribed antisense RNA probe. Whereas the whole probe was protected by Rb-mRNA from nontreated cells, it was truncated in the case of oligonucleotide-treated cells. The size reduction of the probe fits nicely with the idea of mRNA degradation by RNaseH (Walder and Walder 1988) at the site of base-pairing between the mRNA and the oligonucleotide. This targeted degradation of mRNA results in the complete lack of pRb protein synthesis 72 h after application of the oligonucleotides. The protein is absent for about 2 weeks. The effect of pRb depriviation on cell division of normal human fibroblasts is tremendous. Starting 2 days after application and lasting for 2 to 3 weeks the cell doubling time is dramatically shortened, to 18 h from originally 50–60 h. Interestingly, after about 10 days a large number of foci can be detected in the confluent monolayer on plastic. However, these foci of seemingly transformed cells are only of transient nature and are probably spots where fast cell division initiates. After 3 weeks the culture dishes are usually covered with a multilayer of cells (Strauss et al. 1992).

The most efficient single oligonucleotide with regard to growth stimulation was one covering the AUG start codon of the Rb-mRNA. However, the effect on transient focus formation was even higher by combined application of this and another oligonucleotide which is complementary to a downstream region of the coding sequence (Strauss et al. 1992). This is probably due to initiation of translation at a downstream AUG in the first case leading to a truncated protein still having the repressor function. Since we have never clearly seen such a truncated protein in our gels, it is probably synthesized at a low rate, if at all.

4 Malignant Transformation of pRb-Deprived Cells Is the Result of Additional Genetic Event(s)

Loss of function of the Rb gene has been postulated as the direct cause of retinoblastomas and has also been implicated in the development of other types of cancers (Weinberg 1990). The immortalizing gene products of different DNA tumor viruses form complexes with pRB and may thereby inactivate its normal function (Weinberg 1990). We were therefore interested to find out whether deprivation of pRb directly leads to immortalization or transformation or even to tumorigenicity. To this end we used the soft agar growth assay. At different times after treatment with antisense oligonucleotides we seeded 10^4 cells into 5 cm dishes in 0.3/0.6% soft agar. After 4 weeks about twenty little colonies with less than 30 cells and one or two large colonies were detected, whereas no colonies were found on control dishes with non-treated cells. The large colonies from several dishes were picked. Most of the colonies could not be established as cell lines suggesting absence of immortalization. Two colonies could be expanded into a confluent monolayer on a 10 cm dish but could not be passaged afterwards. Upon addition of fresh antisense oligonucleotides one colony could be passaged five times on a 10 cm dish suggesting extension of the life span by repression of pRb synthesis. The cells also had a transformed phenotype – indicated by a rounded morphology and continuous focus formation. Injection of 10^7 cells from this clone into nude mice did not result in tumor formation (Strauss et al. 1992).

We conclude from these experiments that the loss of pRb leads to a dramatic increase in replication and cell division but not automatically to an increase in immortalization or transformation. However, transformed cells seem to arise as the consequence of a mutation as suggested by the frequency of colony formation in soft agar. The survival of the transformed cells obviously require the continuous absence of the Rb protein. Tumor formation probably requires a further genetic event. In order to study the development of a tumor cell after pRb deprivation long-term repression of pRb synthesis is necessary.

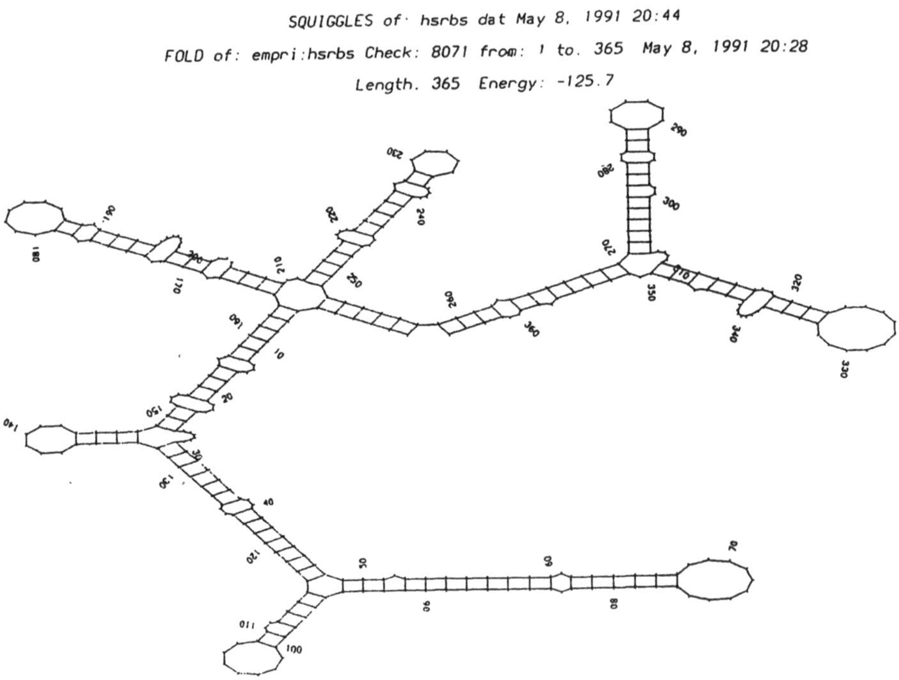

Fig. 2. Predicted secondary structure of the antisense RNA sequence (365 n.t.) which is most active in biological assays as obtained using program FOLD

5 Repression and Modulation of pRb Synthesis by Antisense RNA

In order to optimize the expression of antisense RNA for Rb, we constructed several recombinants carrying different parts of the human Rb1 cDNA in antisense orientation downstream of different promoters (CMV, SV40 early, MMTV, T7). The constructs were transfected into Chinese hamster CHO cells and into mouse NIH3T3 cells. The formation of large colonies or foci was used as an indicator for the function of the recombinants. With all promoters an antisense fragment spanning the first 365 nucleotides of the cDNA was found to be most efficient. Interestingly, computer analysis of the antisense RNA revealed the formation of a high degree of secondary structure (Fig. 2). If this predicted structure really holds true for the in vivo situation, it might be the cause of a high stability of the antisense RNA. The antisense effect could be exerted by partial melting of the secondary structure and base-pairing of a relatively short sequence.

Using the established rodent cell lines as recipients, we have consistently detected focus formation and growth in soft agar at a high frequency (Table 1). Transformed clones efficiently formed tumors in nude mice. This suggests that the loss of pRb can directly cause transformation in established cells which have already undergone the first step(s) to malignancy.

Normal human fibroblasts (5×10^5) were then transfected with pCMVasRb and pSV2neo and selection with Geneticin was applied. Cells transfected with pSV2neo only did not form colonies in the presence of 200 µg/ml of Geneticin. Five colonies of densely growing cells were detected in the case of cotransfection with pCMVasRb after 3 weeks. The colonies were picked, expanded, and analyzed for the presence of pRb by Western blotting of immunoprecipitates (Fig. 3). In all colonies the amount of pRb was dramatically reduced or undetectable. All clones were changed in their morphology and grew very fast as observed previously after treatment with antisense oligonucleotides. We are currently analyzing the transformation properties of the clones.

Table 1. Transformation of NIH3T3 cells by antisense Rb-RNA

Vector[a]	Foci/10^5 cells	Colonies in soft agar[b]
pM6SVT3N/M6SVT7N	1/2	4/5
pM6SVT7N + pT7asRb	8/4	105/94[c]
pM6SVT3N + pT3asRb	7/3	158/125[c]

[a] The pM6SV vectors are retroviral vector plasmids with an endogenous SV40 promoter driving the expression of nuclear RNA polymerases from either phage T3 or phage T7. The asRb recombinants carry the 5'-terminal 365 nucleotides of the Rb cDNA in antisense orientation downstream of either the T3 promoter or the T7 promoter.

[b] For soft agar assays 10^4 cells were seeded in 5-cm dishes. Cells were grown in 0.3% agar for 4 weeks and large colonies were scored.

[c] Cells for this assay are derived from one isolated focus.

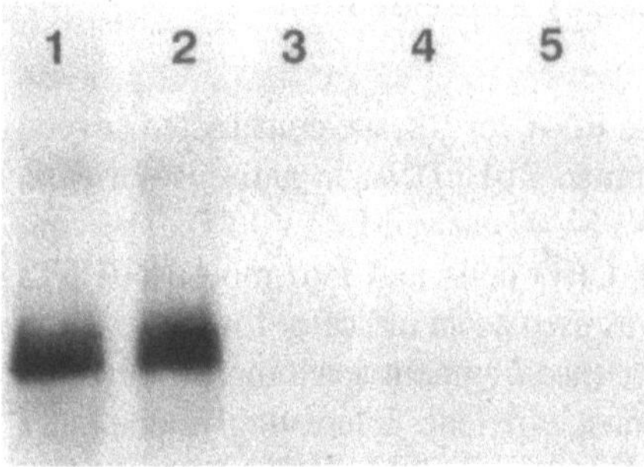

Fig. 3. Detection of pRb by immunoprecipitation and chemiluminescence using the monoclonal antibody 1F8 (Bartek et al. 1992). Protein from one 10-cm dish was immunoprecipitated and run on one lane. *Lanes 1,2* control human fibroblasts IMR-90; *lanes 3–5* three individual clones isolated after cotransfection of pCMVasRb and pSV2neo and selection in 200 μg/ml of Geneticin

We have constructed vectors allowing for expression of ribozymes targeted to the Rb-mRNA (Fig. 4). Constructs with the T7 promoter were expressed in vitro and the resulting transcripts were incubated with Rb-mRNA. Complete cleavage of the target RNA at the predicted site was observed at a 1:1 molar ratio after 30 min (Fig. 5). The biological function of the constructs was confirmed by stimulation of focus formation in CHO cells. However, we were unable to detect the specific cleavage pattern within the cells. Thus, it remains to be clarified how the ribozyme transcripts function in vivo.

We are currently trying to modulate the expression of pRb in two different ways. Firstly, we are applying the regulated expression of the antisense RNA using either the MMTV promoter or the regulated expression of T7 polymerase. Secondly, we are introducing constructs into pRb-deprived cells which allow for regulated expression of pRb which is not affected by the antisense RNA.

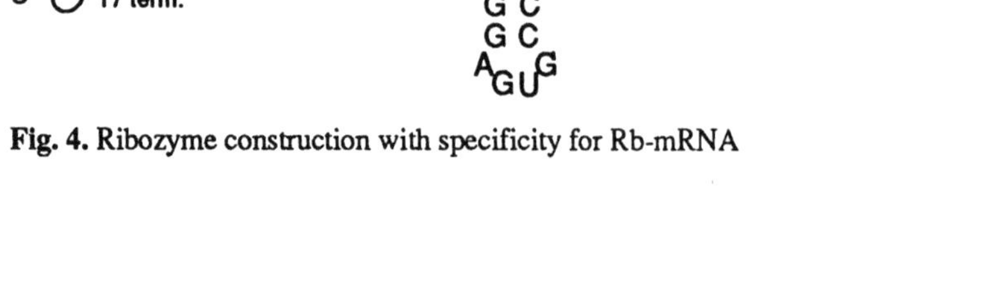

Fig. 4. Ribozyme construction with specificity for Rb-mRNA

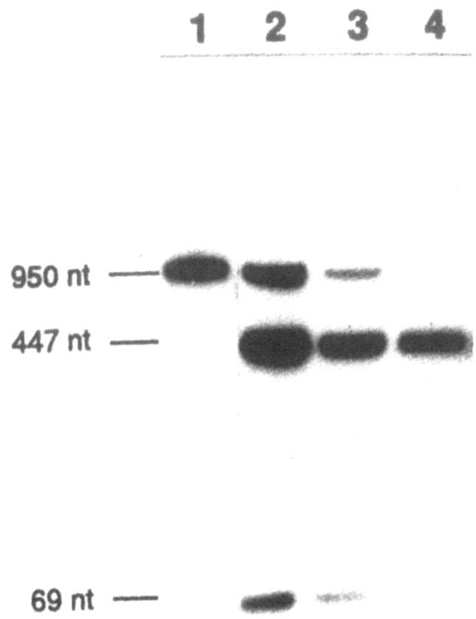

Fig. 5. Cleavage pattern of an Rb-specific ribozyme. An in vitro transcribed and labeled sense RNA fragment of 950 nucleotides (*lane 1*) was incubated with the ribozyme transcript shown in Fig. 4 for 30 min at 37 °C. The ribozyme/target molar ratios were 1:5 (*lane 2*), 1:2.5 (*lane 3*), and 1:1 (*lane 4*) whereby the amount of ribozyme was constant and the amount of target RNA was increased

6 Conclusion

Our results clearly demonstrate that antisense approaches can successfully be used to stimulate cell division by inhibiting the synthesis of the tumor suppressor gene product pRb. The data confirm the proposed function of pRb as a negative regulator of growth and, at the same time, they indicate that the loss of pRb is sufficient to stimulate cell division, whereas complete malignant transformation requires one or two additional genetic events. The high mitotic activity induced by the absence of pRb is necessarily linked to fast DNA replication. This most likely allows for an increased error frequency or results in a reduced activity of the repair machinery. This, in turn, would lead to an increased mutation rate in any gene including proto-oncogenes and tumor suppressor genes. We are currently testing this hypothesis.

The antisense approach in connection with regulation of the level of inactivation is a powerful tool for studies on the regulation of cell division in general. It is especially suited for investigations on the function of repressor proteins. It will be useful to study the role of other tumor suppressor gene products and will help to define their site of action within the pathway of growth regulation.

References

Bacon ThA & Wickstrom E (1991) Walking along human c-myc mRNA with antisense oligodeoxynucleotides: maximum efficacy at the 5'cap region. Oncogene Res 6:13–19

Bartek J, Vojtesek B, Grand RJA, Gallimore PhH & Lane DP (1992) Cellular localization and T antigen binding of the retinoblastoma protein. Oncogene 7:101–108

Ch'ng JLC, Mulligan RC, Schimmel P & Holmes EW (1989) Antisense RNA complementary to 3' coding and noncoding sequences of creatine kinase is a potent inhibitor of translation in vivo. Proc Natl Acad Sci USA 86:10 006–10 010

Fearon ER & Vogelstein B (1990) A genetic model for tumorigenesis. Cell 61:759–767

Helene C & Toulme JJ (1990) Specific regulation of gene expression by antisense, sense and antigene nucleic acids. Biochem Biophys Acta 1049:99–125

Holt JT, Gopal TV, Moulton AD & Nienhuis AW (1986) Inducible production of c-*fos* antisense RNA inhibits 3T3 cell proliferation. Proc Natl Acad Sci USA 83:4794–4798

Hunter T (1991) Cooperation between oncogenes. Cell 64:249–27

Lieber A, Kiessling U & Strauss M (1989) High level gene expression in mammalian cells by a nuclear T7-phage RNA polymerase. Nucl Acids Res 17:8485–8493

Lieber A, Sandig V, Sommer W, Bähring S & Strauss M (1992) Stable high level gene expression in mammalian cells by T7 phage RNA polymerase. Methods Enzymol (in press)

Marshall ChJ (1991) Tumor suppressor genes. Cell 64:313–326

Rosolen A, Whitesell L, Ikegaki N, Kennett RH & Neckers LM (1990) Antisense inhibition of single copy N-myc expression results in decreased cell growth without reduction of c-myc protein in a neuroepithelioma cell line. Cancer Res 50:6316–6322

Strauss M, Hering S, Lieber A, Herrmann G, Grifin BE & Arnold W (1992) Stimulation of cell division and fibroblast focus formation by antisense repression of retinoblastoma protein synthesis. Oncogene 7:769–773

Venturelli D, Travali S & Calabretta B (1990) Inhibition of T-cell proliferation by a MYB antisense oligomer is accompanied by selective down-regulation of DNA polymerase alpha expression. Proc Natl Acad Sci USA 87:5963–5967

Walder RY & Walder JA (1988) Role of RNase H in hybrid-arrested translation by antisense oligonucleotides. Proc Natl Acad Sci USA 85:5011–5015

Weinberg RA (1990) The retinoblastoma gene and cell growth control. TIBS 15:199–203

Weinberg RA (1991) Tumor suppressor Genes. Science 254:1138–1146

Growth Suppressors and the Decision to Replicate the Genome

Biological Phenotypes of Tumor-Derived Human p53 Mutants

C. A. Finlay[1] and R. S. Quartin[1]

1 Introduction

p53 is a cellular protein expressed at low levels in the normal cell (Benchimol et al. 1982; Thomas et al. 1983). p53 was originally discovered as a 53 000 kDa protein that co-immunoprecipitates with the SV40 large T antigen in SV40 transformed cells (Lane and Crawford 1979; Linzer and Levine 1979). In T antigen transformed cells, the level of p53 are approximately 100-fold higher than those found in the non-transformed cell; the elevation of p53 levels results primarily from the increased stability of p53 when bound to the SV40 large T antigen (the half-life of p53 bound to T antigen is extended to 24 h) (Oren et al. 1981). These observations led to the hypothesis that the elevation of p53 levels is critical to the process of cellular transformation. Cotransfection of p53 cDNA or genomic clones with an activated *ras* gene resulted in the transformation of primary rat cells in culture (Jenkins et al. 1984; Parada et al. 1984; Eliyahu et al. 1984), providing direct evidence that deregulated overexpression of p53 results in cellular transformation. The p53 gene was then classified as an oncogene. Further studies showed, however, that the wild-type p53 gene did not cooperate with an activated *ras* gene to transform cells (Eliyahu et al. 1988; Finlay et al. 1988); those cDNA and genomic clones activated for transformation with *ras* each possessed a single nucleotide change resulting in a missense mutation in the murine p53 protein. In fact, overexpression of wild type murine p53 suppresses the formation of transformed foci following transfection of primary rat cells with two cooperating oncogenes (e.g. *ras* plus Ela from adenovirus type 5) (Eliyahu et al. 1989; Finlay et al. 1989), and inhibits the growth of transformed cells in vitro (Baker et al. 1990a; Diller et al. 1990; Mercer et al. 1990; Johnson et al. 1991). The large number of mutations that serve to activate p53 for transformation with *ras* (in a region spanning over 25% of the p53 protein) indicates that these mutations are loss of function mutations and that one mechanism by which mutant p53 expression contributes to cellular transformation is to form oligomeric protein complexes with the endogenous wild-type rat p53 protein and, through a dominant negative mechanism (Herskowitz 1987), inactivate a p53 growth regulatory function (Baker et al. 1990b; Hinds et al. 1990).

To determine whether inactivation of p53 function was important in the development of human tumors, Baker et al. (1989) conducted an detailed analysis of p53 in human colorectal carcinomas. Over 75% of colorectal carcinomas had deletions in a region of chromosome 17p that contains the p53 gene and the remaining allele fre-

[1] Department of Molecular Biology, Lewis Thomas Laboratory, Princeton University, Princeton, NJ 08544–1014, USA.

43. Colloquium Mosbach 1992
DNA Replication and the Cell Cycle

quently possessed a missense mutation. Sequence analysis of a wide variety of human tumors (breast, colon, esophageal, lung, liver) has since demonstrated that missense mutations in p53 (with a deletion of the remaining allele) occur quite frequently in human cancers (for recent reviews, see Hollstein et al. 1991; Finlay 1992), and p53 is now classified as a tumor suppressor gene. These data suggest a model for tumor progression in which a normal cell with a missense mutation in one p53 allele has a growth advantage over cells with two wild type p53 alleles (through a dominant negative mechanism). The increased proliferative capacity of these cells would result in more cell divisions and an increased probability of mutation (allelic loss) occurring at the remaining wild-type allele. The resultant cell would possess a malignant phenotype.

The majority of the mutations observed in human tumors are missense mutations, suggesting mutant p53 proteins are selected for tumor cell growth. These mutations do not occur randomly; the mutations are clustered in four of the highly conserved domains of the p53 protein, suggesting these regions are of functional importance. Approximately 40% of the missense mutations occur at five "hot spot" residues (amino acids 175, 248, 249, 273 and 282) (Hollstein et al. 1991; Finlay 1992). Although mutations at these "hot spots" are seen most often; tumors from different tissues do have different patterns of p53 mutations. This observation may reflect the different concentrations of mutagens in different tissues or may reflect a growth advantage conferred by a specific mutant in a given tissue.

2 Results

The prevalence of elevated levels of mutant p53 proteins in human tumors suggests that overexpression of mutant p53 is selected for tumor development. Indeed, there is evidence from several studies that mutant p53 proteins can provide a growth promoting function in transformed cells in the absence of a wild type allele (Wolf et al. 1984; Chen et al. 1990; Deppert et al. 1990; Shaulsky et al. 1991). Are there properties common to all of the mutant proteins? Do these mutants differ in their ability to transform cells? To address these questions, a biochemical and biological characterization of mutant p53 cDNA clones representing the different "hot spot" mutant p53 alleles was conducted.

The ability of each human p53 mutant to suppress the growth of the Saos-2 cell line (a human osteosarcoma cell line which does not express p53 (Masuda et al. 1987) was assayed by determining the plating efficiency of these cells following transfection with either the wild type p53 gene or any of the mutant p53 genes. Expression of the human wild type protein (with either a proline or an arginine at amino acid 72) results in a 95% inhibition of plating efficiency (Table 1), demonstrating the sensitivity of the Saos-2 cell line to wild-type p53 expression. A single missense mutation in p53 at amino acid 143 (V to A), 175 (R to H), 248 (R to W), 273 (R to H), or 281 (D to G), however, eliminates the ability of p53 to inhibit transformed cell growth. In all instances, transfection with a mutant p53 clone is similar to transfecting with neo alone. In addition, although no cell lines (0/12) expressing the wild-type protein were ob-

Table 1. The effect of wild-type or mutant p53 on the plating efficiency of SAOS-2 cells

Plasmid	Colonies/10^6 Cells	Average	CMV-Bam-Neo
CMV-Bam-Neo	608,1774, 850, 1450	1020	1.00
p53-WT (Pro)	56, 133, 50, 10	62	0.06
P53-WT (Arg)	52, 113, 30, 30	56	0.05
p53-143A	1240, 2056, 1310, 1450	1514	1.48
p53-175H	266, 1449, 800, 810	831	0.81
p53-248W	ND, 1553, 800, 1140	1164	0.86
p53-273H	1084, 1638, ND, 965	1229	0.96
p53-281G	245, 1095, 720, 800	715	0.70

tained from transfections with wild type p53, expression of each mutant allele was observed in 30–50% of the cell lines derived from these transfections (data not shown).

The mutant p53 proteins share additional properties in common and these are summarized in Table 2. In all instances, the half-lives of the mutant p53 proteins are extended, and the levels of the mutant proteins are elevated in the transformed cell lines (Hinds et al. 1990). All of the mutant proteins bind poorly or not at all to the SV40 large T antigen (Bargonetti et al. 1991; R. Quartin and A. J. Levine, unpubl. results), bind poorly to DNA (Bargonetti et al. 1991; Kern et al. 1991) and in general, have reduced transactivation activity (Fields and Jang 1990; Raycroft et al. 1991).

Although the different mutations eliminated p53 suppression activity to similar extents, the "hot spot" mutants differed in several properties (Table 2). First, the ability to cooperate with an activated *ras* gene and transform cells was not equivalent. The mutation observed most frequently in human tumors (175 R to H) reproducibly resulted in the greatest number of transformed foci (approximately two- to sixfold higher than observed with the 281 or the 273 mutation, respectively) (Hinds et al. 1990). Second, only two of the mutants (143 and 175) were found in complexes with the constitutively expressed member of the heat shock family (hsc 70) (Hinds et al. 1990) and were immunoprecipitated with the mutant specific monoclonal antibody PAb 240 (Gannon et al. 1990; R. Quartin and A. J. Levine, unpubl. results).

3 Discussion

Since the definition of the p53 gene as a tumor suppressor gene, there has been rapid progress in the characterization of both wild-type and mutant p53 activities. All of the mutant p53 proteins have lost critical functions that are likely to be related to the ability of the wild-type p53 protein to regulate cell growth (DNA binding and transactivation activity, for example). The diversity of the phenotypes of the mutants (e.g. different conformations, different associations with cellular proteins), however, sug-

Table 2. Summary of the characteristics of mutant human p53 proteins

	Mutant amino acid residue					
	143	*175*	*248*	*273*	*281*	*WT*
Half-life	(h)	(h)	ND	(h)	(h)	(min)
Ability to suppress transformed cell growth	–	–	–	–	–	+
SV40 T-antigen binding	(10–20%) + (80–90%) –	–	–	–	–	+
DNA binding (in vitro)	–	–	–	–	–	+
Transactivation (Gal-4)	–	–	ND[a]	+	ND	+
Relative trans-formation frequency 2	22	ND	8	4	1	
Conformational alteration (Pab240 binding)	+	+	–	–	–	–
HSC70 binding	+	+	–	–	–	–

[a] ND = not determined

gests there may be subtle differences in the ability of mutant proteins to promote cell growth (both in the presence or the absence of a wild-type allele). Further characterization of the growth promoting activity of each mutant in the absence of a wild-type allele is necessary to determine whether a tumor overexpressing a mutant p53 protein has any growth advantage over a non-expressing tumor cell. Such information could prove important for the future diagnosis and prognosis of human cancer.

How could mutant p53 proteins promote growth in the absence of a wild type allele? Perhaps by associating with cellular proteins (which may or may not normally bind to the wild-type p53 protein) and altering their normal functions. For example, a 90 kd cellular protein known to bind to either the wild-type or the mutant p53 proteins (Hinds et al. 1990) has recently been identified as the rat homologue of the mdm-2 gene (Momand et al. 1992). When mdm-2 is overexpressed, 3T3 cells are converted to a tumorigenic phenotype (Fakhazadeh et al. 1991). Whether the ability of mdm-2 to bind to wild-type p53 is related to the growth promoting activity of mdm-2, however, remains to be elucidated.

References

Baker SJ, Fearon ER, Nigro JM, Hamilton SR, Preisinger AC, Jessup JM, vanTuinen P, Ledbetter DH, Barker DF, Nakamura Y, White R & Vogelstein B (1989) Chromosome 17 deletions and p53 gene mutations in colorectal carcinoma. Science 244:217–221
Baker SJ, Markowitz S, Fearon ER, Willson JKU & Vogelstein B (1990a) Suppression of human colorectal carcinoma cell growth by wild-type p53. Science 249:912–915
Baker SJ, Preisinger AC, Jessup JM, Paraskeva C, Markowitz S, Willson JK, Hamilton S & Vogelstein B (1990b) p53 gene mutations occur in combination with 17p allelic deletions or late events in colorectal tumorigenesis. Cancer Res 50:7717–7722
Bargonetti J, Friedman PN, Kern SE, Vogelstein B & Prives C (1991) Wild-type but not mutant p53 immunopurified proteins bind to sequences adjacent to the SV40 origin of replication. Cell 65:1083–1091
Benchimol S, Pim D & Crawford L (1982) Radioimmunoassay of the cellular protein p53 in mouse and human cell lines. EMBO J 1:1055–1062
Chen P-L, Chen Y, Bookstein R & Lee W-H (1990) Genetic mechanisms of tumor suppression by the human p53 gene. Science 250:1576–1579
Deppert W, Buschhausen-Denker G, Patschinsky T & Steinmeyer K (1990) Cell cycle control of p53 in normal (3T3) and chemically transformed (MethA) mouse cells. II. Requirement for cell cycle progression. Oncogene 5:1701–1706
Diller L, Kassel J, Nelson CE, Gryka MA, Litwak G, Gebhardt M, Bressac B, Ozturk M, Baker SJ, Vogelstein B & Friend SH (1990) p53 functions as a cell cycle control protein in osteosarcomas. Mol Cell Biol 10:5772–5781
Eliyahu D, Raz A, Gruss P, Givol D & Oren M (1984) Participation of p53 cellular tumor antigen in transformation of normal embryonic cells. Nature 312:646–649
Eliyahu D, Goldfinger N, Pinhasi-Kimhi O, Shaulsky G, Skurnik Y, Arai N, Rotter V & Oren M (1988) Meth A fibrosarcoma cells express two transforming mutant p53 species. Oncogene 3:313–321
Eliyahu D, Michalovitz D, Eliyahu S, Pinhasi-Kimhi O & Oren M (1989) Wild-type p53 can inhibit oncogene-mediated focus formation. Proc Natl Acad Sci USA 86:8763–8767
Fakharzadeh SS, Trusko SP & George DL (1991) Tumorigenic potential associated with enhanced expression of a gene that is amplified in a mouse tumor cell line. EMBO J 10:1565–1569
Fields S & Jang SK (1990) Presence of a potent transcription activating sequence in the p53 protein. Science 249:1046–1049
Finlay CA (1992) Normal and malignant growth control by p53. In Oncogenes II, (eds.) Benz CC & Liu ET, Kluwer Academic Publishers, Norwell, Massachusetts, in press
Finlay CA, Hinds PW, Tan T-H, Eliyahu D, Oren M & Levine AJ (1988) Activating mutations for transformation by p53 produce a gene product that forms an hsc70-p53 complex with an altered half-life. Mol Cell Biol 8:531–539
Finlay CA, Hinds PW & Levine AJ (1989) The p53 proto-oncogene can act as a suppressor of transformation. Cell 57:1083–1093
Gannon JV, Greaves R, Iggo R & Lane DP (1990) Activating mutations in p53 produce a common conformational effect: A monoclonal antibody specific for the mutant form. EMBO J 9:1595–1602
Herskowitz I (1987) Functional inactivation of genes by dominant negative mutations. Nature 329:219–222
Hinds PW, Finlay CA, Quartin RS, Baker SJ, Fearon ER, Vogelstein B & Levine AJ (1990) Mutant p53 cDNAs from human colorectal carcinomas can cooperate with ras in transformation of primary rat cells. Cell Growth Diff 1:571–580
Hollstein M, Sidransky D, Vogelstein B & Harris CC (1991) p53 mutations in human cancers. Science 253:49–53
Jenkins JR, Rudge K & Currie GA (1984) Cellular immortalization by a cDNA clone encoding the transformation-associated phosphoprotein p53. Nature 312:651–654

Johnson P, Gray D, Mowat M & Benchimol S (1991) Expression of wild-type p53 is not compatible with continued growth of p53-negative tumor cells. Mol Cell Biol 11:1–11
Kern SE, Kinzler KW, Baker SJ, Nigro JM, Rotter V, Levine AJ, Friedman P, Prives C & Vogelstein B (1991) Mutant p53 proteins bind DNA abnormally in vitro. Oncogene 6:131–136
Lane DP & Crawford LV (1979) T antigen is bound to a host protein in SV40-transformed cells. Nature 278:261–263
Linzer DIH & Levine AJ (1979) Characterization of a 54K dalton cellular SV40 tumor antigen in SV40 transformed cells. Cell 17:43–52
Masuda H, Miller C, Koeffler HP, Battifora H & Cline MJ (1987) Rearrangement of the p53 gene in human osteogenic sarcomas. Proc Natl Acad Sci USA 84:7716–7719
Mercer WE, Shields MT, Amin M, Sauve GJ, Appella E, Romano JW & Ullrich SJ (1990) Negative growth regulation in a glioblastoma tumor cell line that conditionally expresses human wild-type p53. Proc Natl Acad Sci USA 87:6166–6170
Momand J, Zambetti GP, Olson DC, George D & Levine AJ (1992) The mdm-2 oncogene product forms a complex with the p53 protein and inhibits p53 mediated transactivation. Cell 69:1237–1245
Oren M, Maltzman W & Levine AJ (1981) Post-translational regulation of the 54K cellular tumor antigen in normal and transformed cells. Mol Cell Biol 1:101–110
Parada LF, Land H, Weinberg RA, Wolf D & Rotter V (1984) Cooperation between gene encoding p53 tumor antigen and ras in cellular transformation. Nature 312:649–651
Raycroft L, Schmidt JR, Yoas K & Lozano G (1991) Analysis of p53 mutants for transcriptional activity. Mol Cell Biol 11:6067–6074
Shaulsky G, Goldfinger N & Rotter V (1991) Alterations in tumor development in vivo mediated by expression of wild-type or mutant p53 proteins. Cancer Res 51:5232–5237
Thomas R, Kaplan L, Reich N, Lane DP & Levine AJ (1983) Characterization of human p53 antigen employing primate specific monoclonal antibodies. Virology 131:502–517
Wolf D, Harris N & Rotter V (1984) Reconstitution of p53 expression in a nonproducer Ab-MuLV-transformed cell line by transfection of a functional p53 gene. Cell 38:119–126

Association of Wild-Type P53 with Cell Differentiation: Induction of Pre-b Cell Maturation in Vitro

V. Rotter[1], G. Shaulsky[1], and N. Goldfinger[1]

1 Introduction

Recent experiments suggest that the p53 protein, which has been shown to act as a dominant oncogene (Eliyahu et al. 1984; Jenkins et al. 1984; Parada et al. 1984. Wolf et al. 1984a) can also function as an antioncogene (Eliyahu et al. 1989; Finlay et al. 1989; Lane and Benchimol 1990; Munroe et al. 1988). This apparent dual activity was resolved when it was found that *mutant* p53 can enhance the malignant process, whereas *wild type* p53 may function as a negative growth regulator.

The notion that cell transformation involves the inactivation of *wild type* p53 was initially deduced from the observation that the p53 gene in several types of human and mouse cell lines and primary tumors, had been rearranged or inactivated by point mutations (Munroe et al. 1988; Wolf and Rotter 1985; Ahuja et al. 1989; Baker et al. 1989; Kelman et al. 1989; Nigro et al. 1989; Takahashi et al. 1989). This hypothesis was further supported by the observation that *wild type* p53 failed to enhance malignant transformation but rather suppressed the transforming activity of other oncogenes (Eliyahu et al. 1988; Hinds et al. 1989). A comparison of the various p53 encoded proteins, based on their ability to transform primary rat embryonic fibroblasts in cooperation with the *ras* oncogene, indicated that *mutant* p53 induced the appearance of morphologically transformed foci (Eliyahu et al. 1984; Jenkins et al. 1984; Parada et al. 1984), whereas *wild type* p53 did not (Eliyahu et al. 1988; Hinds et al. 1989). On the contrary, *wild type* p53 suppressed malignant transformation of primary embryonic fibroblasts induced by cotransfection of *ras* oncogene together with Ela, *mutant* p 53 or *myc* (Eliyahu et al. 1989; Finlay et al. 1989).

A more direct approach to study growth regulation by p 53, was taken by Mercer et al. (1990; 1990a), who showed that the expression of *wild type* p 53 in human gliosarcoma cells induced growth arrest before entering the S phase. Conformational changes of a temperature-sensitive *mutant* p53 into *wild-type* p53 were also found to interfere with cell proliferation of transformed embryonic fibroblasts (Michalovitz et al. 1990). These findings suggested that the activity of *wild type* p53 is associated with both cell proliferation and malignancy.

In a different study, it was found that while *wild type* p53 interfered with proliferation of a colorectal carcinoma that contained a mutated p53 gene, no effect was detectable when it was expressed in a colorectal adenoma that contained a *wild type* p53 gene (Baker et al. 1990). It was suggested that the magnitude of *wild type* p53 suppressive activity varies in cells exhibiting a *wild type* or *mutant* p 53 gene (Baker et

[1] Departments of Cell Biology, The Weizmann Institute of Science, Rehovot 76100, Israel.

43. Colloquium Mosbach 1992
DNA Replication and the Cell Cycle

al. 1990). This notion, however, makes it difficult to estimate the net effect of p53 on cell growth regulation.

Growth modulators were shown to play a key role in cell proliferation. While some factors induce cell proliferation, others induce growth arrest. Proliferation is responsible for self renewal, whereas physiological growth arrest was shown, in several systems to be associated with cell differentiation. The goal of our study was to explore the physiological pathway through which p53 acts and its association with cell growth and differentiation. To that end, our strategy was to study expression of *wild type* p53 in a cell system that lacks any expression of endogenous p53. In the present experiments we studied the role of *wild type* p53 in pre-B cell proliferation and differentiation. We found that introduction of *wild type* p53 into cells at an early phase of their differentiation pathway induces them to advance to a more differentiated stage.

2 Results

2.1 Establishment of L12 Cell Lines Expressing Wild Type P53

To elucidate the mechanism of action of *wild type* p53 and to assess the significance of its expression in vivo, we studied the expression of *wild type* p53 in the p53 non-producer cell line, L12. This cell line which is an Ab-MuLV-transformed lymphoid pre-B cell line was the first example of p53 non-producer cells (Rotter et al. 1980). The p53 gene of these cells had been rearranged by an integration of a Moloney murine leukemia virus into the first p53 intron (Wolf and Rotter 1984). This viral integration interfered with the expression of the mature p53 mRNA (Wolf et al. 1984). While all other Ab-MuLV-transformed p53 producer cells develop lethal tumors in syngeneic mice, L12 cells developed regressor tumors. We found that reconstitution of *mutant* p53 in L12 cells, changed their phenotype and rather than developing regressor tumors they produced lethal tumors (Wolf et al. 1984a). This led us to conclude that *mutant* p53 enhanced the transformed phenotype of these p53 non-producer cells.

In the experiments presented here, we examined the effect of *wild type* p53 expression on cell proliferation in vitro, tumor development in vivo and state of differentiation in vitro of the pre-B Ab-MuLV transformed, L12 cells.

To establish *wild type* p53 producer cell lines, L12 cells were cotransfected with the murine *wild type* p53 and the selectable bacterial *gpt* gene. The expression vector pSVLp53cD consisted of the *wild type* p53 cDNA isolated from a normal murine T-cell library (Arai et al. 1986), subcloned downstream to the SV40 late promoter. Drug-resistant clones obtained 2–3 weeks after gene transfer were single cell cloned and the resulting L12-cD established cell lines were further analyzed.

The p53 protein expressed in these clones was evaluated by its specific binding to anti-p53 monoclonal antibodies. Although mutations may occur in a number of sites, *mutant* p53 proteins could be distinguished from the *wild type* p53 protein by their differential expression of specific antigenic epitopes. While *mutant* p53 forms retained the antigenic epitope recognized by PAb-240, *wild type* p53 lacked this epitope and bound the PAb-246 anti-p53 monoclonal antibodies instead (Yewdell et al. 1986; Gannon et al. 1990). Fig. 1 shows that in all L12-cD-derived clones, the p53 protein expressed, bound specifically the PAb-246 anti-p53 monoclonal antibodies (lanes d); no immunoprecipitation was evident when the *mutant* specific anti-p53 monoclonal antibodies PAb-240, were used (lanes b). When 230-23-8, a different p53 producer Ab-MuLV established cell line, or L12-M8-3A2, an L12 derived cell line established by transfection of *mutant* p53 protein encoded by pSVLp53M8 cDNA, were tested we found, as expected, that the protein was specifically immunoprecipitated with PAb-240 (lanes b), but no binding was evident when PAb-246

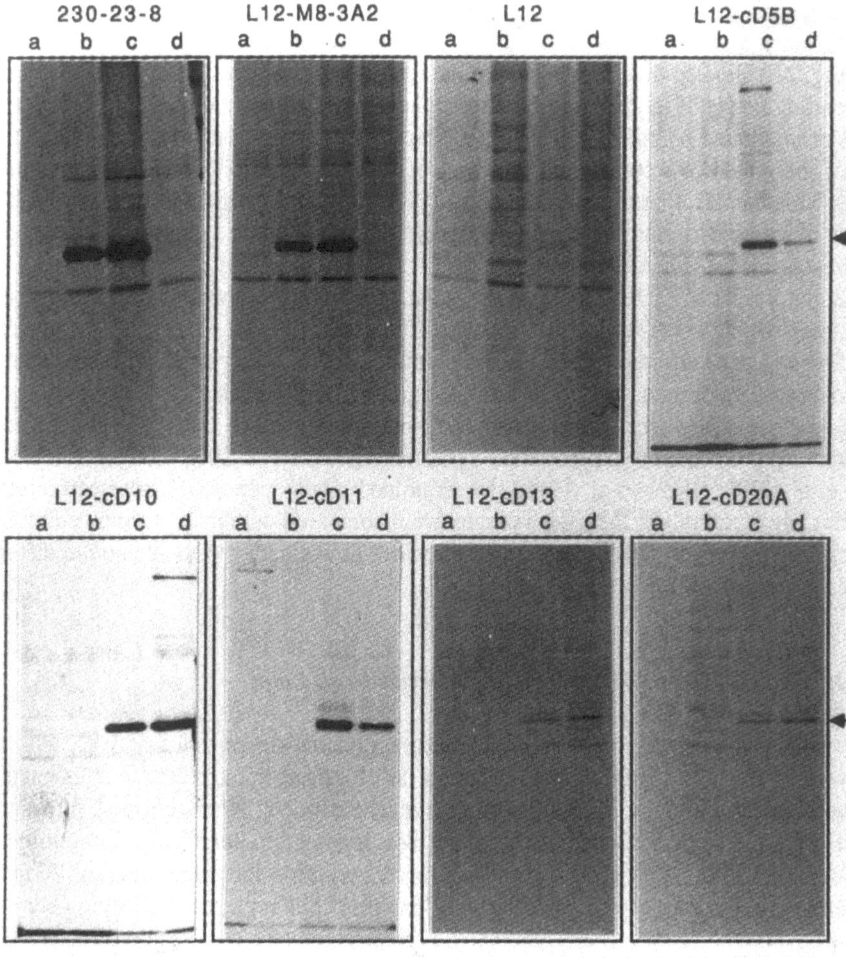

Fig. 1. Expression of wild-type p53 protein in L12-cD derived clones. Equal amounts of TCA-insoluble radioactive material were immunoprecipitated with the following specific antibodies: *a* non-immune serum; *b* monoclonal anti-p53 PAb-240; *c* monoclonal anti-p53 PAb-242 and *d* monoclonal anti-p53 PAb-246. *Arrows* point to the relative position of p53 protein

(lanes d) was used instead. *Wild type* and *mutant* p53 expressed the PAb-242 antigenic determinant (lanes c). In agreement with our previous studies (Rotter et al. 1981; Wolf and Rotter 1984; Wolf et al. 1984, Wolf et al. 1984a), L12 totally lacked p53 expression when analyzed with all available anti-p53 monoclonal antibodies (see Fig. 1). These results strongly suggest that the protein expressed in the L12 derived cell lines was the authentic *wild type* p53 protein that is expected of the p53cD cDNA clone. Based on the specific antigenic determinants expressed in the protein we concluded that no mutations or alterations in the *wild type* p53 protein had occurred upon gene transfer or drug selection. It should be added that measured levels of p53 protein expression were reproducible and stable in culture.

As these cell lines expressed various levels of the *wild type* p53 protein, it was important to assess the number of integrated cDNA copies in the individual clones. For that purpose, genomic DNA was prepared and digested with EcoRV which was expected to cut at a single site within the pSVL derived constructs. Southern blot analysis (Fig. 2A) showed that, in addition to the rearranged endogenous p53 genomic corresponding bands of the parental L12 cells, the various clones exhibited different patterns of integrated p53 cDNA sequences. No correlation was found between the number of integrated p53 copies and the level of p53 expression. Digestion of genomic DNA with BamHI, that released the p53 cDNA, indicated that in addition to variable levels of integrated rearranged p53 sequences, all of the clones contained intact p53 coding sequences (Fig. 2B).

In the present study we report on the isolation of five independent L12-cD derived cell lines which were single cell cloned, all harboring the *wild type* p53 sequences. All five cell lines are maintained stably in culture, representing an exceptional example of transformed cells which proliferate in spite of the constitutive expression of *wild type* p53. However, since it had been previously shown that p53 induces cell growth arrest in several experimental systems (Mercer et al. 1990, 1990a; Michalovitz et al. 1990; Baker et al. 1990), we examined whether these L12-derived clones exhibit any variations in their growth rate when compared with their parental cell line. Our following experiments were aimed therefore at evaluating possible subtle differences in the growth rate of these cell lines.

2.2 In Vivo Cell Growth Patterns of L12-Derived Cell Lines

To evaluate the effect of *wild type* p53 expression on the capacity of L12 cells to grow in vivo we measured the rate of tumor development in mice. These cells have a unique pattern of tumor development in syngeneic mice. When injected into syngeneic $C_{57}L/J$ mice L12 cells induce the development of large size local tumors (Rotter et al. 1980; Wolf et al. 1984a) that regress within a short time interval. Judged by these criteria, it is conceivable that L12 parental cells represent an in vivo benign growing tumor. To determine whether L12 tumorigenicity can be modified by the *wild type* p53 protein, cells from the various lines were injected into syngeneic mice and tumor development was monitored. Tumor development of the L12-cD derived clones was reduced when compared to the parental L12 cells. Expression of *wild type* p53 caused a reduction in the incidence and the size of the developing tumors. The

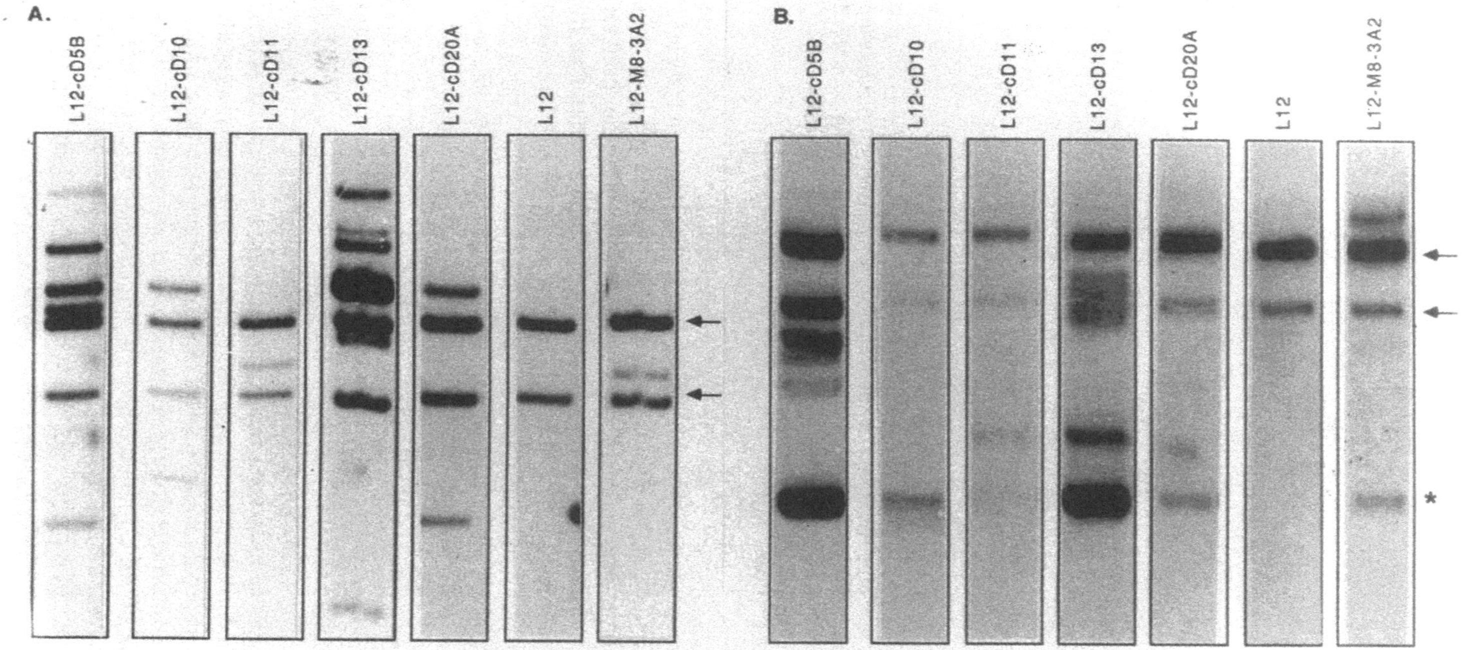

Fig. 2. Southern blot analysis of L12-cD derived clones. Genomic DNA was digested with either EcoRV (**A**) or BamHI (**B**), fractionated on 0.8% agarose gels, transferred to nitrocellulose filters and hybridized to radiolabelled full length p53cD cDNA insert. *Arrows* point to the endogenous p53 sequences, the *asterisk* (*) points to the intact 2 kb p53 cDNA integrated fragment

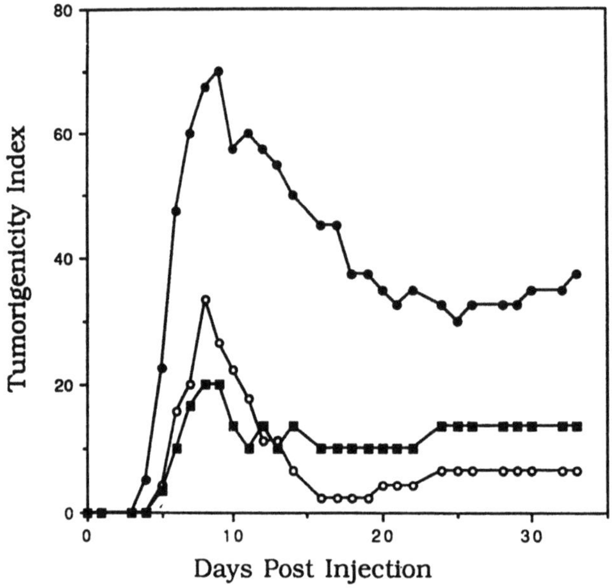

Fig. 3. Evaluation of tumor development of L12-cD derived in vivo. L12-derived clones, were injected subcutaneously into individual syngeneic, male $C_{57}L/J$ mice. Mice were monitored daily for tumor development and tumorigenicity index was calculated as described in experimental procedures. *Full circles* L12-induced tumors; *empty circles* L12-cD5B induced tumors; *full squares* L12-cD10 induced tumors

most striking results were obtained when L12-cD5B and L12-cD10 were compared with L12 parental cells (Fig. 3). A detailed comparison of the different cell lines indicated that this suppression in tumor development was correlated with the levels of p53 protein expressed in the individual L12-cD-derived cell lines as determined by immunoprecipitation (Fig. 1) and immunoblotting (data not shown). Clone L12-cD10 which expressed the highest levels of *wild type* p53 protein exhibited the lowest rate of tumor development (Fig. 3). Clone L12-cD20A which expressed lower levels of p53, (see Fig. 1) showed an in vivo pattern similar to that of the parental L12 cells (data not shown).

2.3 In Vitro Cell Growth Patterns of L12-Derived Cell Lines

To further evaluate differences in in vitro cell proliferation we measured the rate of ^{3}H-thymidine incorporation of the various cell lines. Equal numbers of viable cells were aliquoted into microtiter plates and the levels of ^{3}H-thymidine incorporation was measured at different time intervals. In repeated experiments we consistently found that L12-cD derived clones exhibit a reduced level of ^{3}H-thymidine incorporation, as compared with the parental L12 cell line. ^{3}H-thymidine incorporation was in inverse correlation with the amounts of constitutively expressed *wild type* p53. Indeed, clones L12-cD5B and L12-cD10, expressing rather high levels of p53 (see Fig. 1) exhibited the lowest ^{3}H-thymidine incorporation. Clone L12-cD20A, which expressed very low levels of the *wild type* p53, exhibited a pattern of ^{3}H-thymidine incorporation almost identical to the parental L12 cell line (Fig. 4A). Interestingly, however, although ^{3}H-thymidine incorporation was consistently different, the ratio between ^{3}H-thymidine incorporation of L12 parental cells and that of the L12-derived clones measured at different times points was almost constant. Calculation of the doubling times based on ^{3}H-thymidine incorporation clearly indicated almost identical values for the different cell lines examined (Fig. 4B, dotted bars). The fact that L12 and L12-cD derived cell lines that are of similar doubling time still exhibit significant variations in ^{3}H-thymidine incorporation, raises the possibility that the L12-cD clones are dynamic populations, consisting of dividing cells with a doubling time similar to the parental L12 cells, as well as growth arrested cells. Alternatively, these cell lines have an overall longer G_0/G_1 phase that is not reflected when doubling time is calculated by ^{3}H-thymidine incorporation that is measuring DNA replication. To resolve this, we have estimated the doubling time by counting the number of cells at various time points of growing cell lines. Using this parameter we found that while L12 or L12 expressing *mutant* p53 (L12-M8-3A2) have almost identical doubling times, L12-cD derived clones expressing *wild type* p53 exhibited an increment of up to 60% in the doubling time (L12-cD10, Fig. 4B open bars). Prolongation of doubling time was directly correlated with the levels of p53 synthesis. It should be added that at each time point, viability of cells was close to 100% and no morphological changes were observed. These results suggest, that the L12-cD derived cell lines represent dynamic cell populations consisting of a certain percentage of proliferating cells and an increased percentage of viable, but growth arrested cells. Lower ^{3}H-thymidine incorpo-

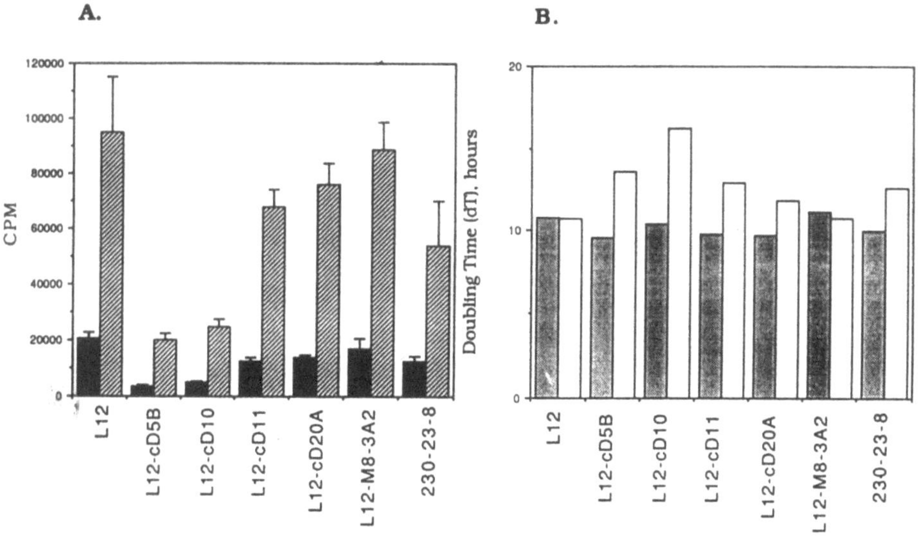

Fig. 4. Evaluation of in vitro cell growth of L12 derived clones. ^{3}H-thymidine incorporation (cpm above background), 48 h (*dotted bars*) and 72 h (*hatched bars*) after plating (A). Doubling times evaluated by ^{3}H-thymidine incorporation (h) calculated from ^{3}H-thymidine incorporation data *(dotted bars)* and from viable counts values (*empty bars*) of individual cell lines are presented (B)

ration may indicate, therefore, that cells have an extended G_0/G_1 phase, thus permitting less DNA replication within a given time interval.

To gain more insight into the effect of *wild type* p53 on the cell cycle of L12 cells, we determined the distribution of DNA content in the population by FACS analysis. A comparison of cell cycle profiles indicated different patterns between L12-cD derived clones and the parental L12 cells. All L12-cD cell lines consistently contained a significantly higher percentage of cells in the G_0/G_1 phase (up to 14%) then in L12 cells (Fig. 5). Again, we found a direct correlation between the levels of p53 expres-

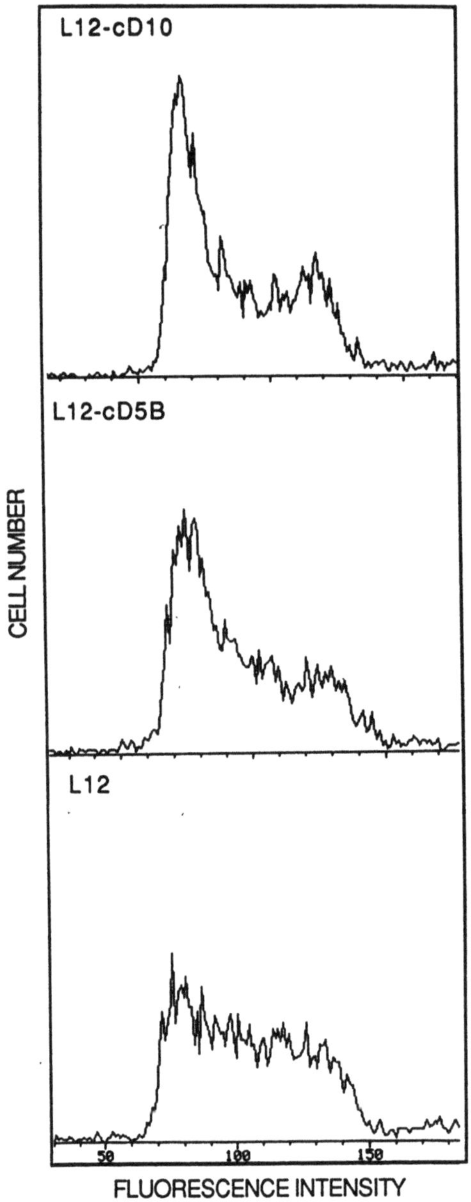

Fig. 5. The effect of wild-type p53 on L12 cell cycle. DNA histograms indicating propidium iodide fluorescence intensity were generated for L12, L12-cD5B and L12-cD10. On the *linear horizontal axis* scale, *70* represents 2N chromosomes and *140* represents 4N chromosomes.

Table 1. Distribution of cells at the various cell cycle phases[a]

Cell line	G_0/G_1 (%)	S (%)	M/G_2 (%)
L12	33.82	32.95	33.23
L12-cD10[b]	47.81	28.89	23.30
L12-M8-2C1[c]	38.86	34.20	26.94
L12-M8-3A2[c]	34.55	38.06	27.39
L12-*gpt*-1A3[d]	38.85	33.72	27.43

[a] Cell cycle was analyzed as described in experimental procedures.
[b] L12 cell transfected with p53cD, coding for wild-type p53, and pSV*gpt* drug resistance marker.
[c] L12 cell transfected with p53-M8, coding for *mutant* p53, and pSV*gpt* drug resistance marker.
[d] L12 cell transfected with pSV*gpt* drug resistance marker only.

sion and the increased numbers of cells accumulating in the G_0/G_1 phase. The relatively increased incidence of L12 cells in the S phase compared to L12-cD cells expressing the *wild type* p53 protein, may explain why ^{3}H-thymidine incorporation was higher in L12 cells. Figure 5 compares clones L12-cD5B and L12-cD10 with the parental L12 cell line. It should be noted that the cell cycle pattern of the L12 cell line is rather unusual in that the ratio of cells in the G_0/G_1 is low compared to those in the S phase. It can be seen from the same figure however, that L12-cD5B and L12-cD10, presented a typical cell distribution pattern with respect to cell cycle phases. The accumulation of cells at the G_0/G_1 phase indicated that the L12-cD populations contain a larger fraction of growth arrested cells. Analysis of cell cycle patterns of L12 derived cell lines established by transfection with either *mutant* p53 encoded by pSVLp53M8, L12-M8-2C1 and L12-M8-3A2 or by the drug resistance pSV-*gpt* gene only, L12-*gpt*-1A3, have shown no significant changes in their cell cycle patterns when compared with the parental L12 cell line (Table 1). This suggests that alterations in the cell cycle patterns observed in L12-cD derived cell lines is most likely a result of expression of *wild type* p53 protein rather then clonal selection. Next we evaluated whether these changes in the cell cycle patterns reflected a more differentiated cell population profile.

2.4 Pre-B Cell Differentiation

The L12 cell line was established by infection of bone-marrow cells with the Ab-MuLV virus (Rosenberg et al. 1976). Cell lines of this type were shown to be pre-B cells exhibiting typical genetic features and specific cell surface markers (Levitt et al. 1980; Maki et al. 1980). The differentiation pathway from early pre-B cells to mature B cells has been well characterized. Early cells in this pathway were found to contain the germ line heavy and light chain immunoglobulin genes. Upon maturation, the heavy chain immunoglobulin is rearranged, thus permitting the synthesis of μ immunoglobulin that accumulates initially within the cytoplasmic compartment (Siden et al. 1979; Siden et al. 1980). Analysis of pre-B cell lines established by the Ab-MuLV showed variable levels and different species of immunoglobulin synthesis. However, approximately 40% of these cell lines produce no immunoglobulin molecules detectable by either metabolic labelling or immunofluorescence (Alt et al. 1981). DNA from the vast majority of Ab-MuLV transformed cell lines, regardless of whether μ positive or immunoglobulin null cell lines, exhibited rearrangements in both heavy chain alleles (Alt et al. 1981). Some Ab-MuLV transformed cells, however, were shown to differentiate in vitro to a more advanced stage, involving rearrangements in the light chain immunoglobulin gene that is leading to the synthesis of κ light chain (Lewis et al. 1982).

In an effort to evaluate possible changes in cell differentiation stage of L12-cD derived cells expressing the *wild type* p53, we have analyzed the expression of immunoglobulin proteins as well as organization of the heavy and light immunoglobulin genes. To that end, cells were metabolically labelled and cell lysates were immunoprecipitated with anti-μ antibodies. Figure 6a, shows that, while no μ immunoglobulin was detected in L12 parental cells, L12-cD5B and L12-cD10 exhibited significantly increased levels of μ immunoglobulin synthesis. The levels of μ immunoglobulin synthesis that varied in the individual L12-cD cell lines, were found to be collinear with the levels of p53 synthesis (data not shown). Analysis of 112-M8-3A2 cell line expressing *mutant* p53 protein, did not reveal any significant μ immunoglobulin synthesis (Fig. 6a).

To further assess immunoglobulin expression in these cells, we have measured the steady state levels of cytoplasmic μ immunoglobulin. For that purpose we permeabilized the cell membranes of the various cell lines and measured the amount of fixed cytoplasmic μ immunoglobulin. FACS analysis of fixed and permeabilized cells stained with fluorescent anti-μ antibodies, showed that L12-cD5B and L12-cD10 exhibited 10–100 folds higher specific fluorescent cytoplasmic densities, compared to the parental L12 cells (Fig. 6b). It should be mentioned that fluorescence intensity was measured in a logarithmic scale. As no cell surface immunoglobulin was detected in any of the L12-cD derived clones (data not shown) we concluded that the μ immunoglobulin measured by immunoprecipitation represented the cytoplasmic μ immunoglobulin which was detectable by FACS analysis. Based on the results generated by these two experimental approaches we concluded that both the rate of synthesis and the steady state levels of μ immunoglobulin were augmented in L12-cD cell lines expressing *wild type* p53 protein.

Rearrangements in the immunoglobulin heavy chain gene in pre-B cells have been shown to precede the synthesis of μ heavy immunoglobulin protein. Therefore, it was of interest to examine whether augmented expression of μ involves rearrangements in the μ heavy immunoglobulin gene. Genomic DNA prepared from the various clones was digested with EcoRI and hybridized with $J_{I,I}$, that probes the J_3–J_4 region of the μ heavy chain immunoglobulin (Alt et al. 1981). Figure 7A shows that the parental L12 pre-B cells exhibited rearrangements in both heavy chain alleles. The typical 6.2 kb EcoRI fragment found in the germ line (Alt et al. 1981) (Fig. 7A NIH-3T3) was not evident in L12 cells; instead, smaller size fragments were defected. No further rear-

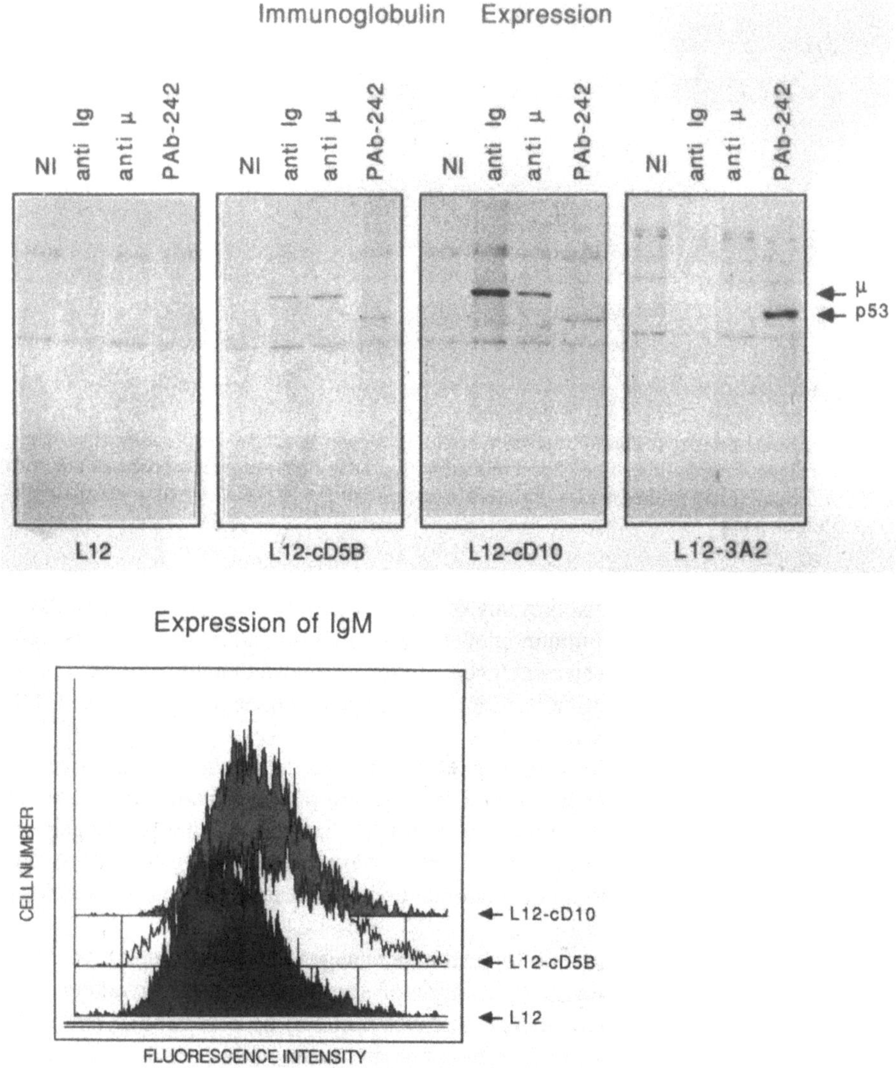

Fig. 6. Effect of wild-type p53 protein on μ heavy chain immunoglobulin expression in L12 cells. Immunoprecipitation of newly synthesized μ heavy chain immunoglobulin. Cells were lysed and immunoprecipitated with non-immune serum (*NI*), goat anti-mouse total immunoglobulin, goat anti-mouse μ antibodies, and anti-p53 PAb-242 monoclonal antibodies (**a**). Cytoplasmic μ immunoglobulin was detected in permeabilized cells stained with FITC conjugated, goat anti-mouse μ chain antibody. Logarithmic distributions of FITC-derived fluorescence intensity were assayed (**b**)

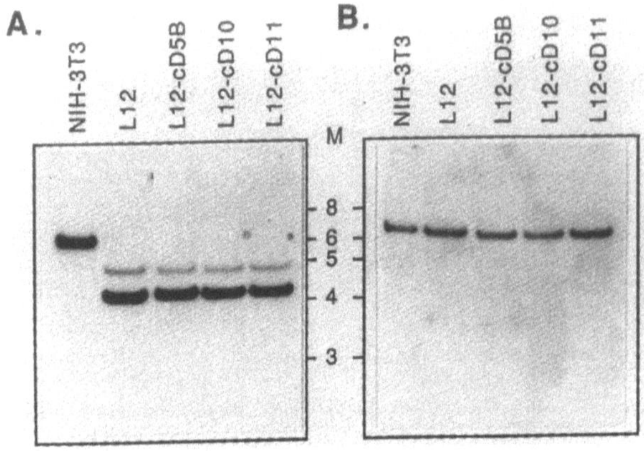

Fig. 7. Analysis of heavy μ and light κ immunoglobulin gene structure. High molecular weight DNA was digested with either EcoRI and probed with μ heavy chain specific probe (**A**) or with EcoRI and BamHI and probed with κ light chain-specific probe (**B**) *M* Molecular weight markers in kilobase pairs

rangements were, however, found in any of the L12-cD cell lines examined. Furthermore, when the κ light chain immunoglobulin was analyzed, no rearrangements were found in the L12 or L12-cD derived clones. Figure 7B shows that all cell lines examined showed the typical germ line, 6.8 kb EcoRI-BamHI fragment (Lewis et al. 1982) of the κ light chain immunoglobulin gene.

It is therefore possible that no further rearrangements occurred and that induction of μ immunoglobulin expression, induced by *wild type* p53 was mediated by a direct or indirect transactivation mechanism. An alternative possibility, that rearrangement is a continuous event which occurs at random in every differentiating cell in this cell population, a situation that does not permit detection of rearrangements in our assay system, can not be excluded.

An additional approach we took to determine changes in cell differentiation stage of the L12 derived cell lines was the analysis of specific cell surface markers. The B220 antigen is a B cell surface marker that was found to be expressed at relatively low densities on bone marrow pre-B cells and in greater amounts on mature B lineage cells (Coffman and Weissman 1981, 1983; Kinkade et al. 1981). A comparison between the parental L12 cells and the L12 derived clones expressing the *wild type* p53 have indicated that the latter expressed accentuated levels of the B220 cell surface marker. FACS analysis of stained cells, showed significantly higher fluorescent densities of B220 on the cell surface of L12-cD cell lines compared with the parental L12 cells (Fig. 8). This suggests that L12-cD derived clones represent a more advanced stage along the differentiation pathway. It is worth mentioning that the B220 cell surface marker is recognized by the RA3-2C2 monoclonal antibody (Coffman and Weissman 1981; Coffman and Weissman 1983) which also recognizes the p53 protein (Rotter et al. 1981). However, apart from the RA3-2C2 antigenic determinant, these two proteins share no other properties. L12 cells that totally lack the nuclear p53 (Rotter et al. 1980; Wolf et al. 1984a; 1984b; Wolf and Rotter 1984) still express

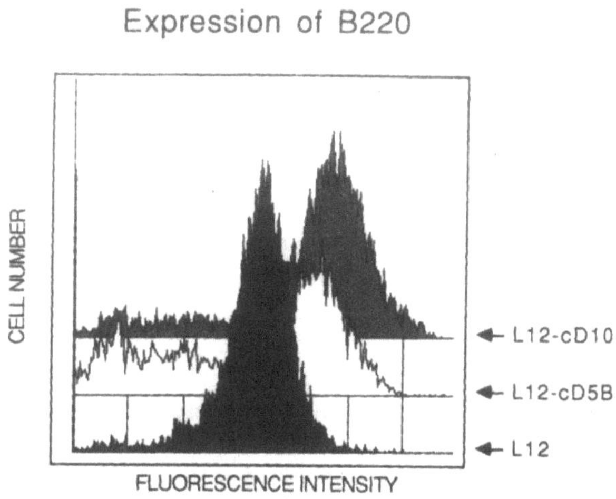

Fig. 8. The effect of wild-type p53 protein on B220 expression in L12 derived cell lines. Exponentially growing intact cells were reacted with a rat anti-mouse B220 antibodies and analyzed by FACScan (Becton-Dickinson) The logarithmic distributions of FITC derived fluorescence intensity of the various cell lines are presented

the cell surface B220 marker. In our experiments, scoring of B220 was performed with either RA3-2C2 or 14.8 (Kinkade et al. 1981), two independently established anti-B220 monoclonal antibodies. Results presented in Fig. 8 show the staining profiles obtained with the 14.8 anti-B220 antibodies.

The observation that L12-cD derived clones were induced to express cytoplasmic μ immunoglobulin, coupled with the fact that these cells expressed augmented levels of B220, suggests that the expression of *wild type* p53 caused these cells to advance to a more differentiated stage, compared with the parental cells from which they were derived.

3 Discussion

The p53 growth regulator is a short-lived protein that exerts its activity in normal cells at low molar concentrations. Overexpression of *wild type* p53 resulted in growth arrest of transformed cells (Mercer et al. 1990, 1990a; Michalovitz et al. 1990; Baker et al. 1990).

In addition to interfering with cell proliferation (Mercer et al. 1990, 1990a; Michalovitz et al. 1990; Baker et al. 1990), the *wild type* p53 protein inhibited malignant transformation of primary embryonic cells induced by transfection with other oncogenes (Eliyahu et al. 1989; Finlay et al. 1989), suggesting that p53 functions as a tumor suppressor gene. Several mechanisms of action of *wild type* p53 were proposed in these experimental systems. In all cases, however, the possibility that the endogenously expressed p53 affects the activity of exogenous *wild type* p53 was considered. In cases where *mutant* p53 was shown to cooperate with the ras oncogene in transforming primary embryonic rat fibroblasts, it was suggested that endogenous *wild type* p53 was inactivated because of the formation of a complex with the transfected *mutant* p53 which permitted the malignant transformation (Finlay et al. 1989; Levine

1990). In the case of colorectal tumors, again, the cell type was a determining factor in the inhibition of cell proliferation mediated by *wild type* p53 (Baker et al. 1990). In cases where the host cells expressed *mutant* p53, *wild type* p53 was inhibitory, whereas there was no effect in cells containing *wild type* p53. In all these experimental models which were studied, the activity of the transfected gene was superimposed on the activity of the endogenously expressed p53. Under these circumstances, the measured effects reflected the combined activity generated by the exogenous and endogenous p53 proteins (Finlay et al. 1989; Baker et al. 1990; Levine 1990).

In an effort to characterize the physiological activity of *wild type* p53, we used a strategy that consisted of studying the expression of this *wild type* gene in a cell system free of any endogenously expressed p53 protein. Studying L12 cells offered two main advantages. First, because these cells lack the expression of any endogenous p53 protein, introduction of *wild type* p53 was expected to represent the authentic unadulterated function of the p53 protein. Secondly, the fact that L12 cells are pre-B cells that can differentiate under in vitro conditions permitted studying the involvement of *wild type* p53 in cell differentiation.

The observation that *wild type* p53 affected the growth rate of cells, as measured by the development of tumors in vivo and by an extended doubling time in vitro, coupled with the fact that cells expressing *wild type* p53 protein acquired a different cell cycle pattern, indicated that L12-cD derived clones contained a higher percentage of growth arrested cells. These cell lines represent therefore profiles of more differentiated cell populations.

The conclusion that *wild type* p53 triggered cell differentiation was based on the observation that expression of this protein induced the synthesis of μ immunoglobulin in the L12 nonproducer cells and on the fact that cells expressing *wild type* p53 exhibited a higher level of the B220 cell surface marker. It is not clear, however, whether *wild type* p53 directly induced the expression of these two genes or whether one of them or yet a third gene was initially transactivated by *wild type* p53 leading in turn, to further cell differentiation.

The fact that the μ immunoglobulin synthesis induced in L12 cells following *wild type* p53 expression, was not accompanied by further rearrangements strongly suggests that p53 functions by a transactivation mechanism that facilitates the expression of the already rearranged heavy chain immunoglobulin gene. Activation of the μ heavy chain immunoglobulin gene expression was shown to be mediated by a large number of cellular proteins. These cellular proteins were shown to function through their direct binding to specific *motifs* found in the heavy chain enhancer sequences (Beckmann et al. 1990; Libermann et al. 1990; Weinberger et al. 1986; Nelsen et al. 1990; Araki et al. 1988). The possibility that *wild type* p53 functions in a similar way is strongly supported by the fact that this DNA binding protein (Steinmeyer and Deppert 1988; Kern et al. 1990) which is actively transported into the nucleus (Addison et al. 1990; Shaulsky et al. 1990; Shaulsky et al. 1990a) was shown to function as a transcription factor (Fields and Jang 1990; Raycroft et al. 1990). *Wild type* p 53 activity may be exerted by either direct binding to enhancer sequences in the μ heavy chain promoter or, alternatively by facilitating the activity of known immunoglobulin transactivators (Weinberger et al. 1986; Araki et al. 1988; Beckmann et al. 1990; Libermann et al. 1990; Nelsen et al. 1990).

Pre-B cells represent an early developing cell population capable of undergoing self renewal involving proliferation, and at the same time capable of entering a differentiation pathway leading to a cell maturation which was not found to be coupled with terminal cell growth arrest (Alt et al. 1981; Lewis et al. 1982). We propose that the *wild type* p53 expressed in these pre-B cells pushes them to advance a more mature stage; they then have a lower renewal capacity than the parental cells, but are on an accelerated differentiation pathway. The net balance of these cells still represents a growing population, which nevertheless exhibits a profile of more differentiated cells. L12 pre-B cells were induced to differentiate only when expressing *wild type* p53 but not when expressing *mutant* p53 or drug resistance. This strongly suggests that differentiation of these pre-B cells is a specific result of *wild type* p53 activity rather then clonal selection which might have occurred upon gene transfer and drug selection.

The apparent disagreement between our present study, where we report on the establishment of stably growing clones expressing *wild type* p53 and previous reports, showing unequivocally that p53 is a growth arrest gene, can be explained by the fact these experimental models are concerned with the expression of *wild type* p53 in different microcellular environments. It is plausible that in transformed gliosarcoma (Mercer et al. 1990b) or fibroblastic cells (Michalovitz et al. 1990) that are probably more advanced in their differentiation pathways, expression of *wild type* p53 induced terminal differentiation, leading to a net effect of growth arrest. It can therefore be assumed that p53 induces cell differentiation at either an earlier or a later phase on the route of cell differentiation. Manifestation of growth arrest induced by *wild type* p53 is most likely in inverse correlation with the cell differentiation state. When introduced at an early phase of cell differentiation, *wild type* p53 causes almost no growth arrest, whereas, at a more advanced phase cell growth arrest is predominant.

It is possible that L12 cells, which are pre-B cells can differentiate in vivo to a certain extent. Development of regressor tumors upon injection of L12 cells may be the manifestation of a differentiation process that could be accelerated following expression of *wild type* p53. Conversely, however, it is possible that when these cells express *mutant* p53 instead, cell differentiation is blocked, and thus cells cannot differentiate but they develop aggressive proliferating tumors instead. In agreement with this notion, previous reports (Mora et al. 1980; Ben-Dori et al. 1987; Reisman and Rotter 1989) as to modulations in p53 expression in differentiating cell populations, can be interpreted by the fact that *wild type* p53 is upregulated upon differentiation, this being the case in normal embryonic cells at 14–16 day of gestation (Mora et al. 1980). However, when *mutant* p53 is measured, for example, in Friend Leukemic cells (Ben-Dori et al. 1987), p53 must be downregulated to permit their cell differentiation.

In conclusion, we suggest that *wild type* p53 functions as a physiological cell differentiation factor. Introduction of *wild type* p53 into cells at an early phase of their differentiation pathway induces them to advance to a more differentiated stage. However, when cells are at a more advanced phase of their differentiation, expression of *wild type* p53 pushes them to terminal differentiation that is accompanied by growth arrest.

4 Experimental Procedures

4.1 Gene transfer

The p53-cD cDNA clone isolated from a normal T cell library, consists of the full length *wild type* p53 and an additional 95 base pairs containing an SV40 splice acceptor sequence upstream of the coding region, derived from the original pcD-p53 plasmid (Arai et al. 1986). The p53-M8 cDNA clone isolated from a Meth A λgt10 library, codes for a mutant p53 protein that has an alternatively spliced C-terminus (Wolf et al. 1985). The mammalian expression vector used in these experiments was pSVL (Pharmacia). This vector consists of SV40 late promoter, VP1 intron and SV40 late polyadenylation signal. The various cDNA clones were inserted into the unique BamHI site of the polylinker (Shaulsky et al. 1990a). The recombinant plasmid pSVLp53cD (20 μγ) or pSVLp53M8 (20 μγ) and selective marker, pSV*gpt* (Mulligen and Berg 1981) (5 μγ), were introduced into exponentially growing cells (2×10^7) by electroporation (Sudgen et al. 1985). Following electroporation, the cells were grown in RPMI-1640 supplemented with 10% heat inactivated Fetal Calf Serum (FCS) and 2×10^{-5} β-mercaptoethanol. After 24 h incubation, the cells were transferred to selectable RPMI-1640 medium containing 2 μγ/ml mycophenolic acid, 150 γλ/m xanthine and 15 μλ/ml hypoxanthine (Mulligen and Berg 1981). Growing clones detected 7 to 10 days after electroporation were single cell cloned by limiting dilution. Clonality was verified by immunofluorescent staining against p53 and morphology.

4.2 Immunoprecipitation

Cells were metabolically labelled for 1 h at 37 °C in methionine-less DMEM, supplemented with 10% heat-inactivated dialyzed FCS and 0.125 mCi/ml of ^{35}S-methionine (Amersham). Cells were lysed in a buffer containing 50 mM Tris-HCL pH 8. 5 mM EDTA, 150 mM NaCl, 0.5% Nonidet-40 and 1 mM PMSF, sonicated and precleared on 5% protein A-Sepharose CL-4B (Sigma). Equal amounts of TCA-insoluble radioactive material were incubated for 12 h at 4 °C with the following specific anti-p53 antibodies; monoclonal anti-p53 PAb-240 (Ganon et al. 1990) monoclonal anti-p53 PAb-242 (Yewdell et al. 1986) and monoclonal anti-p53 PAb-246 (Yewdell et al. 1986). The antibodies used were ascitic fluids from the peritoneal cavity of hybridoma-bearing syngeneic Balb/c mice which were spun to remove tissue debris, diluted in PBS, and used without further purification. To immunoprecipitate newly synthesized μ heavy chain immunoglobulin, exponentially growing cells were washed twice with PBS and metabolically labelled for 4 h with ^{35}S-methionine. Cells were lysed, precleared by ultracentrifugation (40 K, 30 min) and immunoprecipitated with goat anti-mouse total immunoglobulin or goat anti-mouse μ antibodies (Bio-Makor, Rehovot, Israel). The immune complexes were collected on protein A-Sepharose CL-4B and washed four times with a buffer containing 5% sucrose, 1% Nonidet-40, 0.5 M NaCl, 50 mM Tris-HCL pH 7.4 and 5 mM EDTA. Immunoprecipitates were separated on 10% SDS-polyacrylamide gels.

4.3 Genomic DNA Preparation and Analysis

Cells were washed with PBS, suspended in lysis buffer (0.5% SDS, 50 mM Tris-HCl pH8, 5 mM EDTA) and proteinase K (Boehringer-Mannheim) (0.2 μγ/ml) and incubated for 24 h at 37 °C. Cell lysates were extracted twice with an equal volume of TE-equilibrated phenol, followed by one extraction with an equal volume of chloroform-isoamyl alcohol (24:1). The DNA was precipitated with an equal volume of isopropyl alcohol, centrifuged, washed twice with 80% ethanol and dissolved in TE. For p53 specific sequences, genomic DNA was digested with either EcoRV or BamHI, fractionated on 0.8% agarose gels, transferred to nitrocellulose filters and hybridized to radiolabelled full length p53cD cDNA insert. For the analysis of heavy μ and light κ immunoglobulin genes structure, genomic DNA was prepared and digested either with EcoRI or with a combination of EcoRI and BamHI and probed with either a 2.0 kb EcoRI-BamHI insert of $J_{I, I}$ that hybridizes to the J_3–J_4 region of the heavy chain immunoglobulin, or a 0.5 kb PstI insert of Bg4 that hybridizes with the J_k region of the light chain immunoglobulin (probes were a kind gift from Prof. E. Canaani and A. Rosner, Department of Chemical Immunology, The Weizmann Institute of Science).

4.4 In vivo Doubling Time

In order to achieve equal exponential growth conditions, cells were plated at 2 x 10^5 cells/ml and incubated without any selectable drugs at 37 °C for 24 h. Out of the growing culture, an inoculum was removed and re-plated at 2 x 10^5 cells/ml in fresh medium as above. Twenty-four hours later, in vitro growth parameters were examined.

For ^{3}H-thymidine incorporation into DNA, cells were plated in 96 well plates at 5 x 10^3 cells/ml. Twelve to 72 h later, cells were pulse labelled with 10 μXι/ml ^{3}H-thymidine (Nuclear Research Center-Negev, Be'er Sheva, Israel) for 4 h and harvested onto GF/C filters using an automatic cell harvester (Dynatech). Dried filters were counted by liquid scintillation and radiolabelled nucleotide incorporation was calculated as a mean of 11 repetitions above background. Viable cell count was determined at 12–18 h intervals after plating by staining triplicate samples with 0.05% Eosin-Y in PBS. In all cases, cell viability was over 98%.

4.5 Cell Cycle Analysis

Exponentially growing cells were pulse labelled with 10 μM Bromodeoxyuridine (BrdU) (Sigma) for 10 min, washed twice with PBS and fixed with 70% ethanol at 0 °C for 30 min. HCl was added to a final concentration of 2N and cells were further incubated for 30 min at room temperature, washed once with 0.1 M $Na_2B_4O_7$ pH 8.5 and resuspended in PBS containing 0.5% Tween 20. Cells were stained with an FITC-conjugated mouse anti-BrdU antibody (Becton-Dickinson no. 7583) according to the manufacturers reco mMended procedure, resuspended in PBS containing 50 μγ/ml propidium iodide (Sigma) and at least 5 x 10^3 cells (as gated by light scatter) were analyzed by FACScan (Becton-Dickinson using the Consort 30 program). DNA histograms indicating propidium iodide fluorescence intensity were generated.

4.6 Detection of Cytoplasmic μ Immunoglobulin

Cells were permeabilized and fixed with pre-cooled (–20 °C) 70% ethanol for 30 min at 0 °C. Cells were then stained for 30 min at 0 °C with a 1:10 dilution of FITC conjugated, goat anti-mouse μ chain antibody (Sigma Immunochemicals). Following two washes with PBS, cells were resuspended in PBS containing 50 μγ/ml propidium iodide and at least 5 x 10^3 cells (as gated by light scatter) were analyzed by FACScan (Becton-Dickinson). The logarithmic distributions of FITC derived fluorescence intensity of the various cell lines are presented.

4.7 Expression of the B220 Cell Surface Marker

Exponentially growing cells were washed twice with PBS and reacted with a rat anti-mouse B220 antibody (Kinkade et al. 1981) (B220 clone 14.8, American Tissue Culture Collection) for 30 min at 0 °C. Cells were washed twice with PBS and reacted with an FITC conjugated, rabbit anti-rat IgG (Bio-Makor, Rehovot, Israel) for 30 min at 0 °C. Cells were washed twice and analyzed by FACScan (Becton-Dickinson).

4.8 Tumorigenicity in mice

L12-derived clones, from the earliest passage available, were grown in RPMI-1640 medium without the selective drugs for 48 h prior to injection. Cells were washed and resuspended in PBS, 2 x 10^6 cells were injected subcutaneously into individual syngeneic male $C_{57}L/J$ mice. The mice were monitored daily for tumor development and were graded from 0 to 5, according to tumor size and the overall clinical status: 0, no detectable tumors; 1, very small palpable tumors; 2, visible tumors, smaller than 0.5 in diameter; 3, tumors larger than 0.5 cm in diameter, no obvious paralysis; 4, tumors larger than 0.5 cm in diameter with partial or complete paralysis; 5, death. At each survey, the individual scores were summed up. The sum was divided by the maximal possible score (which stands for 100% death), and multiplied by 100 to represent the tumorigenicity index.

Acknowledgements. This work was supported in part by a grant from the Leo and Julia Forchheimer Center for Molecular Genetics, and the Ebner foundation, the Weizmann Institute of Science. V. R. holds the Norman and Helen Asher Professorial Chair in Cancer Research and a Career Development Award from the Israel Cancer Research Fund. The authors would like to thank Prof. D. Zipori and Dr. D. Reisman for fruitful discussion and criticism. Ms. M. Baer prepared and edited the manuscript.

References

Addison C, Jenkins, JR & Sturzbecher HW (1990) The p53 nuclear localization signal is structurally linked to a $p34^{cdc2}$ kinase motif. Oncogene 5:423–426

Ahuja H, Bar-Eli M, Advani SH, Benchimol S & Cline MJ (1989) Alterations in the p53 gene and the clonal evolution of the blast crisis of chronic myelocytic leukemia. Proc Natl Acad Sci USA 86:6783–6787

Alt F, Rosenberg N, Lewis S, Thomas E & Baltimore D (1981) Organization and reorganization of immunoglobulin genes in A-MuLV-transformed cells: rearrangement of heavy but not light chain genes. Cell 27:381–390

Arai N, Nomura D, Yokota K, Wolf D, Brill E, Shohat O & Rotter V (1986) Immunologically distinct p53 molecules generated by alternative splicing. Mol Cell Biol 6:3232–3239

Araki K, Maeda H, Wang J, Kitamura D & Watanabe T (1988) Purification of a nuclear *trans*-acting factor involved in the regulated transcription of a human immunoglobulin heavy chain gene. Cell 53:723–730

Baker SJ, Markowitz S, Fearon ER, Willson JKV & Vogelstein B (1989) Chromosome 17 deletions and p53 gene mutations in colorectal carcinomas. Science 244:217–221

Baker SJ, Markowitz S, Fearon ER, Willson JKV & Vogelstein B (1990) Suppression of human colorectal carcinoma cell growth by wild-type p53. Science 249:912–915

Beckmann H, Su LK & Kadesch T (1990) TFE3: A helix-loop-helix protein that activates transcription through the immunoglobulin enhancer mE3 motif. Genes Dev 4:167–179

Ben-Dori R, Resnitzky D & Kimchi A (1987) Changes in p53 mRNA expression during terminal differentiation of murine erythroleukemia cells. Virology 161:607–611

Coffman RL & Weissman IL (1983) Immunoglobulin gene rearrangement during pre-B cell differentiation. Mol Immunol 1:31–38

Coffman RL & Weissman IL (1981) B220: a B cell specific member of the T200 glycoprotein family. Nature 289:681–683

Eliyahu D, Goldfinger N, Pinhasi-Kimhi O, Shaulsky G, Skurnik Y, Arai N, Rotter V & Oren M (1988) Meth A fibrosarcoma cells express two transforming mutant p53 species. Oncogene 3:313–321

Eliyahu D, Michalovitz D, Eliyahu S, Pinhasi-Kimhi O & Oren M (1989) Wild-type p53 can inhibit oncogene-mediated focus formation. Proc Natl Acad Sci USA 86:8763–8767

Eliyahu D, Raz A, Gruss P, Givol D & Oren M (1984) Participation of p53 cellular tumor antigen in transformation of normal embryonic cells. Nature 312:646–649

Fields S & Jang SK (1990) Presence of a potent transcription activating sequences in the p53 protein. Science 249:1046–1049

Finlay CA, Hinds PW & Levine AJ (1989) The p53 proto-oncogene can act as a suppressor of transformation. Cell 57:1083–1093

Gannon JV, Greaves R, Iggo R & Lane DP (1990) Activating mutations in p53 produce a common conformational effect: a monoclonal antibody specific for the mutant form. EMBO J 9:1595–1602

Hinds P, Finlay C & Levine AJ (1989) Mutation is required to activate the p53 gene for cooperation with the ras oncogene and transformation. J Virol 63:739–746

Jenkins JR, Rudge K & Currie GA (1984) Cellular immortalization by a cDNA clone encoding the transformation-associated phosphoprotein p53. Nature 312:651–654

Kelman Z, Prokocimer M, Peller S, Kahn Y, Rechavi G, Manor Y, Cohen A & Rotter V (1989) Rearrangements in the p53 gene in Philadelphia chromosome positive chronic myelogenous leukemia. Blood 74:2318–2324

Kern SE, Kinzler KW, Baker SJ, Nigro JM, Rotter V, Levine AJ, Friedman P, Prives C & Vogelstein B (1990) Abnormal DNA-binding and phosphorylation of mutant p53 protein in vitro. Oncogene (in press)

Kinkade PW, Lee G, Watanabe T, Sun L & Scheid MP (1981) Antigens displayed on murine B lymphocyte precursors. J Immunol 127:2262–2268

Lane D and Benchimol S (1990) p53: Oncogene or anti-oncogene Genes Dev 4:1–8

Levine AJ (1990) The p53 protein and its interactions with the oncogene products of the small DNA tumor viruses. Virology 177:419–426

Levitt D & Cooper MD (1980) Mouse pre-B cells synthesize and secrete m heavy chains but not light chains. Cell 19:617–625

Lewis S, Rosenberg N, Alt F & Baltimore D (1982) Continuing kappa-gene rearrangement in a cell line transformed by Abelson murine leukemia virus. Cell 30:807–816

Libermann TA, Lenardo M & Baltimore D (1990) Involvement of a second lymphoid-specific enhancer element in the regulation of immunoglobulin heavy-chain gene expression. Mol Cell Biol 10:3155–3162

Maki R, Kearney JK, Paige C & Tonegawa S (1980) Immunoglobulin gene rearrangement in immature B cells. Science 209:1366–1369

Mercer WE, Amin M, Sauve GJ, Appella E, Ulirich SJ & Romano JW (1990) Wild type human p53 is antiproliferative in SV40-transformed hamster cells. Oncogene 5:973–980

Mercer WE, Shields MT, Amin M, Sauve GJ, Appella E, Romano JW & Ullrich SJ (1990b) Negative growth regulation in a glioblastoma tumor cell line that conditionally expresses human wild-type p53. Proc Natl Acad Sci USA 87:6166–6170

Michalovitz D, Halevy O & Oren M (1990) Conditional inhibition of transformation and of cell proliferation by a temperature-sensitive mutant of p53. Cell 62:671–680

Mora PT, Chandrasekaran K, McFarland W (1980) An embryo protein induced by SV40 virus transformation of mouse cells. Nature 288:722–724

Mulligen RC & Berg P (1981) Selection for animal cells that express the *E. coli* gene coding for xanthine guanine phosphoribosyltransferase. Proc Natl Acad Sci USA 78:2072–2076

Munroe DG, Rovinski B, Bernstein A & Benchimol S (1988) Loss of a highly conserved domain on p53 as a result of gene deletion during Friend-virus-induced erythroleukemia. Oncogene 2:621–624

Nelsen B, Kadesch T & Sen R (1990) Complex regulation of the immunoglobulin μ heavy-chain gene enhancer: μB, a new determinant of enhancer function. Mol Cell Biol 10:3145–3154

Nigro JM, Baker JS, Preisinger AC, Jessup JM, Hostetter K, Cleary K, Bigner SH, Davidson N, Baylin S, Devilee P, Glover T, Collins FS, Weston A, Modali R, Harris CC & Vogelstein B (1989) p53 gene mutations occur in diverse human tumor types. Nature 342:705–708

Parada LF, Land H, Weinberg RA, Wolf D & Rotter V (1984) Cooperation between gene encoding p53 tumor antigen and ras in cellular transformation. Nature 312:649–651

Raycroft L, Wu H & Lozano G (1990) Transcriptional activation by wild type but not transforming mutants of the p53 anti oncogene. Science 249:1049–1051

Reisman D & Rotter V (1989) Two promoters that map to 5'-sequences of the human p53 gene are differentially regulated during terminal differentiation of human leukamic cells. Oncogene 4:945–953

Rosenberg N & Baltimore D (1976) A quantitative assay for transformation of bone marrow cells by Abelson murine leukemia virus. J Exp Med 143:1453–1463

Rotter V, Witte ON, Coffman R & Baltimore D (1980) Abelson-murine leukemia virus-induced tumors elicit antibodies against a host cell protein, p50. J Virol 36:547–555

Shaulsky G, Ben-Ze'ev A & Rotter V (1990) Subcellular distribution of the p53 protein during the Cell Cycle of Balb/c 3T3 Cells. Oncogene 11:1707–1711

Shaulsky G, Goldfinger N, Ben-Ze'ev A & Rotter V (1990a) Nuclear localization of p53 is mediated by several nuclear localization signals and plays a role in tumorogenesis. Mol Cell Biol 10:6565–6577

Siden E, Alt FW, Shinfeld L, Sato V & Baltimore D (1981) Synthesis of immunoglobulin μ chain gene products precedes synthesis of light chains during B lymphocyte development. Proc Natl Acad Sci USA 78:1823–1827

Siden EJ, Baltimore D, Clark D & Rosenberg NE (1979) Immunoglobulin synthesis by lymphoid cells transformed in vitro by Abelson murine leukemia virus. Cell 16:389–396

Steinmeyer K & Deppert W (1988) DNA binding properties of murine p53. Oncogene 3:501–507

Sugden B, Marsh K & Yates J (1985) A vector that replicates as a plasmid can be efficiently selected in B-lymphocytes transformed by EBV. Mol Cell Biol 5:410–413

Takahashi T, Nau MM, Chiba I, Birrer MJ, Rosenberg RK, Vinocour M, Levitt M, Pass H, Gazdar AF & Mina JD (1989) p53: A frequent target for genetic abnormalities in lung cancer. Science 246:491–494

Weinberger J, Baltimore D & Sharp PA (1986) Distinct factors bind to apparently homologous sequences in the immunoglobulin heavy-chain enhancer. Nature 332:846–848

Wolf D & Rotter V (1984) Inactivation of p53 gene expression by an insertion of Moloney murine leukemia virus-like DNA sequences. Mol Cell Biol 4:1402–1410

Wolf D & Rotter V (1985) Major deletions in the gene encoding the p53 antigen cause lack of p53 expression in HL-60 cells. Proc Natl Acad Sci USA 82:790–794

Wolf D, Admon S, Oren M & Rotter V (1984) Abelson murine leukemia virus-transformed cells that lack p53 protein synthesis express aberrant p53 mRNA species. Mol Cell Biol 4:552–558

Wolf D, Harris N & Rotter V (1984a) Reconstitution of p53 expression in a non-producer Ab-MuLV-transformed cell line by transfection of a functional p53 gene. Cell 38:119–126

Wolf D, Harris N, Goldfinger N & Rotter V (1985) Isolation of a full length cDNA clone coding for an immunological distinct p53 molecule. Mol Cell Biol 5:127–132

Yewdell JW, Gannon JV & Lane DP (1986) Monoclonal antibody analysis of p53 expression in normal and transformed cells. J Virol 59:444–452

The Growing Biological Scenario of Growth Arrest

C. Schneider[1], G. Del Sal[1], C. Brancolini[1], S. Gustincich[1], G. Manfioletti[1]
and M. E. Ruaro[1]

1 Introduction

Cell proliferation is governed by a fine and dynamic equilibrium between the "in cycle" and the "out of cycle" compartments. Positive (oncogenes) and negative (antioncogenes) effector elements are deputed to regulate this equilibrium: the balance between the positive and negative circuits thus represents a continuous integration of all acting elements.

A vast array of elements in the positive circuit is now known through the study of oncogenes (Bishop 1991). Positive elements within the core of the cell-cycle regulating machinery have also been discovered in the last few years (Nurse 1991). Just as oncogenes have been defined through the activating mutations of the respective protooncogenes (gain of function), genetic lesions (loss of function) have defined the tumor suppressor genes. These are thus defined as "dispensable" components of the negative circuit that enable a cell to receive and process growth inhibitory signals from its surrounding. As a result cell proliferation potential is increased and deregulated (Weinberg 1991; Marshall 1991) as typically observed in transformed cells. Pioneering work on cell fusion in the late sixties (Harris 1970) first established the existence of this class of "tumor suppressor genes", based on the fact that fusion of tumor cells with normal cells almost invariably resulted in the outgrowth of nontumorigenic hybrids.

Another class of proliferation suppressor genes would not obey the above definition since their loss is incompatible with life, but they are still considered as bona fide suppressors of proliferation. These genes may act as transducers of survival signals that are themselves negative signals in terms of proliferation control.

Animal cells have evolved a complex and multifaceted system of proliferation control, which has to obey the harmonious rules imposed by tissue organization and architecture. Proliferation restrictions do not come solely from nutritional factors but mainly from: (1) adhesion to extracellular matrix and factors adsorbed to it, (2) cell-cell adhesion mediated through the plasma membrane, (3) exchange of diffusible molecules through cell junctions.

Proliferation restrictions can commit cells to one of the following programs: (1) reversible withdrawal from cell cycle (Go), (2) terminal differentiation, (3) suicidal death (through apoptosis or senescence)

---

[1] ICGEB (International Centre for Genetic Engineering and Biotechnology) and CIB (Consortium for Interuniversitary Biotechnology) Laboratories, Area di Ricerca, 99 Padriciano, 34012 Trieste, Italy.

43: Colloquium Mosbach 1992
DNA Replication and the Cell Cycle

Our work in recent years has concentrated on the study of the genetic program of growth shutdown leading to reversible cell cycle withdrawal (Go). When cultured fibroblasts are exposed to growth restricting conditions such as diminished growth factors in the medium (serum deprivation) or high cell density (contact inhibition) they enter the Go phase. As a response to these external growth restricting conditions fibroblasts have been shown to increase the expression of a particular set of genes that has been called *gas* (growth arrest specific) (Schneider et al. 1988). When growth-arrested cells are reintroduced into cell cycle, expression of *gas* genes is rapidly down-regulated.

Gas genes are strictly associated with the "out of cycle" compartment and their downregulation could represent a prerequisite for entering a new cycle. Their possible involvement as elements of the negative circuit could thus be proposed. Altogether *gas* genes have represented the first attempt to dissect the biology of the growth arrest and we shall now examine what is known about them.

2 Results

2.1 Gas 3

Gas 3 has been reported (Manfioletti et al. 1990) to encode a 22 kd integral membrane glycoprotein with one N-glycosylation site. Through data base search, two groups (Spreyer et al. 1991, Welcher et al. 1991) have independently reported a strict homology between the mouse *gas* 3 sequence and a rat cDNA. This clone was isolated because it shows an axon regulated expression: it is abundantly expressed in differentiated quiescent Schwann cells in the peripheral nervous system (PNS) and following sciatic nerve injury is rapidly down-regulated. After nerve injury Schwann cell proliferation is resumed: this represents the first example of an equivalent "in vivo" regulation of a *gas* gene.

Gas3 (PMP-22) protein has been shown to be synthesized by Schwann cells and to be a major component of PNS myelin; this indicates it could be a differentiation-specific gene involved in myelin formation (Sinpes et al. 1992).

Our evidence argues against this role since its mRNA is expressed in a regulated manner in fibroblasts and is also highly expressed in other non-neuronal tissues such as lung and intestine of the mouse. We also obtained recent evidence that Gas3 protein is expressed in the mouse lung (T. Dragani, pers. comm.). Previously a drop in *gas* 3 mRNA expression was reported in chemically induced tumors of the mouse lung (Re et al. 1992).

We would thus favour a role for Gas3 protein that is not limited to a specific differentiative function (such as myelin formation in Schwann cells) but could be more generally linked to growth control also in other systems.

In fact it has recently been shown that a point mutation Gly->Asp in the last membrane domain of Gas3-PMP22 seems to be responsible for the *tr* (trembler) phenotype in mice (both the gene responsible for *tr* and Gas3-PMP22 map in fact in the same chromosomal locus) (Suter et al. 1992). This autosomal dominant disease which manifests as a Schwann cell defect, is characterized both by severe hypomyelination of PNS and continuing Schwann cell proliferation throughout life.

The reported evidence indicates that Gas3-PMP22 would have a double role in the association between the differentiative pathway of Schwann cells and their cessation of proliferation. Whether Gas3 function is mediated though receptor like action for an extracellular diffusible growth regulator (which could be produced by the migrating axon or other cells) or as a receptor for cell-cell interaction (either homophilic or heterophilic) it is not known yet. It is possible that the signal converged through Gas3 could be a "survival signal" necessary for surviving in a particular differentiated state. It should be added here, that the *gas3* functionally homologue gene in the central nervous system (PLP, proteolipid protein), is similarly involved in myelination and growth control. It has in fact been shown that oligodendrocytes from mutant mice deficient in PLP (*jimpy*) proliferate more rapidly than normal cells (Skoff 1982) but most of these cells ultimately die (Knapp et al. 1986).

2.2 Gas 2

The biosynthesis of Gas2 protein in NIH 3T3 cells exactly reflects the pattern of its mRNA expression. Since the protein half-life has been determined to be longer than 12 hrs, the total amount of protein does not change appreciably when growth arrested cells are reintroduced into the growth cycle. However, there is a remarkable difference in the phosphorylation level between the Gas2 protein present in serum-starved and growth-stimulated cells (Brancolini et al. 1992). Indeed, Gas2 is highly phosphorylated at serine residue(s) between 5 and 15 min after serum addition (Brancolini et al. in prep).

We have studied in vitro the timing of activation of the kinase(s) which phosphorylates Gas2. This occurs within 5–10 min after the addition of serum, in accordance with the in vivo studies. Probably a fast switch, as generated by phosphorylation rather than by changes in its amount, might be a more convenient system to regulate its activity.

Gas2 does not show any homology with previously cloned genes, but the intracellular localization of its protein product has shed light on its hypothetical function. Gas2 is present in serum-starved cells at the level of the cell border, where it colocalizes with the microfilaments network system; it is also detectable, albeit at lower intensity, along the stress fibers (Brancolini et al. 1992).

An interesting point regards the localization of Gas2 in cells which normally do not express it, such as exponentially growing NIH 3T3 cells. A microvillar apparatus is present, on the cell surface, in about 10% of the cells. If Gas2-GST (glutathione S-transferase) bacterially produced fusion protein is injected into these cells it clearly localizes in the microvillar apparatus. These data strengthen the idea that Gas2 is a component of the cortical actin filament system.

Work in progress on the analysis of Gas2 function has shown that overexpression of deleted derivatives of the carboxy terminus of the *gas2* gene in NIH 3T3 fibroblasts is able to modify the architecture of the microfilament network system and profoundly alter cellular shape. Therefore, we focus our attention on the active role of

Gas2 in the organization of the microfilament system and cell shape. A possible link between growth control and cell morphology has been recently demonstrated for the polyphosphoinositide/actin binding protein profilin which can interact with the RAS pathway in *Saccharomyces cerevisiae* (reviewed in Goldschmidt-Clermont and Janmey 1991).

2.3 Gas 6

The complete cDNA sequence of the human and murine Gas6 proteins (Manfioletti et al. in prep.) has indicated a clear homology with both bovine and human vitamin K-dependent protein S, (43% identity) which are involved in the negative cascade of blood coagulation (reviewed in Esmon 1989).

The highest homology (53% identity) between Gas6 and protein S is located in the first amino acid residues in the amino terminus (aa 1-87). This region of protein S contains the γ-carboxy glutamic acid (GLA) domain which is required for the calcium-dependent membrane binding, a characteristic property that could be conserved between Gas6 and protein S.

Another common feature between Gas6 and protein S relates to the presence of four epidermal growth factor-like EGF repeat domains which are located in both cases in the central part of the proteins. An interesting difference is the absence in Gas6 of the thrombin-sensitive site which is present in protein S between the GLA domain and the first EGF-like domain. Lower homologies are detected within the carboxy terminus of the proteins (40% identity).

A different pattern of expression of *gas6* and protein S genes is detected in various human tissues. Protein S is predominantly expressed in the liver; on the other hand, Gas6 is expressed in a more promiscuous way, being detectable in the brain, lung, spleen, intestine, and at a lower level in liver. This suggests that even though Gas6 and protein S show similar structural motifs, reflecting a common evolutionary origin, they might serve different functions.

Recently it has been demonstrated that protein S is mitogenic for smooth muscle cells at physiological concentrations (Gasic et al. 1992). Therefore, it may not be misleading to hypothesize that Gas6 too might be involved in the control of cellular proliferation or cell survival.

2.4 Gas 1

gas1 is the only *gas* gene which is regulated at the transcriptional level. The kinetics of appearance of *gas*1 either in asynchronously growing cells after serum starvation or in cells arrested by density inhibition is inversely correlated with the expression of c-*myc*. This relation is also evident when the expression of *gas*1 is followed during the Go-> S progression of serum-starved cells. In this case *gas*1 expression level falls to the minimum level within 3 h after serum stimulation.

The same behavior described for the *gas*1 mRNA is observed by Western blot analysis for the corresponding protein (Gas1).

Sequence analysis of *gas*1 cDNA indicates that *gas*1 is a plasma membrane glycoprotein with two putative transmembrane domains. Although sequence comparison does not show homology with any known protein, the presence in the extracellular region of sequences such as RGD and Ser-Gly, suggests that Gas1 may be involved in cell-cell or cell-extracellular matrix interactions. In fact the RGD sequence represents the recognition signal of cell adhesion receptors of the integrin family (Ruoshlahti and Pierschbacher 1987), while the sequence Ser-Gly residues are potentially involved in the glycosaminoglycan (GAG) conjugation (Saunders et al. 1989).

Since both *gas*1 mRNA and protein levels are decreased after cell cycle entry, we asked the question whether its ectopic expression, obtained by automated microinjection, could affect DNA synthesis in murine fibroblasts. Overexpression of Gas1 causes a blockage in the cell cycle progression either within the Go->S transition of quiescent cells stimulated to reenter cell cycle or in asynchronously growing cells. Interestingly, the expression of elements of the early serum response such as c-*fos* and c-*jun*, which are involved in the control of the Go->S progression (Kovary and Bravo 1991), is not affected by *gas*1 overexpression (Del Sal et al. submitted).

Transformed cells maintained in low serum are unable to accumulate Gas1 at the same levels as observed in the untransformed cells. On the other hand, they are also unable to stop cell proliferation in response to restricted growth conditions. Nevertheless when Gas1 is overexpressed in cycling transformed cells, they become arrested in cell cycle progression as efficiently as untransformed cells.

The evidence that both normal and transformed cells respond to the Gas1 overexpression in the same way by operating a blockage in cell cycle progression suggests the existence of a constitutive negative pathway leading to growth shutdown (DGS = default growth shutdown) (Del Sal et al. 1992 submitted).

3 Discussion

Among the *gas* genes analyzed, *gas*1 is the only one that has been demonstrated to block cell cycle progression (Del Sal et al. submitted). In addition its regulation seems to occur at the transcriptional level (Manfioletti et al. 1990), while the other *gas* genes seem to be all regulated at the post-transcriptional level (Ciccarelli et al. 1990; Manfioletti et al 1990). It is thus possible that *gas*1 represents an earlier event in the growth shutdown program than the other *gas* genes. Moreover, the absence of any nuclear located-*trans*-activating products among the known *gas* genes reinforces the fact that "early" *gas* genes have still to be discovered. It is possible that the analysis of *gas*1 transcriptional regulatory elements can lead us backward into the early events that control DGS program. The hypothesized DGS program has in fact two requirements in order to be operative: (1) a lower active level of positive elements (such as c-*myc*), (2) a higher functional level of the negative elements (such as *gas*1) that are themselves maintained repressed by the positive elements. In contrast, the "in cycle" compartment has the inverse requirements in order to suppress the DGS program: (1) higher functional level of positive elements (such as c-*myc*) to neutralize DGS and thus (2) a lower active level of negative elements (such as *gas*1).

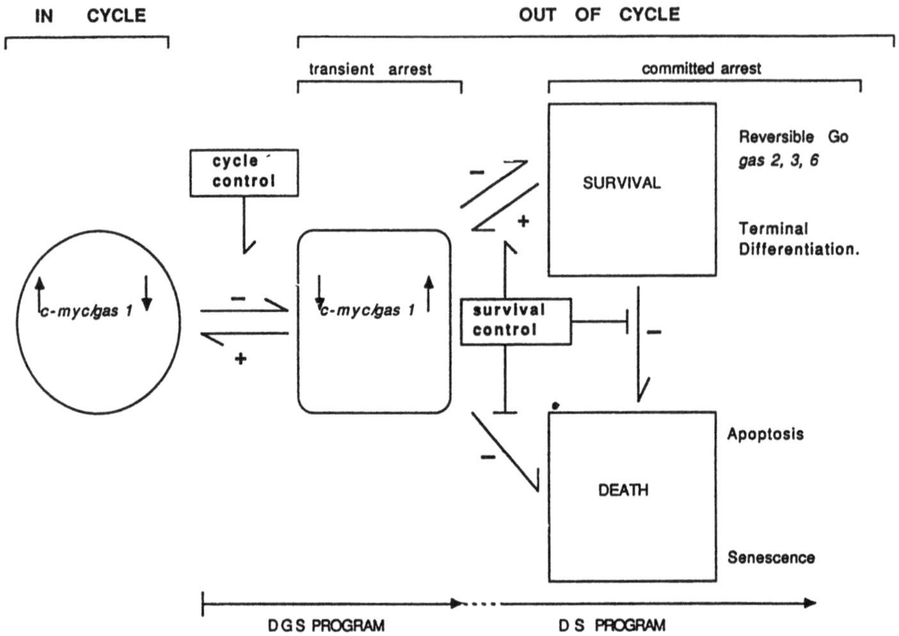

Fig. 1. Model for cell proliferation control. (–) Represents the negative circuit elements constituting the default programs for cycle control (DGS/default growth shutdown) and survival control (DS/default suicide). (+) Represents elements of the antidefault program needed both for "in-cycle commitment" or for "out of cycle survival". Their functions converge in the neutralization of the default programs of growth shutdown (i.e. *c-myc*) or suicide (i.e. *bcl-2*)

The transition between an established "in cycle" compartment to an established "out of cycle" compartment might thus happen through a growth shutdown transition state (i.e. c-*myc* ↓ -*gas*1 ↑) in which the DGS pathway operates a cell cycle progression block that is compatible with cell survival and is reversible (see Fig. 1). At this stage three main programs (Fig. 1) can be followed: (1) to stay in the same differentiation state maintaining the option to go back into "cycle", (2) to terminally differentiate expressing a new set of genes, at the same time surviving as in (1) but losing the possibility to return into "cycle", (3) to die either quickly (apoptosis) or slowly (senescence). The first two programs are both compatible with cell survival, the basic difference being that in the first case cell regeneration is maintained.

The last option, that is to die, has been also proposed to represent a constitutive program of cell death when survival factors are absent (Raff 1992). Thus, just as a cell seems to need signals from other cells in order to proliferate and at the same time to neutralize the DGS pathway, similarly it needs signals from other cells in order to survive and neutralize the intrinsic default suicide (DS) pathway (Hengartner et al. 1992). The posttranscriptionally regulated *gas* genes could serve a role as elements in survival control. Survival control elements are required to escape the constitutive programmed cell death and they are necessary for the out of cycle maintenance. From this it follows that their downregulation represents a necessary condition to return into cycle. Gas3 could thus represent the in vivo counterpart of what has been proposed since its expression is high in differentiated Schwann cells and becomes downregulated when they enter into the cycle. Survival and differentiation factors/elements in this model are not separable in function since they are required to maintain the viability of a cell out of the cycle.

The understanding of growth arrest through the study of gas genes has led us to the extreme view that cells are programmed either to leave the cell division cycle and stay alive or to kill themselves. Cell division would thus represent the dream of a cell (F. Jacob, La logique du vivant), death being its intrinsic programmed fate.

Acknowledgements. This work was supported by funds from the Associazione Italiana per la Ricerca sul Cancro as part of the special project "Oncosuppressor Genes".

References

Bishop JM (1991) Molecular themes in oncogenesis. Cell 64:235–248

Brancolini C, Bottega S & Schneider C (1992) Gas2, a growth arrest-specific protein, is a component of the microfilament network system. J Cell Biol 117:1251–1261

Ciccarelli C, Philipson L & Sorrentino V (1990) Regulation of expression of growth arrest-specific genes in mouse fibroblasts. Mol Cell Biol 10:1525–1529

Del Sal G, Ruaro ME, Philipson L & Schneider C (1992) The growth arrest specific gene, *gas*1, is involved in growth suppression. Cell 70:595–607

Esmon CT (1989) The roles of protein C and thrombomodulin in the regulation of blood coagulation. J Biol Chem 264:4743–4746

Gasic GP, Arenas CP, Gasic TB & Gasic GJ (1992) Coagulation factors X,Xa and protein S as potent mitogens of cultured aortic smooth muscle cells. Proc Natl Acad Sci USA 89:2317–2320

Goldschmidt-Clermont PJ & Jamney PA (1991) Profilin a weak CAP for actin and RAS. Cell 66:419–421

Harris H (1970) Cell fusion. Harvard University Press, Cambridge

Hengartner MO, Ellis RE & Horvitz RH (1992) *Caenorhabditis elegans* gene ced-9 protects cells from programmed cell death. Nature 356:494–499

Kovary K & Bravo R (1991) The Jun and Fos protein families are both required for cell cycle progression in fibroblasts. Mol Cell Biol 11:4466–4472

Knapp PE, Skoff RP & Redstone DV (1986) Oligodendroglial cell death in jimpy mice; an explanation for the myelin deficit. J Neurosci 6:2813–2822

Manfioletti G, Ruaro ME, Del Sal G, Philipson L & Schneider C (1990) A growth arrest-specific (*gas*) gene codes for a membrane protein. Mol Cell Biol 10:2924–2930

Marshall CJ (1991) Tumor suppressors genes. Cell 64:313–326

Nurse P (1990) Universal control mechanism regulating onset of M phase. Nature 344:503–508

Raff MC (1992) Social controls on cell survival and cell death. Nature 356:397–400

Re CF, Manenti G, Borello MG, Colombo MP, Fisher HJ, Pierotti MA, Della Porta G & Dragani TA (1992) Multiple molecular alterations in mouse lung tumors. Mol Carcinogen 5:155–160

Roushlati E & Pierschbacher MD (1987) New perspectives in cell adhesion: RDG and integrins. Science 238:491–497

Saunders S, Jalkanen M, O'Farrell S & Bernfield M (1989) Molecular cloning of syndecan, an integral membrane proteoglycan. J Cell Biol 108:1547–1556

Schneider C, King R & Philipson L (1988) Genes specifically expressed at growth arrest of mammalian cells. Cell 54:787–793

Skoff RP (1982) Increased proliferation of oligodendrocytes in the hypomyelinated mouse mutant-jimpy. Brain Res 248:19–31

Snipes GJ, Suter U, Welcher AA & Shooter ME (1992) Characterization of a novel peripheral nervous system myelin protein (PMP-22/SR13). J Cell Biol 117:225–238

Spreyer P, Kuhn G, Hanemann CO, Gilen C, Shaal H, Kuhn R, Lemke G & Muller HW (1991) Axon-regulated expression of Schwann cell transcript that is homologous to a 'growth arrest-specific' gene. EMBO J 10:3661–3668

Suter U, Welcher AA, Ozceli T, Jackson SG, Kosaras B, Franke U, Billings-Gagliardi S, Sidman LR & Shooter ME (1992) Trembler mouse carries a point mutation in a myelin gene. Nature 356:241–244

Weinberg RA (1991) Tumor suppressor genes. Science 254:1138–1146

Welcher AA, Suter U, De Leon M, Snipes GJ & Shooter EM (1991) A myelin protein is encoded by the homologue of a growth arrest-specific gene. Proc Natl Acad Sci USA 88:7195–7199

MIX
Papier aus verantwortungsvollen Quellen
Paper from responsible sources
FSC® C105338

If you have any concerns about our products,
you can contact us on
ProductSafety@springernature.com

In case Publisher is established outside the EU,
the EU authorized representative is:
Springer Nature Customer Service Center GmbH
Europaplatz 3, 69115 Heidelberg, Germany

Printed by Libri Plureos GmbH
in Hamburg, Germany